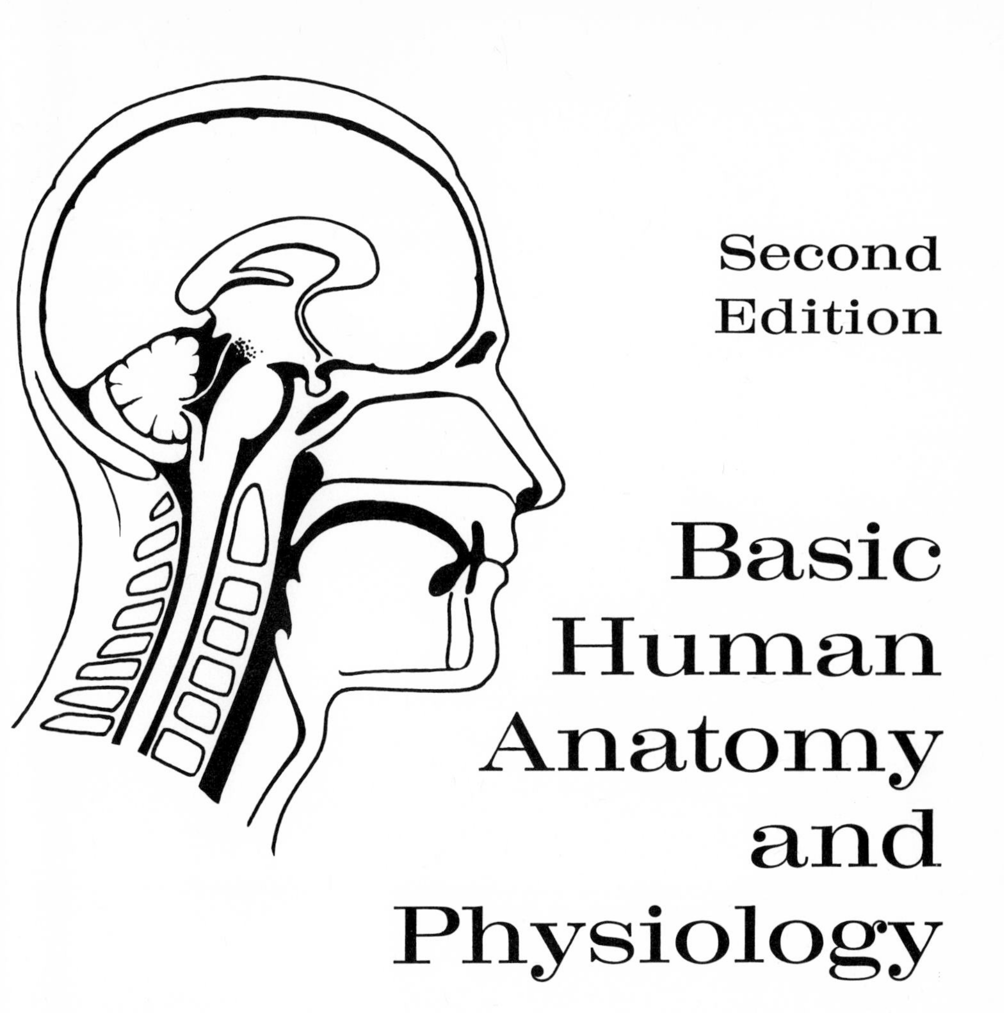

Second Edition

Basic Human Anatomy and Physiology

Charlotte M. Dienhart, Ph.D.
Assistant Professor of Anatomy
Division of Basic Health Sciences
Emory University, Atlanta, Georgia

illustrated by STEVEN P. GIGLIOTTI

W. B. Saunders Company
Philadelphia · London · Toronto

W. B. Saunders Company: West Washington Square
Philadelphia, Pa. 19105

1 St. Anne's Road
Eastbourne, East Sussex BN21 3UN, England

833 Oxford Street
Toronto, M8Z 5T9, Canada

Listed here is the latest translated edition of this book together with the language of the translation and the publisher.

French (2nd Edition) – Les Editions HRW Ltee.,
Montreal, Canada

Basic Human Anatomy and Physiology ISBN 0-7216-3081-2

Print No.: 9 8 7 6 5

In memory of my parents

PREFACE

The second edition of *Basic Human Anatomy and Physiology* has, like the first edition, the primary purpose of providing for paramedical personnel a clear understanding of the principles of the human body's structure and function. The author urges the reader to keep in mind that the scope of the book has been consciously tailored to serve the needs of those who do not wish to pursue detailed knowledge of the subject matter. However, for those seeking additional knowledge, in this revision specific references have been included at the end of each chapter, and at the end of the book there is a list of general references in anatomy and physiology.

Some of the changes incorporated in this revision have been suggested by those who were kind enough to review the book critically and to make suggestions for improvement. An entirely new chapter on the integumentary system has been added, as well as material on control mechanisms, acid-base balance, water balance, cellular variations, and fertilization and implantation. It is hoped that these and other changes will make the book more useful.

The author is most grateful for the generally positive response the first edition has received. It is especially gratifying to hear this response from students; after all, the book has been written for them. Many favorable comments have been made concerning Mr. Gigliotti's fine work, which indeed greatly enhances the text. Special thanks go to my friends and colleagues, Dr. A. F. Baradi, Dr. Ellen Fuller, and Dr. Stephen W. Gray for generously agreeing to criticize constructively some of the new material. The author alone, however, is responsible for any errors, imperfections, or omissions.

Finally, it has been a distinct pleasure to work with Mr. Robert E. Wright and the staff of W. B. Saunders Company. Their patience, understanding, and expert assistance are gratefully acknowledged.

CHARLOTTE M. DIENHART

CONTENTS

Chapter 1

INTRODUCTION

TERMINOLOGY – ITS IMPORTANCE

Before one studies the structure (anatomy) and function (physiology) of the human body, it is necessary to know the meanings of certain words used in describing body parts and activities. Just as it is essential to know the vocabulary when studying a foreign language, so too the student studying a science must know the vocabulary pertaining to that science. This mastery of scientific language will be easier for those who have already taken a course in the biological sciences – familiarity comes with usage. In any event, the student should have available a medical dictionary and use it as a valuable study aid. The following pages will contain definitions and explanations of those basic words and terms which are most often used in the study of the human body.

GENERAL BODY PLAN

Anatomical Position. Whenever one describes the location or relationship of body parts, it is necessary to think of the whole body in a certain fixed position. This position is called the *anatomical position* and is shown in Figure 1–1. The body is erect, facing forward. The arms are at the sides with the palms of the hands facing forward.

Body Surfaces. The front surface of the body is its *anterior* or *ventral* surface. Moreover, the palms of the hands and the soles of the feet are the ventral surfaces of these extremities. *Dorsal* or *posterior* surfaces are the back of the body, the backs of the hands, and the tops of the feet.

Directional Terms. *Superior* or *cranial* means above or toward the head. The opposite term is *inferior* or *caudal,* meaning below or toward the feet. Another pair of terms is *medial,* meaning toward the midline or middle, and *lateral,* meaning away from the midline or more toward the periphery. *Central* and *peripheral* are opposite terms meaning toward the center and away from the center, respectively. *Proximal* means close to the origin or source of attachment, whereas *distal* refers to a point farther away from the origin or attachment.

Remember that all the foregoing terms refer to relationships of structures to each other *when the body is in the anatomical position*. Later it will be seen that these terms are also useful in describing relationships of internal body parts.

The following are some examples of how these terms can be used: The chest is *superior to* the abdomen, but *inferior to* the neck. The sternum (breast bone) is *medial to* the ribs. The knee is *proximal to* the ankle, but *distal to* the hip.

Planes or Sections. A cut lengthwise through the body dividing it into right and left portions gives a *longitudinal* or *sagittal* section. If this cut is exactly in the midline, it is called a *midsagittal* section. If one were to face the body from the side and make a lengthwise cut, a *frontal* or *coronal* section would result, forming ventral and dorsal portions. Dividing the body by a horizontal cut into upper and lower portions gives a *transverse* or *horizontal* section. These planes or sections apply to all the structures inside the body as well as to the body as a whole (Fig. 1–2).

Body Cavities. The cavities, or spaces, of the body contain the internal organs or *viscera*. The two main cavities are called the *ventral* and *dorsal*. The ventral is the larger cavity and is subdivided into two portions by the diaphragm, a dome-shaped respiratory muscle. The upper, *thoracic* or chest cavity contains the heart, lungs, trachea, esophagus, large blood vessels, and nerves. The *abdominopelvic* cavity lies below the diaphragm. In its abdominal portion are the liver, gallbladder, stomach, pancreas, spleen, kidneys, and intestines. The lower or pelvic part contains the urinary bladder, rectum, and reproductive organs.

The smaller of the two main cavities is called the *dorsal* cavity. As its name implies, it contains organs lying more posteriorly in the body. The upper or *cranial* cavity houses the brain. The vertebral portion houses the spinal cord (Fig. 1–3).

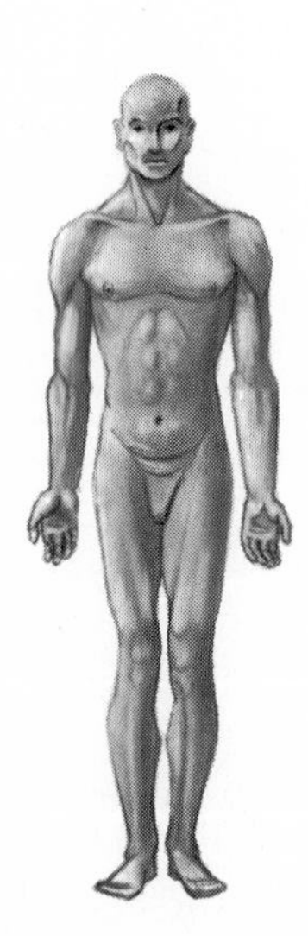

Figure 1–1. The anatomical position.

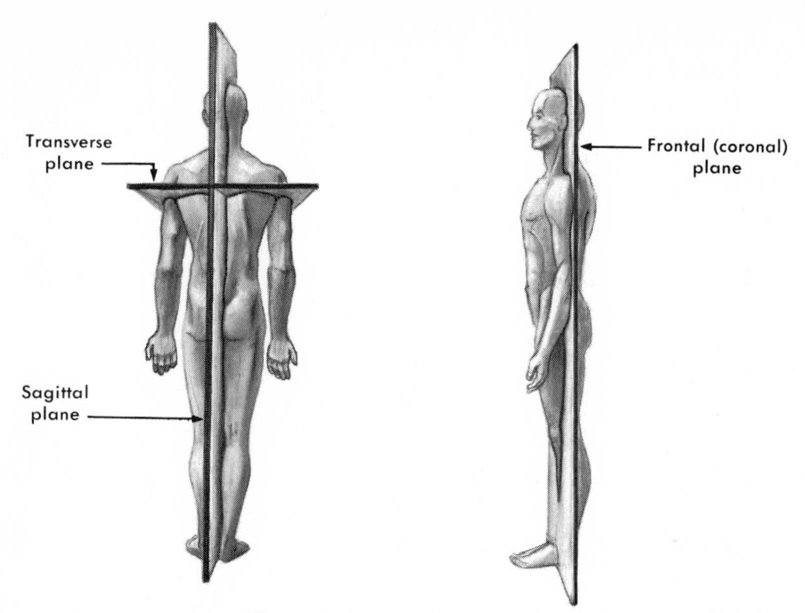

Figure 1–2. Planes of the body.

STRUCTURAL UNITS

One of the most striking facts about the human body is that it begins from the union of just two cells—the egg and the sperm. Through the process of growth and development these two cells give rise to all the cells, tissues, organs, and systems that make up the adult human body.

The Cell. The structural and functional unit of any living organism is the cell. The human body is composed of countless millions of these tiny units. Although the existence of cells has been known for many years, only recently have scientists been able to discover the great complexity and intricate organization of the human cell. If one excludes viruses, the cell is regarded as the smallest unit of living matter which can live independently and reproduce itself.

Body cells vary in size from those which can be seen only through a high-power microscope to the ovum (egg cell) which is barely visible to the naked eye. Although most diagrams show a typical cell as being round, body cells actually are found in a variety of shapes—elongated, square, star-shaped, oval, etc. The basic structure and functions of all human cells are similar, but certain types of cells perform special additional functions which will be pointed out later.

The cell can be thought of as a highly organized "molecular factory." That is, it carries on within itself a tremendous number of very complex and intricate functions at the molecular level. The cell contents are enclosed by a thin envelope called the *cell membrane,* which will be described later in more detail. The two main parts of the cell contents are the *nucleus* and the *cytoplasm.* The nucleus is the "control center" of the cell—it controls growth, repair, and reproduction. The agents of heredity, the chromosomes, are contained in the nucleus. The species of an organism (man, dog, cow) is determined by the

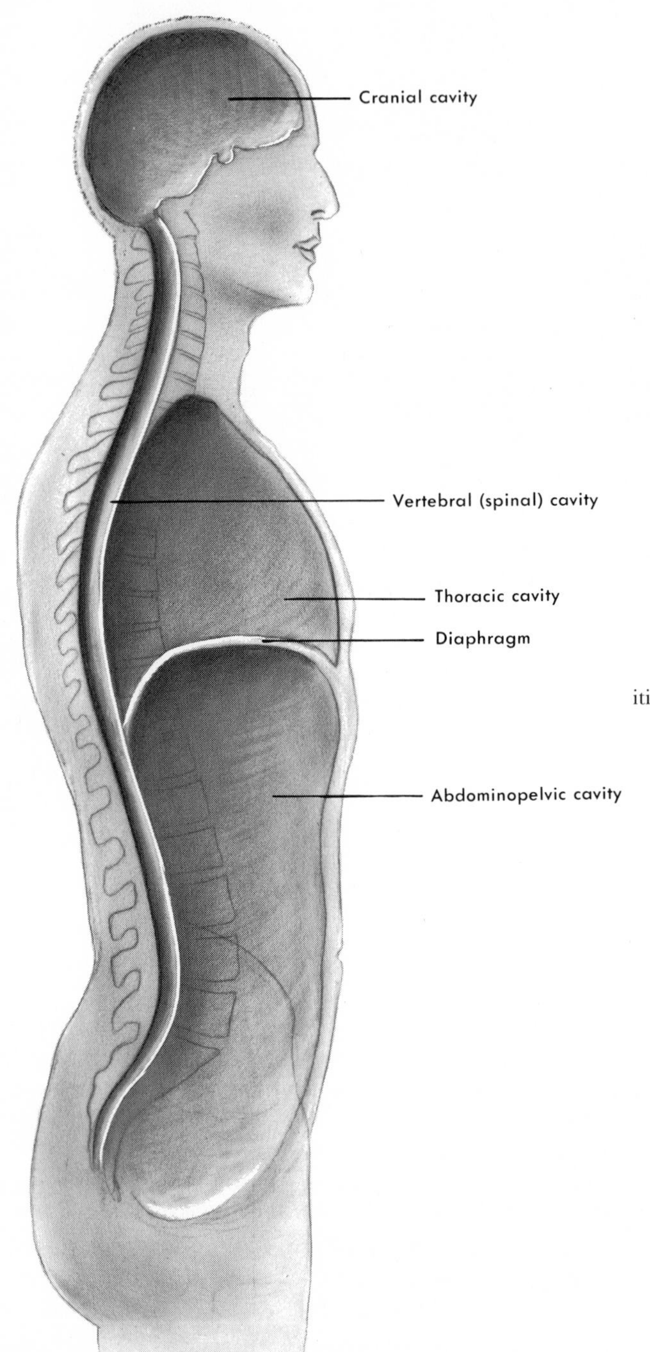

Figure 1–3. The body cavities.

chromosomes in each cell. A human cell normally has 46 chromosomes, 22 somatic pairs and 2 sex chromosomes.

The cytoplasm makes up all of the cell except the nucleus. It contains many different structures called organelles (little organs) which carry out cell metabolism. One such structure, called a *mitochondrion*, is a tiny, rod-shaped or spherical body that is barely visible under the microscope. These mitochondria are often referred to as the "power plants" of the cell since they supply it with most of its usable energy. The more metabolically active the cell, the more mitochondria it contains. They range in numbers from a few to two or three thousand per cell. Other structures found in the cytoplasm are the centrosome, microsomes, Golgi apparatus, and endoplasmic reticulum (Fig. 1–4).

The great increase in our knowledge concerning the structure and functions of the different parts of the cell is due largely to improved techniques of study, such as the development of the electron microscope. This additional knowledge in turn has led to a greater understanding of the entire body in healthy and diseased states.

The process of cell division that cells undergo when they reproduce is called *mitosis*. It consists of a series of steps or phases in which both nucleus and cytoplasm of the parent cell undergo an exact division so that two identical daughter cells are formed. The ability to undergo mitosis varies with cell types. In parts of the body where cells ordinarily have a short life span, such as the outer layer of the skin, cells are constantly dividing to replace those which are dying. On the other hand, mitosis seldom, if ever, occurs in normal adult muscle or brain. One characteristic of malignant growths (cancers) is wild or uncontrolled cell division.

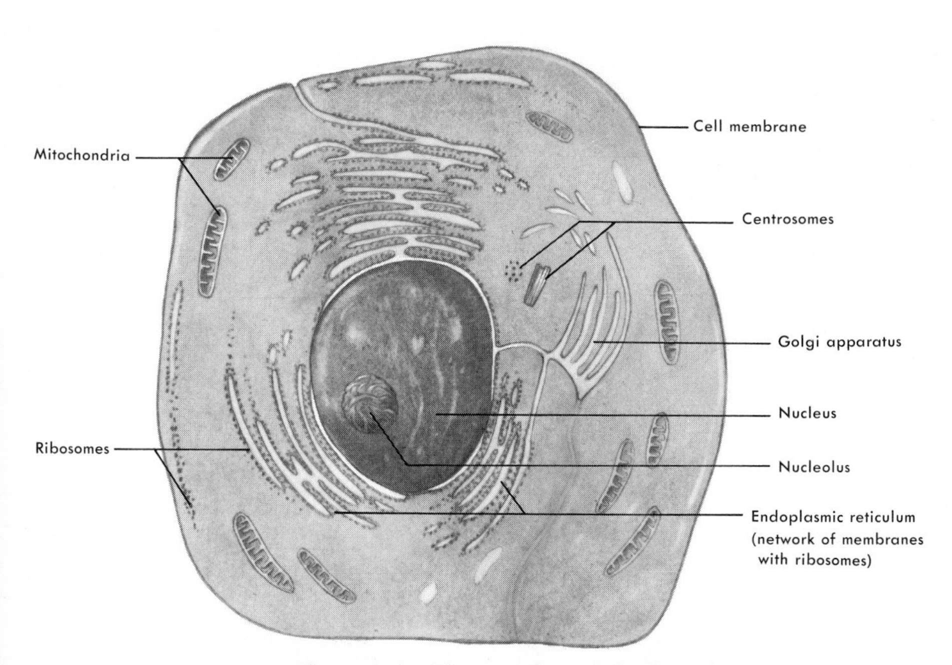

Figure 1–4. Diagram of a typical cell.

Tissues. The fertilized egg cell develops into a mass of cells by continued mitosis. A group of these cells develops certain features or characteristics that set it apart from other cell groups. Such a group of like cells, together with the material between them that holds them together, is called a *tissue.* There are four basic tissues: epithelial, connective, muscle, and nerve. Some authorities list vascular tissue, or blood, as a fifth type; here it will be included as a special kind of connective tissue. In this chapter only the first two types, epithelial and connective tissues, will be discussed, since they are so widespread in the body. The others, including blood, will be described in their respective chapters.

Epithelial tissue, or epithelium, is highly specialized to protect, absorb, and secrete. As a protector, epithelium forms the outer or covering layer of the skin and lines hollow organs or structures such as the air passageways, the stomach, and the intestines. In addition, some of these lining cells are able to absorb materials from substances that pass through these organs. Other epithelial cells, found in glands, produce and release a secretion, such as saliva, sweat, and digestive juices.

The epithelium that lines organs or covers surfaces is subjected to much wear and tear; fortunately, this is a highly cellular tissue with a great capacity to renew or regenerate itself by mitosis. Most types of epithelium are named according to the shape and arrangement of the cells. *Simple epithelium* consists of one layer of cells, and according to cell shape may be squamous (plate-like), cuboidal (cube-shaped), or columnar (like columns) epithelium. *Stratified epithelium* is composed of two or more layers (strata) of these cells. *Glandular epithelium* may be simple (unbranched) or compound (branched) and either tubular (tube-shaped) or alveolar (flask-shaped). Figure 1–5 shows the different types of epithelium.

In addition, certain epithelial cells may have added features according to the functions they perform. For example, to protect the air passages from particles in air inhaled from outside the body, fine hair-like processes are found along the free edges of the lining cells. These are not really hairs but very tiny projections of the actual cell substance itself called *cilia.* Cilia beat with a wave-like motion, much as a field of wheat ripples in the wind. Thus foreign particles may be swept up and out of the air passages.

With the exception of the cornea of the eye, all epithelial surfaces of the body that are exposed to the air are dry surfaces. They are protected by a covering of horny material called *keratin,* which varies in thickness over different parts of the body. Made by the cells themselves, keratin is thickest on the soles of the feet and palms of the hands and thinnest on the eyelids. The hair and nails are made of keratin also, although these structures are harder and coarser because of a slight difference in chemical composition. In Chapter 2, the skin and its appendages are described in greater detail.

Connective tissue, as its name indicates, serves to support other tissues or to connect them together. It can perform this function because most of it is composed of a relatively large amount of intercellular material which accounts for its strength. Its cells, which produce the intercellular material, are relatively few in number.

In general, connective tissue is named according to the kind and arrangement of the intercellular substance, with the predominantly cellular types simply

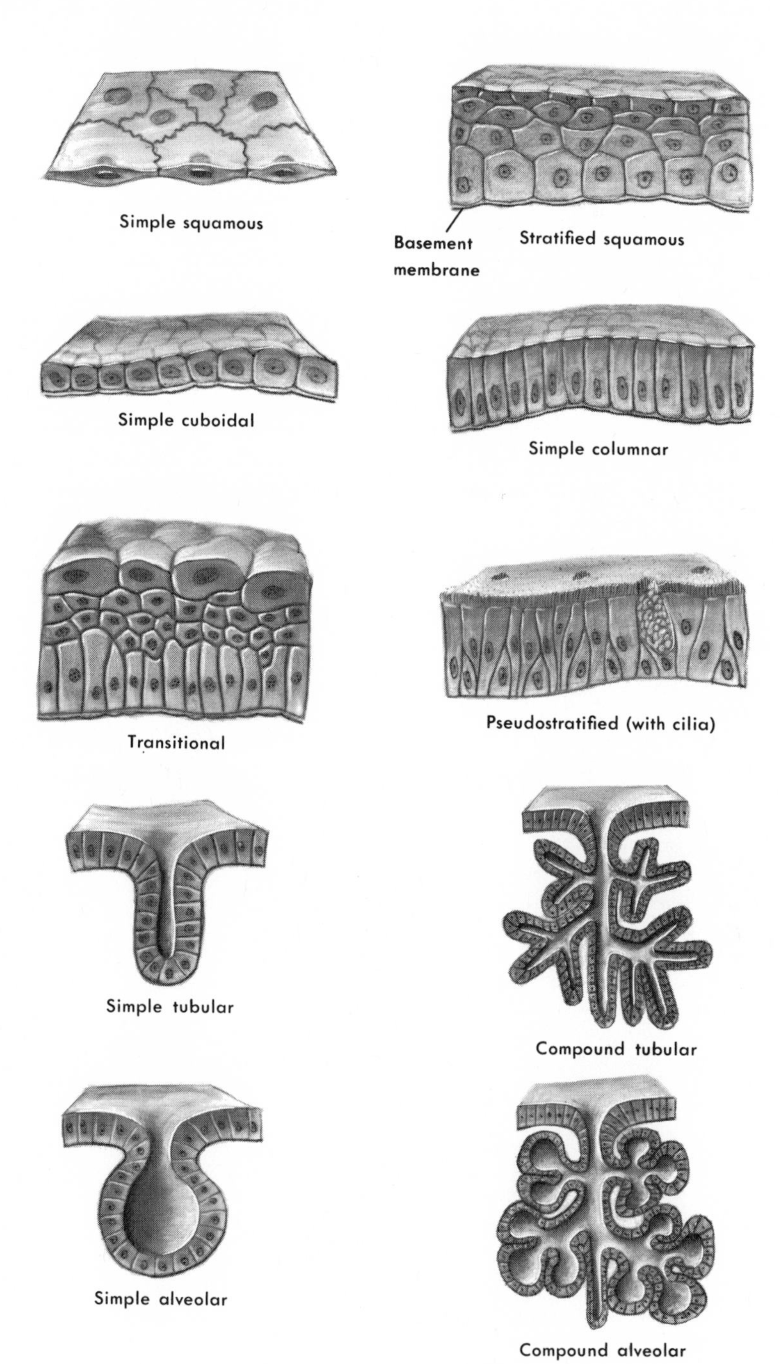

Figure 1–5. Types of epithelium.

given specific names. The mother cells of connective tissue have great ability to develop and change along many different pathways. Thus there are many different types of cells in these tissues (Fig. 1–6).

Loose ordinary connective tissue is both supporting and cellular. It is called "loose" because the intercellular material is delicate and jelly-like in consistency and "ordinary" because it is widely distributed in the body. It is found underneath epithelial tissues, such as under the epithelial layer of the skin and the lining of the digestive tube, around the secretory parts of glands, and as packing material between muscles and internal organs. There are various types of cells scattered throughout this sticky, soft intercellular material. The most numerous are called *fibroblasts*; others include macrophages (histiocytes), foreign body giant cells, mast cells (source of heparin), fat cells, plasma cells, and blood leukocytes (white blood cells).

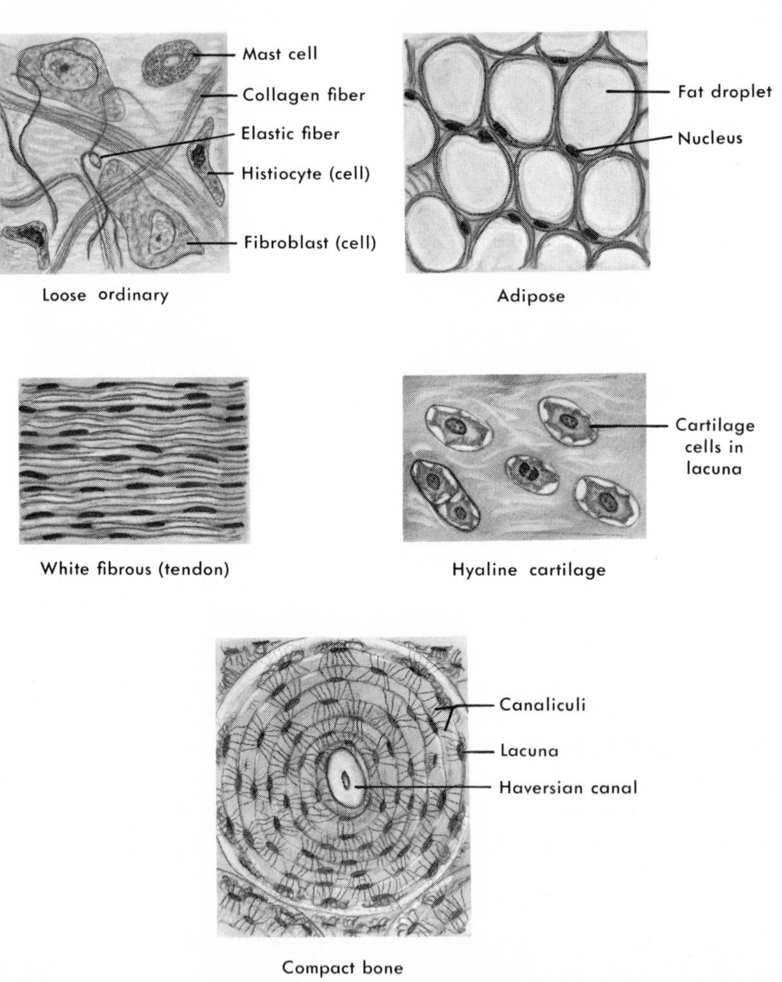

Figure 1–6. Types of connective tissue.

Adipose or *fat tissue* is much like loose connective tissue except that it is highly cellular, being composed of large masses of fat cells closely packed together. Each cell consists of a small droplet of fat which occupies most of the cell. This droplet pushes the nucleus to one side, giving the cell a signet ring appearance. In these cells, then, the body stores its fat. Adipose tissue is found more concentrated in certain parts of the body called fat depots, such as underneath the skin of the abdomen and buttocks. Besides being a reserve food supply, adipose tissue acts as protective padding and provides insulation.

Dense fibrous connective tissue is characterized by having varying numbers and kinds of fibers running through the intercellular material which give it strength. Fibroblasts, which make the fibers, are scattered throughout this intercellular substance. The fibers may be arranged regularly, that is, all in the same direction, or irregularly, in several different directions. An example of the regular type is a tendon, in which strength is required in only one direction. The deeper part of the skin is a good example of irregular dense fibrous tissue; here strength is required in different directions. Dense connective tissue is also found as deep fascia and as coverings or capsules for many organs.

Cartilage is much like dense connective tissue, except that it will bear a certain amount of weight before bending. It can do this because its intercellular substance has the consistency of a firm gel, rather like that of plastic. The cells (chondrocytes) lie in small spaces in the intercellular substance. These spaces are called lacunae, and each lucuna contains one or more chondrocytes. Cartilage is found in the nose, the ear, the larynx, the trachea, and over the ends of bones forming some of the joints. During fetal life, most of the skeleton is first formed from cartilage.

Bone or osseous tissue is the hardest of all connective tissues since its intercellular substance is calcified by the deposition of calcium salts. This calcification makes bone very strong and thus capable of bearing great weight without bending or breaking. As in cartilage, bone cells (osteocytes) lie in lacunae. These are connected by tiny tunnels or canals (canaliculi) which carry nourishment from the blood to the cells.

Membranes. Membranes are sheets of epithelial and connective tissues found in certain areas of the body. The cutaneous membrane is actually the skin that covers the entire body. The outer layer, or *epidermis*, is stratified squamous epithelium, whereas the underlying layer, or *dermis*, is connective tissue.

Mucous membrane lines all body passageways which open to the exterior, such as the digestive and urinary tracts. It has an outer epithelial layer resting on a connective tissue layer and secretes a sticky substance called *mucus*.

Serous membrane lines the closed cavities of the body and also covers the organs in them. The closed cavities are the pleural cavity (lungs), the pericardial cavity (heart), and the peritoneal cavity (abdominal organs). A serous membrane then is actually a double-layered membrane. The outer or parietal layer lines the cavity, and the inner or visceral layer covers the organs. This epithelial and connective tissue membrane secretes a watery serous fluid.

Some membranes are composed entirely of fibrous tissue. These are *synovial membranes* lining joint cavities and tendon coverings, *perichondrium* covering the outer surface of cartilage, and *periosteum* covering the outer surface of bones.

Fascia is a special kind of connective tissue membrane. Superficial fascia is found as a continuous layer of loose connective tissue lying underneath the skin. It attaches the dermis to underlying structures and contains varying amounts of fat. Deep fascia is the tough fibrous tissue covering muscles, blood vessels, and nerves. It does not contain fat.

Organs. It has been explained that masses of cells which are specialized for certain functions form the basic tissues of the body. When two or more of these tissues combine their functions, they form a structure in the body called an *organ*. The heart, stomach, thyroid gland, liver, and brain are examples of organs. Each is composed of certain types of epithelial, connective, muscle, nerve, and vascular tissues all arranged and coördinated so that they function as a unit.

Systems.. A *system* is a group of organs that is specialized to perform a major body function. For example, the circulatory system is composed of the heart and blood vessels which provide all body cells with nourishment by circulating the blood through all body parts. The skeletal system, composed of all the bones in the body, forms the body's framework.

In this book the body will be studied by systems—this is called systemic anatomy and physiology. The following systems will be discussed:

1. Integumentary system
2. Skeletal system
3. Muscular system
4. Nervous system
5. Circulatory system
6. Respiratory system
7. Digestive system
8. Urinary system
9. Endocrine system
10. Reproductive systems

It is well to keep in mind that the body functions as a unit, that is, as the sum of its individual systems. Therefore, although we will study the body one system at a time, it is important always to think of the body as a single unit or organism. This means that the systems are interdependent, and whatever affects one of them will also affect the body as a whole.

ANATOMICAL VARIATIONS

The beginning student of anatomy tends to think that the structures forming the body always follow a rigid, unvarying plan. Although it is true that what is usually taught pertains to the typical or usual finding, minor variations are almost always present. These variations can be explained mostly on the basis of the prenatal development of the individual, that is, development before he was born. Usually these minor differences do not interfere with normal function, and their presence may not be discovered unless surgery or an autopsy is performed.

Less frequently, more marked variations or deviations from the normal are present. These are called *anomalies* and are also considered to result from abnormalities in growth and development. If an anomaly is serious enough, as in the case of a heart defect that prevents its proper functioning, it may cause the death of the person, even before he is born. On the other hand, some anomalies cause no difficulty or symptoms at all.

VARIATIONS IN CELL GROWTH

Abnormal growth and development of cells, and thus of tissues and organs, can lead to several conditions. In some cases, these abnormalities occur during prenatal life; in others, the problem arises after development is complete.

Aplasia, Hypoplasia, and Atrophy. A tissue or an organ may be absent because it was never formed during prenatal development; this condition is called *aplasia*. Some cases of aplasia, such as in the absence of a vital organ or blood vessel, are incompatible with life. In the event that a missing organ is one of a pair, such as a kidney, then life can be sustained.

The term *hypoplasia* refers to a structure that is underdeveloped or smaller than normal. In this case, compatibility with life even when a vital organ is involved depends on whether or not there is sufficient functioning tissue to carry on normal processes. The causes of aplasia and hypoplasia are not known; both are *congenital* defects (present at birth) and are the result of abnormal influences during prenatal life.

Atrophy refers to a decrease in the size of an organ that originally developed normally; that is, it is an *acquired* condition. Atrophy may be either physiological or pathological in nature. Physiological atrophy normally occurs to some extent in all persons as part of the aging process; structures such as muscle, brain, and reproductive organs begin to decrease in volume.

Pathological atrophy involves a decrease in size that is not associated with aging but with some disease condition. One such condition is *disuse atrophy*, which is the result of prolonged inactivity of a body part or organ. A limb which has been placed in a plaster cast or had its nerve supply severed (denervation) for a long period will show evidence of atrophy.

Any condition in which pressure is exerted on an organ for any length of time will result in *pressure atrophy*. This may stem either from a simply mechanical force on the cells themselves or from a lack of blood supply to the organ because of pressure on its blood vessels. In the latter case, one could designate this as *vascular atrophy*.

An endocrine gland (see Chapter 10) may become overworked and thus develop *exhaustion atrophy*, in which case its cells would atrophy and the gland would be unable to produce its secretions. This in turn would lead to atrophy of the target organs which depend on these secretions for their function.

Hypertrophy and Hyperplasia. An increase in the over-all size of a tissue or organ due to an increase in the volume of *existing* cells is called *hypertrophy*. If the increase is caused by the formation of *new* or additional cells, then the result is termed *hyperplasia*. The former occurs in tissues which normally do not produce new cells, such as muscle and nerve. In epithelial tissues, however, which have a great capacity to reproduce, hyperplasia as well as hypertrophy may occur.

It should be noted that hypertrophy may be a normal process in some tissues as a result of extensive use. The increased muscular development in an athlete is an example of this. On the other hand, a heart which has been weakened by disease may show pathological hypertrophy of its muscle tissue because it is attempting to compensate for its reduced efficiency.

Anaplasia. The basic alteration occurring in cells when they develop into malignant tumors (cancers) is called *anaplasia*. This change is considered irreversible and is probably the chief diagnostic sign of malignancy. Anaplastic cells are often quite bizarre in appearance, showing wide variation in size, shape, and internal structure. Thus, by microscopically examining cells taken from a tumor, a trained pathologist can usually diagnose whether or not the tissue is malignant by noting the presence or absence of cellular anaplasia.

THE LIFE SUBSTANCE AND ITS PROPERTIES

Perhaps the most intriguing thing about the materials that make up the body is that each cell, be it epithelial, connective, or any other, has a certain composition and properties, the total of which we call *life*. The name *protoplasm* has been given to this life substance, which is understood to mean the material from which the cell is made and upon which all its vital functions depend.

Every living cell is made of different but definite amounts of many chemical elements. The three elements in greatest abundance are oxygen, carbon, and hydrogen, which make up about 90 per cent of the total. Some of the others, found in lesser amounts, are nitrogen, calcium, phosphorus, potassium, iron, sodium, sulfur, and magnesium.

When two or more elements are combined in a certain ratio they form a compound. Water is a compound formed from hydrogen and oxygen; it makes up about 65 per cent of the life substance.

Metabolism is a general term which includes all of the physical and chemical changes that occur in living matter. It refers, for example, to food and oxygen being brought into the cell and changed or broken down so that energy can be released. The cell uses this energy for growth and repair by forming new units of protoplasm from the building materials in the food. Metabolism also includes the excretion of waste products by the cell.

Reproduction is an essential characteristic of any living substance: Any material that cannot reproduce itself in kind cannot be called living.

Irritability is the property that allows living material to react to stimuli from its surroundings. Heat, cold, and pain are all stimuli that cause living organisms to respond in some way. Light, for example, is a stimulus that causes some lower organisms to be drawn toward it. Some organisms will attempt to move away if a toxic or painful substance approaches them.

Conductivity is the ability to conduct or lead a stimulus from one point to another. Although possessed by all body cells to a certain degree, this property is most highly developed in nerve cells.

Contractility, or the ability to reduce in length upon stimulation, is also a property of all protoplasm. Muscle cells have this property most highly developed. These cells also possess elasticity which allows them to return to their normal length after contracting.

HOMEOSTASIS

In order to carry on its many complex activities, the cell must continually take materials into itself and excrete its waste products. The cell membrane is a unique structure that is especially constructed for the purpose of regulating exactly the interchange of substances between the outside world and the cell. This interaction between the cell and its environment serves to maintain each of the substances inside the cell within the concentration range necessary for life. This balanced state is called *homeostasis*, which literally means "the same position."

Studies have shown that the cell membrane is extremely thin—about ten millionths of a millimeter in thickness. It is composed of a double layer of lipid (fat) molecules (smallest units of a particle) covered by two protein layers. Very tiny pores or holes are located at intervals, giving the membrane a sieve-like appearance.

Selective permeability is the term used to describe the cell membrane's ability to allow certain substances to pass through it while rejecting others. Some of this can be explained by the pores which act as filters to permit the entry of particles only up to a certain size. In addition, fat-soluble materials presumably are dissolved in the lipid layers of the membrane and so gain entrance. However, it is known that cells admit other materials that are too large for the pores and are not fat soluble, so obviously the membrane has additional mechanisms to handle these.

It has also been shown that different kinds of cells have different degrees of permeability to the same substance. For example, the red blood cell will let water pass into it 100 times more easily than will a tiny, one-celled animal, the amoeba.

It is a basic principle that energy or force is needed to move material from one place to another. In the case of the transport of substances across the cell membrane, if the force is supplied by the environment outside the cell, it is referred to as passive transport. If the energy originated from the cell's own metabolism, the movement is due to active transport. Obviously, active transport calls for more work on the part of the cell than does passive transport.

There is another way in which cells are able to bring substances into themselves. This is by a process of ingestion of solids or liquids called *phagocytosis* or *pinocytosis*, literally, "eating" and "drinking" by the cell. The cell membrane forms pockets or invaginations that draw material from the outside toward the cell interior. The membrane envelops the substance, and then pinches off so that the material remains in the cytoplasm (Fig. 1–7).

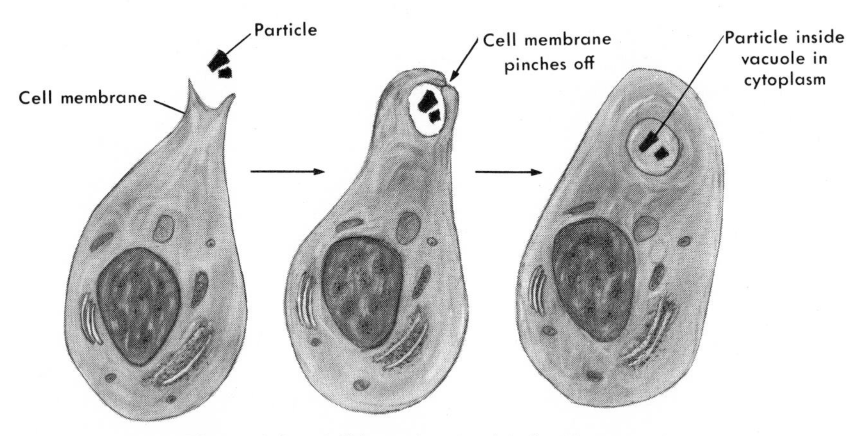

Figure 1–7. Cell ingesting particle by phagocytosis.

BASIC PHYSIOLOGICAL PROCESSES

There are three mechanisms by which substances cross barriers such as cell membranes—filtration, diffusion, and osmosis. *Filtration* is the passage of water or water and dissolved substances across a barrier because of a difference in pressure on the two sides of the barrier. The size of the particles and the size of the holes in the barrier or filter determine which substances will pass. If one dissolves sugar in water, adds some powdered charcoal, and pours the mixture through filter paper into a container, the sugar solution will pass through but the undissolved charcoal will not (Fig. 1–8). The filtration pressure in this example comes from the force of gravity on a column of water; this is called *hydrostatic* pressure. In the body, filtration takes place through the walls of the tiny blood capillaries. Here the pressure responsible is the blood pressure, which is created by the pumping action of the heart. This forces water and certain dissolved substances from the blood through the capillary walls (filter) and into the surrounding tissues. Some materials, such as the blood proteins, are too large and ordinarily do not pass through vessel walls.

Diffusion is the tendency of a substance to distribute itself equally throughout an available space. This happens because the molecules of the substance are in constant motion, whether it be a liquid, a solid, or a gas. Furthermore, a substance tends to move from an area of high concentration to an area of lower concentration. If an open bottle of strong ammonia is placed in a room, the odor of ammonia will eventually permeate the room, thus indicating that it has diffused throughout the area. In the body diffusion of gases (oxygen and carbon dioxide) takes place between the air in the lungs and the blood in the lung capillaries (Fig.

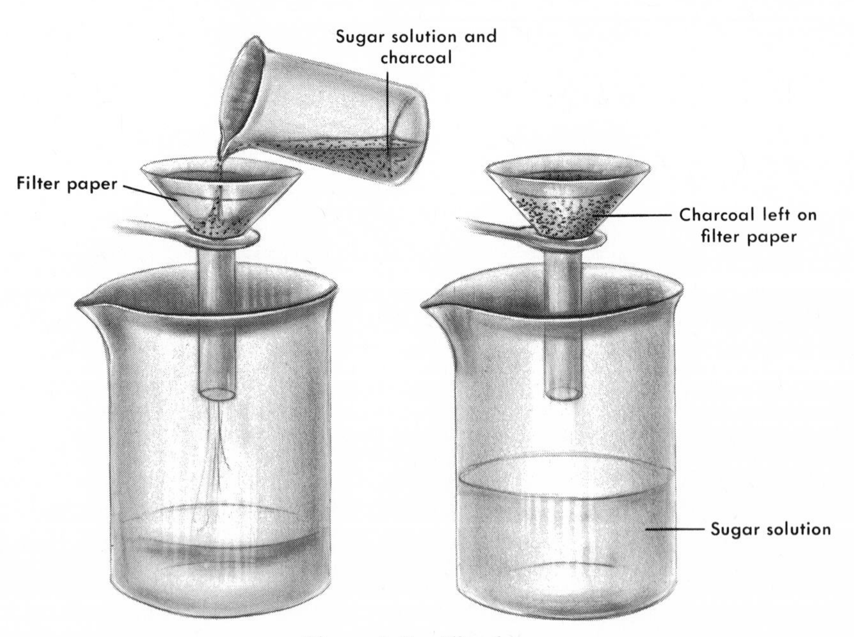

Figure 1–8. Filtration.

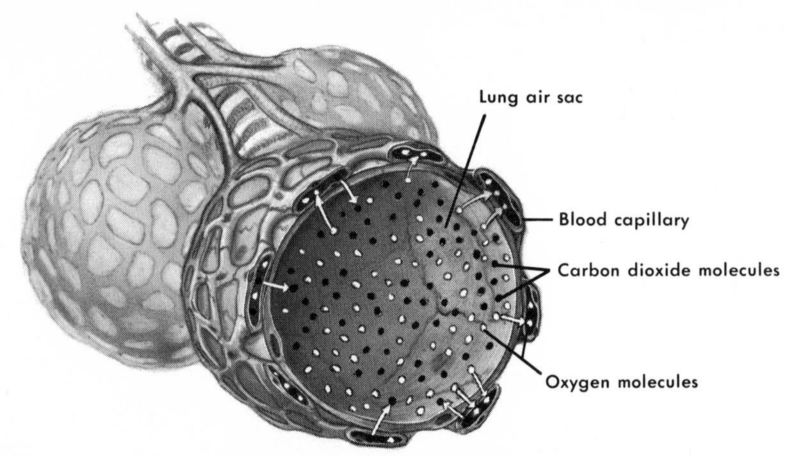

Figure 1–9. Diffusion. Oxygen molecules diffuse out of the air in the lung into the surrounding blood capillaries.

1–9). Another example would be to drop a lump of sugar in a glass of water. In time, the sugar would dissolve and be equally distributed throughout the water. This phenomenon could be verified by simply tasting the water. Here is an example of the diffusion of a solid in a liquid.

Osmosis is actually a special kind of diffusion. It refers specifically to the diffusion of *water* through a semipermeable membrane separating two solutions of unequal concentration. Although water passes through the membrane in both directions, more water will pass toward the side of the more highly concentrated solution, causing an increase in volume on that side of the membrane (Fig. 1–10). In theory, this process will continue until the concentrations (but of course NOT the volumes) of both solutions are equal, that is, until an *equilibrium* is reached. In Figure 1–10, the pressure responsible for moving water into compartment A can be measured. It is called *osmotic pressure,* and its magnitude depends on the con-

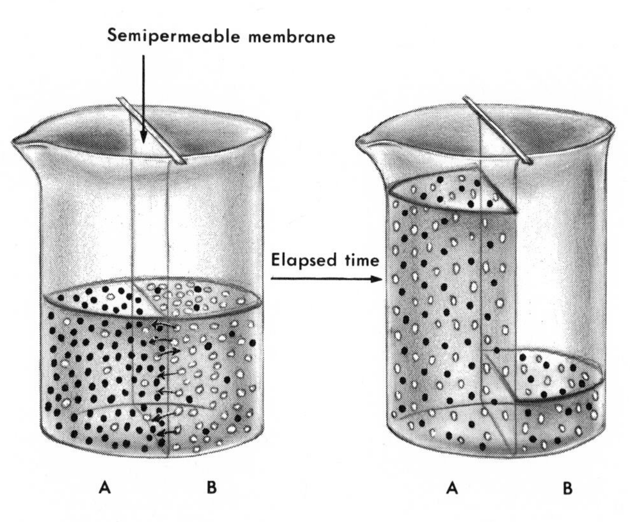

Figure 1–10. Osmosis. Two solutions of unequal concentration separated by a semipermeable membrane. Black dots: molecules of dissolved solute (sugar, for example). Circles: Molecules of water. In time, more water will pass from side B to side A.

centrations of the two solutions. Osmosis, then, represents the migration of water from the lower to the higher concentration and may be thought of as a special kind of mixing process.

The phenomenon of osmosis can be illustrated by using the red cells of the blood. Normally, the concentration of sodium chloride (salt) inside the red blood cell is equal to that in the fluid part of the blood or the plasma in which the cells are contained (about 0.85 per cent). If a drop of blood is mixed with pure water, the mixture becomes clear and transparent. Furthermore, if the mixture is placed under the microscope, no red cells can be seen. What has happened to them? Putting the blood in water diluted the plasma so that its salt concentration was no longer equal to that inside the red cells. Thus, the conditions under which osmosis occurs were created and water passed through the red cell membranes. After a time, the excess water caused the cells to swell and finally burst, or lyse, when the membranes reached their elastic limit. When blood cells are involved, this process is called *hemolysis (hemo* refers to blood). To prevent hemolysis when using salt solutions for intravenous feeding, a 0.85 per cent concentration of salt is used; this solution is referred to as *physiological saline.*

The opposite of hemolysis is called *crenation.* This occurs if blood is mixed with a salt solution that is stronger than 0.85 per cent. Water will then *leave* the cells, causing them to shrivel up, or crenate. Figure 1–11 illustrates hemolysis and crenation.

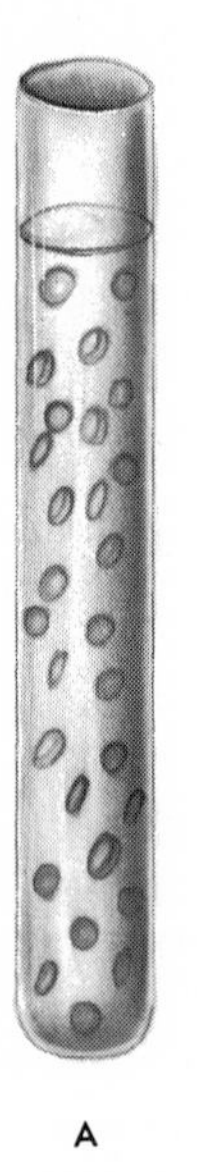

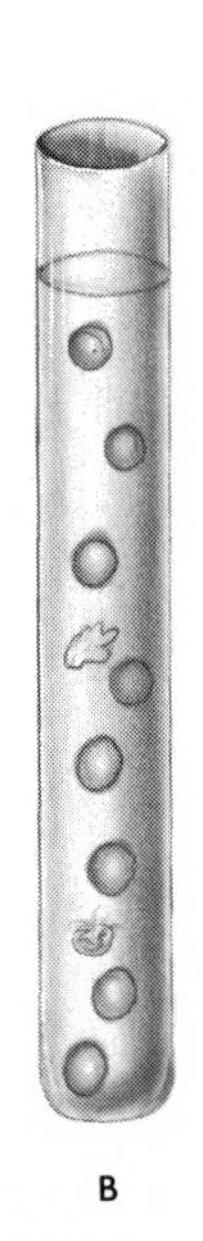

Figure 1–11. Hemolysis. *A*, Blood cells in physiological saline (0.85 per cent sodium chloride) retain normal size and shape. *B*, Blood cells in distilled water. Some cells have already burst or hemolyzed; others are becoming swollen. *C*, Blood cells in strong salt solution (5 per cent). Note shriveled or crenated appearance of cells.

OUTLINE SUMMARY

1. General Body Plan
   a. Anatomical position – all terms used refer to body in this position
   b. Body surfaces and directional terms – most are pairs of opposites
      1) Ventral (anterior); dorsal (posterior)
      2) Superior (cranial); inferior (caudal)

        3) Proximal; distal
        4) Central; peripheral
        5) Medial; lateral
    c. Planes or sections
        1) Sagittal – longitudinal division into right and left portions
        2) Frontal – lengthwise division into dorsal and ventral portions
        3) Transverse – crosswise division into upper and lower portions
    d. Body cavities – contain viscera (organs)
        1) Ventral – divided by diaphragm into thoracic and abdominopelvic
        2) Dorsal – cranium contains brain; vertebral (spinal) column contains spinal cord
2. Structural Units
    a. The cell – basic structural and functional unit
        1) Nucleus – controls growth, repair, reproduction (by mitosis)
        2) Cytoplasm – responsible for cell metabolism; contains mitochondria, centrosomes, Golgi apparatus, endoplasmic reticulum
    b. Tissues – groups of like cells; four basic kinds
        1) Epithelial – covers surfaces and lines cavities; forms secreting cells of glands; types – simple and stratified squamous, cuboidal, columnar, transitional
        2) Connective – supports and connects (includes vascular tissue); types – loose ordinary, adipose, dense fibrous, cartilage, bone; cells – fibroblasts, macrophages, mast cells, fat cells, plasma cells, and white blood cells
        3) Muscle – discussed in Chapter 4
        4) Nerve – discussed in Chapter 5
    c. Membranes – sheets of epithelial and connective tissues
        1) Cutaneous – forms skin
        2) Mucous – lines passageways opening to exterior; secretes mucus
        3) Serous – lines closed cavities; two-layered (parietal and visceral)
        4) Fibrous – synovial, perichondrium, periosteum
        5) Fascia – superficial and deep
    d. Organs – combination of two or more tissues
    e. Systems – groups of organs to perform major body functions (digestion, respiration, etc.)
3. Anatomical Variations
    a. May be minor differences or more marked changes called anomalies
    b. Variations in cell growth
        1) Aplasia – an organ is absent
        2) Hypoplasia – smaller than normal
        3) Atrophy – decrease in size
        4) Anaplasia – basic cell alterations
4. The Life Substance and Its Properties
    a. Protoplasm – composed of specific amounts of many chemical elements which carry out the vital functions of the cell
    b. Properties or characteristics
        1) Metabolism – includes all physical and chemical changes
        2) Reproduction – essential property of living matter
        3) Irritability – reaction to surrounding stimuli
        4) Conductivity – lead stimulus from one point to another
        5) Contractility – reduce in length on stimulation

5. Homeostasis
    a. Homeostasis – state of balance by exact regulation of the interchange of substances between cell and its environment
    b. Cell permeability – selective
        1) Unique structure of cell membrane – fat and protein layers; pores
        2) Energy for transportation of substances – active and passive transport
        3) Ingestion of materials by phagocytosis and pinocytosis
6. Basic Physiological Processes
    a. Filtration – passage of water and dissolved substances across a membrane due to hydrostatic pressure
    b. Diffusion – tendency of a substance to distribute itself evenly throughout a space
    c. Osmosis – passage of water through a semipermeable membrane which separates two solutions of unequal concentration as a result of osmotic pressure

REVIEW QUESTIONS

1. Select the correct answer for each:
   The elbow is (proximal or distal) to the wrist.
   The heart is (cranial or caudal) to the stomach.
   The arms are (central or peripheral) to the trunk.
   The lungs are (medial or lateral) to the trachea.
2. Name the body cavity in which each of these organs is located: liver, esophagus, brain, lungs, pancreas, urinary bladder, spinal cord, rectum, small intestines, heart, trachea, stomach, spleen, kidneys.
3. Where in the cell are the chromosomes found? How many does each human cell have? What are mitochondria and where are they found?
4. What name would be given to one layer of cube-shaped epithelial cells? Make a drawing of a compound alveolar gland.
5. What are cilia and how do they function?
6. What are chondrocytes? Osteocytes?
7. What kind of membrane lines the mouth? The abdominal cavity? The knee joint?
8. True or false: Hyperplasia of skeletal muscles in athletes is a normal process. Explain.
9. What three chemical elements make up most of the protoplasm?
10. What is the most essential difference between living and nonliving material?
11. What is the difference between passive and active transport?
12. What is hemolysis? What would happen to red blood cells if they were put into a 5 per cent salt solution? Why?

ADDITIONAL READING

Bourne, G. H.: Division of Labor in Cells. 2nd ed. New York, Academic Press, 1970.
Brachet, J.: The living cell. Sci. Amer. (Sept.), 1961.
Caspersson, T. O.: Cell Growth and Cell Function. W. W. Norton & Co., Inc. 1950.
Holter, H.: How Things Get into Cells. Sci. Amer. (Sept.), 1961.
Langley, L. L.: Homeostasis. New York, Reinhold Publishing Corp. 1965.

Chapter 2

THE SKIN (INTEGUMENTARY SYSTEM)

INTRODUCTION

By definition, the skin or cutaneous membrane, along with its accessory structures (hair follicles, glands, etc.), may be considered a system—the integumentary system. As such, it can be described in terms of its structure and function, as can the other body systems. In addition, the skin possesses several features which make it a unique and versatile part of the body. The intact skin serves as the first line of the body's defense against invasion by foreign bodies and other harmful substances.

GENERAL STRUCTURE (Fig. 2-1)

The skin is composed of two main layers, the *epidermis* and *dermis*. The surface layer or epidermis consists of sheets of epithelial cells, the number depending upon whether the skin is "thin" or "thick." Deep to the epidermis, to which it is firmly attached, lies the dermis. Composed of connective tissue, the dermis contains most of the accessory structures or appendages of the skin.

EPIDERMIS

The outermost flattened cells of the epidermis are in fact dead cells, forming a layer of horny material called *keratin*. This acts as a protective covering, because if living cells were exposed directly to air they would dry out and hence not survive. The thickness of the keratin layer varies in different areas of the body.

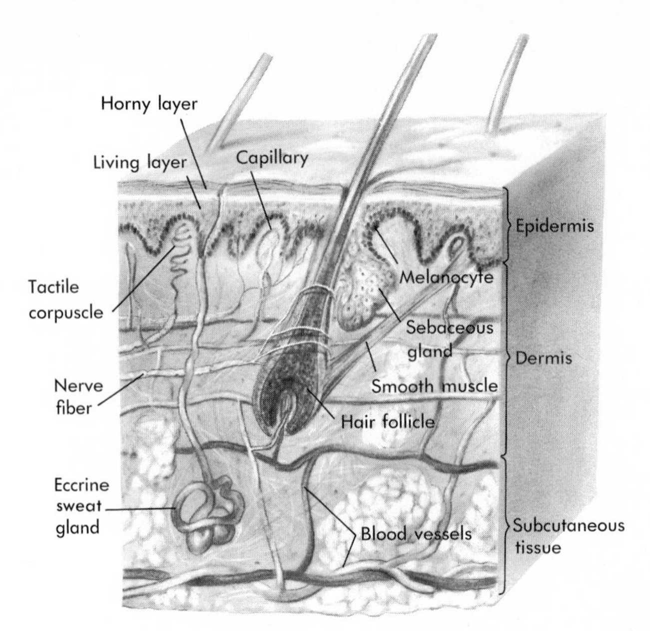

Figure 2–1. Idealized section of skin. The fat-rich subcutaneous layer supports the dermis, in which are blood vessels, smooth muscles and specialized glands. The epidermis is composed of two layers: a horny layer of dead keratin-filled cells, below which is a layer of living cells. Melanocytes, cells producing the pigment granules that determine skin color, are located at the base of the epidermis.

It will be remembered (Chapter 1) that epithelial cells have a great capacity to regenerate. Indeed, this feature assures a constant renewal of the epidermis, as new cells are constantly being formed in the deeper layers and pushed toward the surface. By this process, dead cells are transformed into keratin and are eventually sloughed off the surface of the skin (desquamation) as a result of exposure and abrasion. The regenerative powers of the epidermis are very important, since the many scrapes and scratches suffered by the average person are healed by this self-repairing mechanism. If one suffers a slight cut in the skin which does not bleed, it means that the lesion has not penetrated the dermis, since the epidermis, like all epithelium, is avascular.

Everyone is aware that the palmar surfaces of the fingers have distinctive patterns, no two people having the same pattern. Hence a person's fingerprints are distinctly his own and, after their formation during fetal life (fourth month), never change. In addition, patterns differ between the sexes, among the races, and indeed between a person's right and left hands. A closer examination of these patterns reveals that they consist of alternate ridges and furrows in the epidermis which follow the contours of the dermis beneath.

Pigmentation or skin color is caused by the presence of *melanin*, a pigment produced by special cells in the epidermis called *melanocytes*. The amount of melanin present determines the differences in skin color among the different races. The skins of all races contain some melanin; an individual in any race that has a congenital (present at birth) lack of melanocytes is called an *albino*.

Since the skin of the white race contains the least melanin pigment, blood in capillaries present in the outer part of the dermis shows through and imparts a pink tinge to the skin. Upon exposure to sunlight (or any ultraviolet light), the melanin content of white skin increases, thus producing a darker color or "tan." In some people, tiny patches of melanin appear in the skin as freckles. Besides imparting color to the skin, melanin acts protectively when there is excessive exposure to ultraviolet light. It is known that constant overexposure of the skin to such light may have a carcinogenic (cancer-producing) effect by causing as yet unknown abnormal changes in the skin cells.

DERMIS

The bulk of the skin is composed of the dermis. It consists mainly of irregularly arranged fibrous connective tissue as well as blood vessels, nerve endings, hair follicles, and glands. The dermis lies on a layer of delicate connective tissue containing various amounts of fat; this is the *subcutaneous tissue* or *superficial fascia.* If one picks up a fold of skin between the fingers and then releases it, it will return to its normal position. This elastic property, which allows the skin to mold itself over all body parts, is imparted by elastic fibers in the dermis.

THIN SKIN AND THICK SKIN

The term *thin skin* refers to the thickness of the epidermis, which varies in different areas of the body. Thickness ranges from that of the epidermis covering the eyelids, which is the thinnest, to that covering the back, which is the thickest. All body surfaces except the palms of the hands and soles of the feet are covered by thin skin. Some of the epidermal layers present in thick skin are absent from or are much thinner in thin skin. Hair follicles, absent in thick skin, are present in thin skin, as are sweat glands, although the latter are more numerous in thick skin. Thin skin also differs from thick in that its surface does not present ridges or furrows. An unusually thick layer of keratin covers thick skin.

SKIN APPENDAGES

Skin appendages include hair follicles, nails, sebaceous (oil) glands, and sweat glands. While most of these structures are found in the dermis, they actually develop from epidermal cells which then invade the dermis. There are no sebaceous glands in thick skin.

Hair and Nails. All parts of the body covered by thin skin contain hair follicles, which are epidermal downgrowths into the dermis developed during fetal life. Shortly before birth, the fetus becomes covered by *lanugo,* a coat of very soft, fine hair. A few months after birth, these hairs are replaced by a new growth of coarser hair.

Although man has about the same number of hair follicles as other primates, in most areas of the body they lack color and do not grow much. Thus man does not develop a furry coat all over his body. However, the presence of these hair follicles is extremely important in the repair of damaged epidermis, which will be discussed later.

Soft keratin covers the surface of the epidermis, whereas the nails and hair are composed of hard keratin. Although basically similar, there is a minor chemical difference between the two types. Hard keratin is more permanent; it does not slough off, so hair and nails must be cut from time to time. Hair growth is a continuous process, but periods of active growth alternate with periods of less activity. Contrary to popular opinion, shaving or cutting hair does not accelerate its growth rate. This is logical, since it is difficult to see how cutting dead, keratinized hair-ends could influence the activity of the germinal cells, which are located either deep to or in the dermis.

Most hair follicles are slanted away from the perpendicular. A small bundle of smooth muscle fibers, the *arrector pili*, is attached to the sheath of the follicle and reaches upward to the dermis. These muscles are innervated by the sympathetic nervous system (Chapter 4) and, on contraction due to cold temperatures or emotions, cause the hairs to "stand on end" and produce "goose bumps."

Glands. *Sebaceous* or oil *glands* are located between the hair follicle and the arrector pili muscle. The ducts of most of these glands open into the follicle, although some open directly onto the skin surface. *Sebum*, a fatty substance produced by these glands, lubricates the skin and hair; whether it serves other functions is not clear. It is evident that sebaceous glands enlarge greatly during puberty, probably under the influence of androgenic (male) hormones. This sometimes gives rise to an unsightly skin condition known as *acne*.

Sweat glands are simple tubular in type and are especially numerous in thick skin. Some of their ducts open directly onto the skin surface (eccrine type), while others are associated with hair follicles (apocrine type). The former produce most of the sweat in humans, and their secretion of varying amounts of water helps to regulate body temperature. Dogs have eccrine sweat glands, and only on their paw pads, so their bodies are cooled by evaporation of water through panting.

FUNCTIONS OF SKIN

In addition to providing a barrier against harmful substances, the skin provides a nearly waterproof covering which prevents both drying of tissues and overabsorption of water when the body is immersed. Already noted is the fact that skin pigmentation helps to prevent damage from too much ultraviolet light. By conserving heat upon exposure to cold and by cooling the body by sweating when the ambient (air) temperature is high, the skin helps to regulate body temperature.

Nerve endings in the skin are responsible for the important sense of touch which conveys information from the external environment. Because of its remarkable regenerative powers, the skin is able to heal itself after suffering minor damage. Some chemical substances can penetrate the skin, and this property allows therapeutic compounds to be applied topically. On the other hand, some noxious substances such as heavy metals (e.g., lead) and other poisons also can be absorbed through the skin.

CLINICAL IMPORTANCE

Close scrutiny of the skin is an important aspect of any physical examination, as skin condition can provide the examiner with valuable information concerning general health. Signs worth noting include changes from normal color, the presence of blemishes or eruptions, lack of turger (elasticity), and excessive sweating, dryness, or oiliness. Such conditions may reflect poor nutrition, allergic conditions, the presence of infection, and the like. In addition, there are many diseases of the skin itself; these are considered under the medical specialty called *dermatology*.

Skin Grafting. Some skin wounds, such as burns, involve large areas of the body surface, and the normal self-healing properties of skin are unable to repair such extensive damage. In such cases, skin grafts may be used to aid in repair. The procedure involves transplanting skin flaps from another area of the body to the wounded area, the blood vessels of which will supply blood to the flap.

If the skin from the donor site consists mostly of only the epidermal layers, repair will take place chiefly as a result of growth of epithelial cells from the hair follicles located in the dermis of the damaged area. When the donor flap must consist of dermis as well as epidermis, the cut edges of the area must be sewn together, since no hair follicles remain to provide epithelial cells for repair.

Skin Care. Because of the skin's obvious importance to good health, proper care is essential for maintaining the integrity of this vital organ. Fortunately, in healthy individuals, this involves little more than periodic cleansing, proper diet, and attention to minor cuts and scratches.

In the case of ill persons, however, especially those who are bedridden for extended periods, additional measures must be taken to prevent skin breakdown. Constant pressure on skin areas from the mattress and bed clothing leads to a decrease in blood supply to those areas, resulting in a decrease in oxygen supply to the tissues. Tissue death (necrosis) may follow and sores ("bedsores") which are very difficult to heal may develop.

The aging process also takes its toll on the skin as well as on other organs. Loss of elasticity, decreases in sweat and oil secretions and the like all demand that extra attention be paid to maintain the skin in as healthy a condition as possible.

OUTLINE SUMMARY

1. General Structure of Skin
    a. Composed of two layers
        1) Epidermis—surface epithelial layer covered by soft keratin; contains melanin
        2) Dermis—underlying connective tissue layer; contains blood vessels, nerve endings, hair follicles, glands
    b. Thin skin—covers most of body; has hair follicles
    c. Thick skin—on palms of hands and soles of feet; no hair follicles
2. Skin Appendages
    a. Hair and nails
        1) Composed of hard keratin

        2) Hair follicles are epidermal downgrowths
        3) Arrector pili—smooth muscle attached to follicle
    b. Glands
        1) Sebaceous (oil) glands—lubricate skin and hair
        2) Sweat glands—open directly onto skin (eccrine) or with hair follicle (apocrine)
3. Functions of Skin
    a. Body defense
        1) Barrier to foreign substances
        2) Forms waterproof covering
        3) Pigment protects against ultraviolet light
        4) Helps to regulate body temperature
        5) Great regenerative powers help in wound healing
    b. Other functions
        1) Allows absorption of therapeutic as well as harmful substances
        2) Nerve endings for reception of external stimuli
4. Clinical Importance
    a. Skin examination provides information about general health
    b. Dermatology—specialty devoted to diseases of skin
    c. Skin grafting—used to repair extensive damage
    d. Skin care—especially important in bedridden and older persons

REVIEW QUESTIONS

1. Describe the differences between the epidermis and the dermis.
2. If you were asked to give a *subcutaneous* injection, where would you place the tip of the needle?
3. Why doesn't the epidermis bleed when it is scratched?
4. What is keratin?
5. Sketch a hair follicle, showing its relationships to the arrector pili muscle and glands.
6. What feature would you select as the most important property of the skin?
7. Why are hair follicles necessary in the replacement of large areas of epidermis?
8. Discuss the importance of good skin care, in both healthy and debilitated persons.

ADDITIONAL READING

Montagna, W.: Structure and Function of the Skin. New York, Academic Press, 1962.

Montagna, W., and Billingham, R. E.: Advances in the Biology of Skin. Vol. 5. Wound Healing, New York, Pergamon Press, 1964.

Chapter 3

THE SKELETAL SYSTEM

COMPOSITION AND FUNCTION

The human adult skeleton is composed of 206 bones which are the organs of the skeletal system. Besides forming the framework of the body, the skeleton serves as a means of attachment for the skeletal muscles. It also protects delicate structures such as the brain, heart, and lungs.

The *marrow* or internal soft tissue present in some bones is responsible for *erythropoiesis*, or the formation of red blood cells, as well as the formation of certain kinds of white blood cells. Erythropoietic marrow is found only in the cranial bones, the vertebrae, sternum, hip bone, and the upper ends of the thigh and arm bones. Bones also act as a storehouse for calcium and phosphorus, releasing these minerals to the blood when they are needed.

FORMATION AND DEVELOPMENT

Bone formation *(ossification)* begins about the eighth week of embryonic life. This first bone that forms is called *immature bone* and is similar to the bone tissue that takes part in the repair of fractures. It has relatively more cells and fibers and less mineral and cement substance than *mature bone* which makes up most of the adult skeleton. Almost all the immature bone formed during embryonic and fetal life is later replaced by mature bone.

Bones are first laid down as fibrous membrane structures which are later replaced by cartilage models before becoming ossified. However, the flat bones of the face, the cranial vault and part of the clavicle do not go through the cartilage stage; this is called *membranous* bone formation. Ossification of all bones is complete by about the twenty-fifth year of life.

TYPES OF BONES

Bones whose length is greater than their breadth are called *long bones*. Examples are the bones of the arms and legs. *Short bones* are more cube-shaped; they are found in the wrist and ankle. *Flat bones* have a layer of compact bone on each surface separated by a marrow space. These bones are found in the vault of the skull, and the marrow spaces are called the *diploë*. The sternum (breast bone), ribs, and shoulder blade are also flat bones.

Irregular bones are those of mixed shapes that do not fit into any of the previously mentioned types. Examples are the face bones, the spinal column, and the hip bones. Small bones, such as the kneecap, which develop in tendons are called *sesamoid bones*. *Wormian bones* are accessory bones found between the bones of the skull where their edges come together.

BONE MARKINGS

Bones have holes, projections, and depressions; these are all referred to as *markings*. Small, smooth, flat areas are called *facets*. A small, rounded projection is a *tubercle*; if large, it is called a *tuberosity*. A *spine* is a sharp process; other projections are called *ridges* or *crests*. A *foramen* is a hole or opening in a bone; if it has length, it is a *canal* or a *meatus*.

A depression may be called a *pit*, *groove*, or *fossa*. The rounded projections of bones are called *condyles* or *epicondyles*. Cavities inside bones, especially those of the skull, are called *sinuses*. The enlargement at the end of a bone beyond its con-

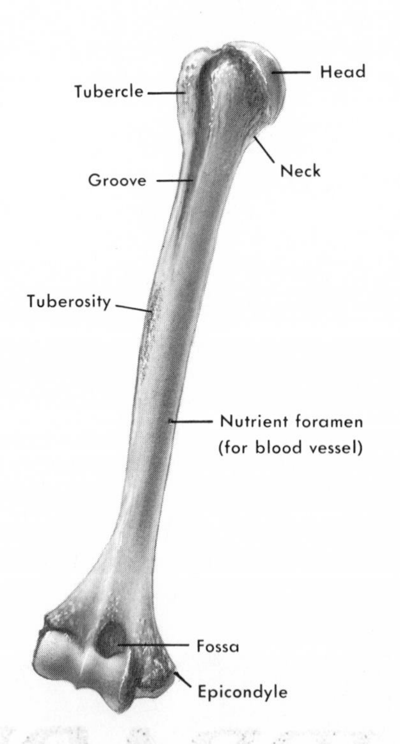

Figure 3–1. A typical long bone (humerus).

stricted part or *neck* is the *head* of the bone. Figure 3–1 shows a typical long bone with some of its markings.

DIVISIONS OF THE SKELETON

The 206 bones of the complete adult skeleton can be placed in two divisions, the *axial skeleton* and the *appendicular skeleton.* The axial skeleton, containing 80 bones, is composed of the skull, vertebrae, and thorax. The distribution of bones is as follows:

Skull	29 bones
Vertebrae	26 bones
Ribs and sternum	25 bones
Total	80

The appendicular skeleton has 126 bones and is composed of the extremities or appendages. The distribution is:

Upper extremities	64 bones
Lower extremities	62 bones
Total	126

Figure 3–2 shows the entire skeleton.

AXIAL SKELETON (80 BONES)

Skull (29 Bones). The skull can be divided into the *cranium* (eight bones) and the *face* (15 bones) plus the six ear ossicles ("little bones").

The cranium forms the bony box that encloses the brain. It is composed of four single bones (frontal, occipital, sphenoid, ethmoid) and two sets of paired bones (parietal, temporal). Figure 3–3 shows two views of the skull in which all but one of these bones (the ethmoid) can be seen.

The cranial bones are united by jagged, saw-toothed, immovable joints called *sutures.* The frontal bone is joined to the two parietal bones by the *coronal* suture; the two parietal bones meet each other at the *sagittal* suture. The occipital bone meets the parietal bones at the *lambdoidal* suture; the parietal bones are joined to the temporal bones by the *squamous* sutures.

At birth parts of the cranium have membrane instead of bone covering the brain tissue. This membrane enables the skull to be molded during the birth process and allows for its growth after birth. These unossified areas are called *fontanels* and are often known as the soft spots on a baby's skull (Fig. 3–4). There are six fontanels, and they are all usually closed by the second year of life.

The *frontal* bone forms the anterior region of the cranium, the forehead, and the roof of the bony orbit (eye socket). The two *parietal* bones form the upper sides and roof of the skull. The *occipital* bone forms the posterior portion and part of the base of the skull. There is a large hole in this bone called the *foramen magnum* through which the spinal cord becomes continuous with the brain.

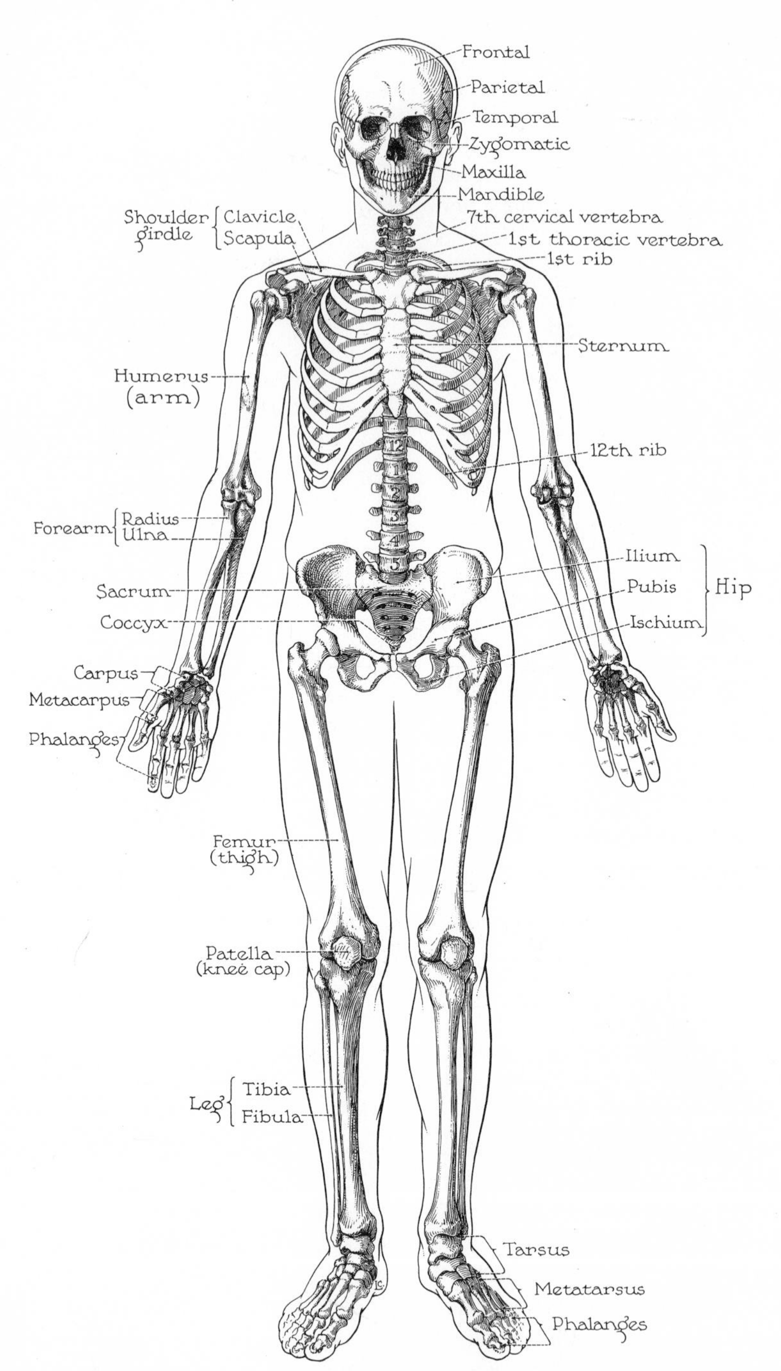

Figure 3–2. Anterior view of the human skeleton. (From King, B. G., and Showers, M. J. Human Anatomy and Physiology, Ed. 5. Philadelphia, W. B. Saunders Co., 1963.)

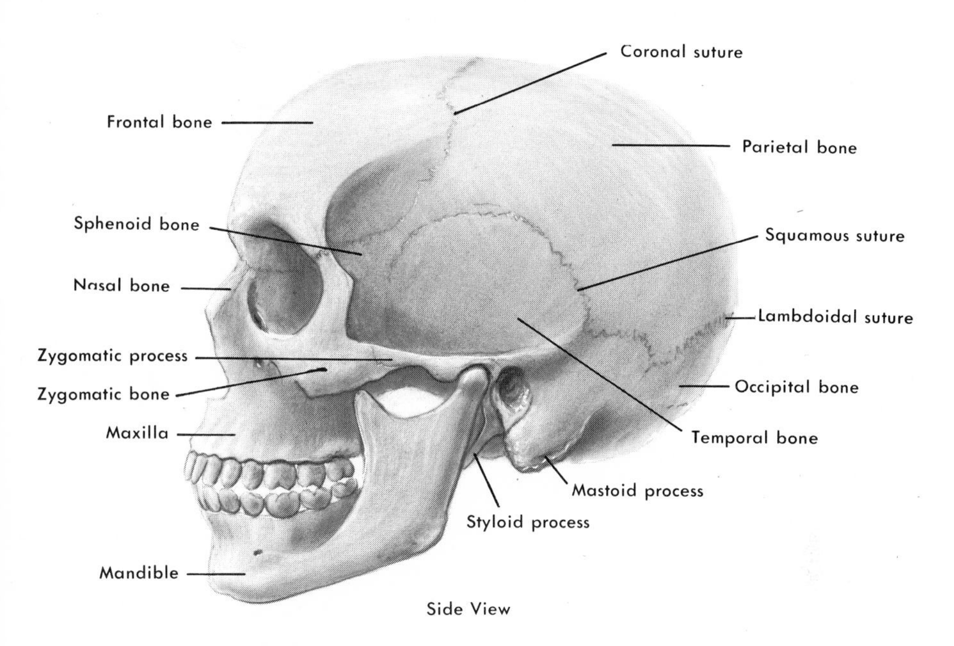

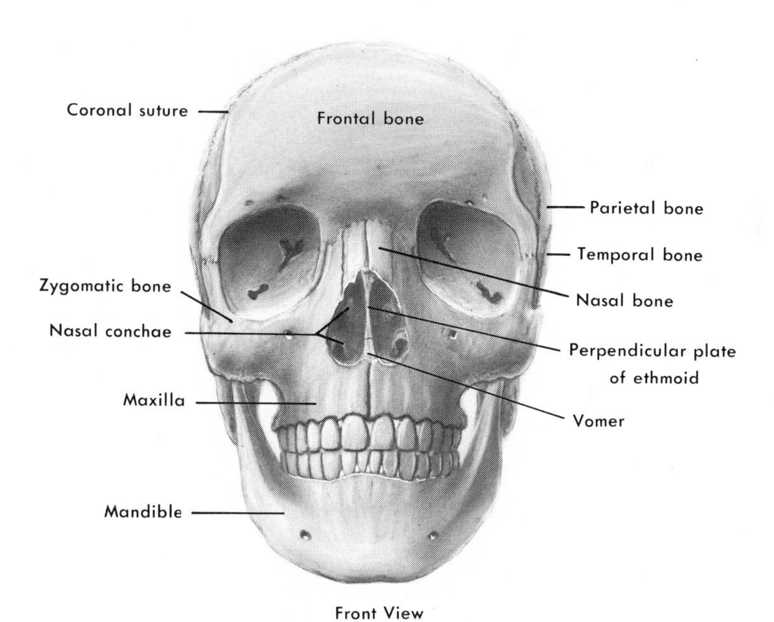

Figure 3-3. The skull.

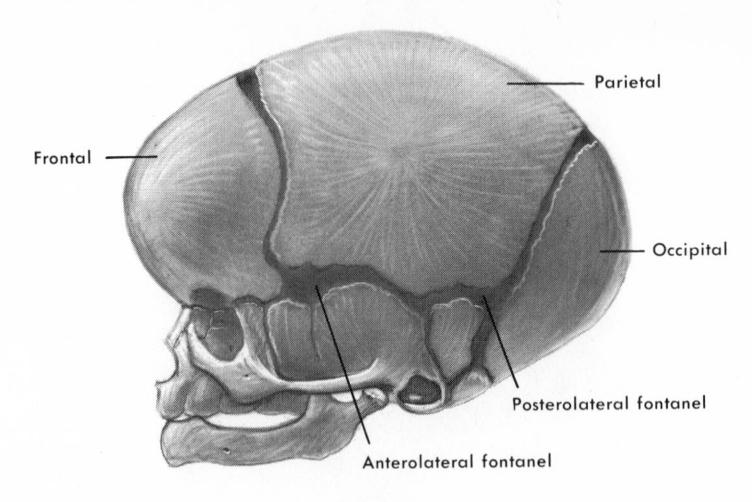

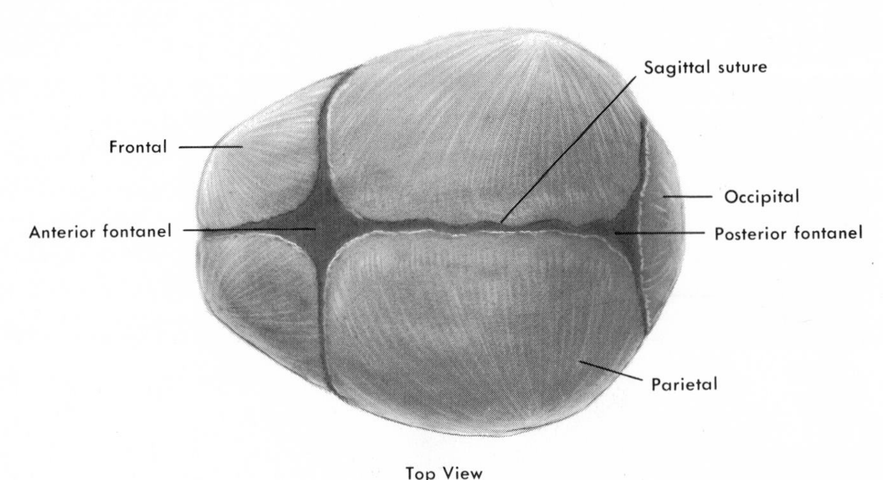

Figure 3–4. The fetal skull.

The two *temporal* bones form the lower sides of the skull (the ear region). Each of these bones houses the inner ear, the opening into which is called the *external auditory meatus*. Behind the meatus is the *mastoid process* which contains the *mastoid air sinuses*. A part of the temporal bone that arches out anteriorly is called the *zygomatic process*. This process joins with the zygomatic bone of the face to form the *zygomatic arch*.

The *ethmoid* is an irregular bone located above the nose. A side view of it can be seen in a sagittal section of the skull (Fig. 3–5). The thin, flat *perpendicular plate* of the ethmoid joins with a bone of the face, the *vomer*, to form the bony *nasal septum*. The sharp superior portion of the perpendicular plate is called the *crista galli*. At right angles to the perpendicular plate is the *cribriform* or *horizontal plate*

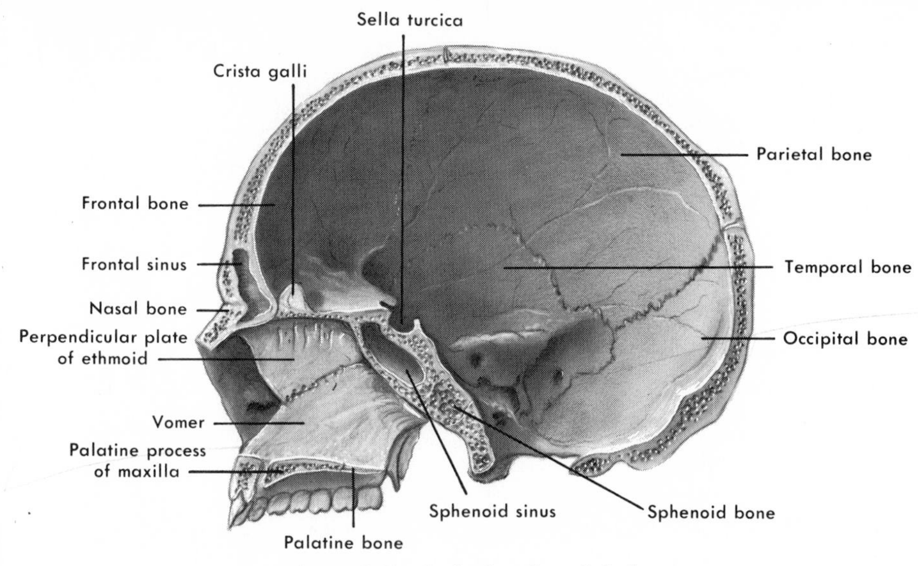

Figure 3–5. Sagittal section of skull.

which has perforations for the passage of nerves carrying the sense of smell from the nasal cavity to the brain. On either lateral wall of the nasal cavity are two scroll-shaped projections of the ethmoid, the *superior* and *middle nasal conchae* (Fig. 3–6).

The *sphenoid* is a bat-shaped bone forming part of the floor of the cranium (Fig. 3–7). It is composed of a body and two sets of *wings*, the greater and the lesser. Parts of the wings form parts of the walls of the bony orbits. On the superior surface of the body is a saddle-like fossa called the *sella turcica* (Turkish saddle) which houses the pituitary gland. The bat's "feet" project from the inferior

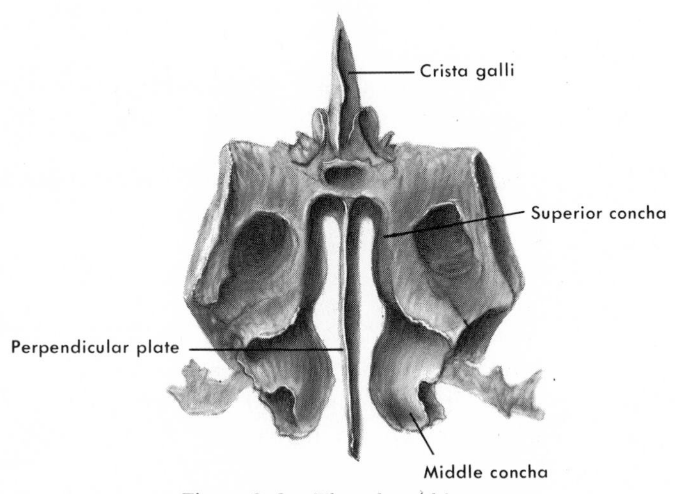

Figure 3–6. The ethmoid bone.

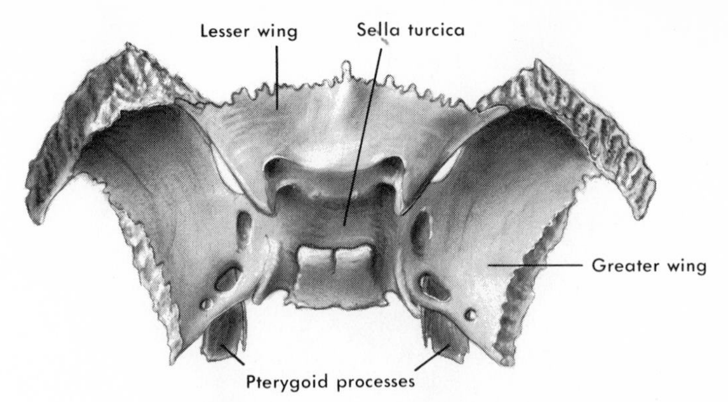

Figure 3–7. The sphenoid bone.

surface of the bone and are called the *pterygoid processes*. There are several small foramina in the sphenoid for the passage of nerves. Because it articulates (meets with) all the other cranial bones at some point, the sphenoid is called the "key" bone of the cranium (Fig. 3–8).

The bones of the face complete the skull. Of the 15 bones, three are single and six are paired:

Single: mandible, vomer, hyoid (strictly speaking, the hyoid is not a face bone, but it will be included here with them)
Paired: maxilla, zygomatic, lacrimal, nasal, inferior nasal concha, palatine

The *mandible* or lower jaw bone has a central portion or body which forms the chin. The upper border is called the *alveolar process* and contains sockets for the lower teeth. On either side of the body is an upward curved portion called the *ramus*. The rounded condyles of the rami articulate with the *mandibular fossae* of the temporal bones to form the hinged jaw joint. Note that the mandible is the only movable bone in the skull.

The *vomer* is a thin bone shaped like a plowshare. It has already been described as forming the lower part of the nasal septum. The *hyoid* bone is shaped like a **U** and lies in the neck just under the mandible. It does not articulate with any of the skull bones but is attached to the *styloid* processes of the temporal bones by ligaments.

The two *maxillae* are fused in the midline to form the upper jaw. The alveolar process contains sockets for the upper teeth. Each maxilla forms part of the floor of the orbit and the anterior part of the *hard palate*. Because the maxillae articulate at some point with each of the other face bones (except the mandible), they are known as the "key" bones of the face.

The *zygomatic bone* forms the prominence of the cheek. By joining with the zygomatic process of the temporal bone, it helps to form the zygomatic arch. The zygomatic bone has a broad connection with the maxilla and is also united with the frontal and sphenoid bones.

The *lacrimal* bones are thin and scale-like, resembling fingernails. They are situated in the anterior part of the medial wall of each orbit. Each has an opening which helps to form the *nasolacrimal* canal. Two small *nasal* bones are joined in the midline to form the bridge of the nose.

The *inferior nasal conchae* are two irregular scroll-shaped bones, each situated on the lateral wall of the nasal cavity. Note that these are similar to the superior and middle nasal conchae, which are parts of the ethmoid bone. The irregular **L**-shaped *palatine* bones form the posterior portion of the hard palate, part of the orbits, and part of the nasal cavities.

There are three tiny bones in the middle portion of each ear. These are called the *ear ossicles* and are named the malleus (hammer), incus (anvil), and stapes (stirrup).

There are four pairs of *paranasal sinuses* in the skull. These sinuses are spaces in bones that are lined with mucous membrane and communicate with the nasal cavities. They are located in the frontal, maxillary, sphenoid, and ethmoid bones. Their functions are to lighten the bones of the skull and to give resonance to the voice.

Vertebral Column (26 Bones). The bones of the vertebral column are superimposed on one another in a series which provides a flexible supporting column for the trunk and head (Fig. 3–9). The bones are separated by fibrocartilaginous *intervertebral discs*. The vertebrae show a uniform plan of structure with

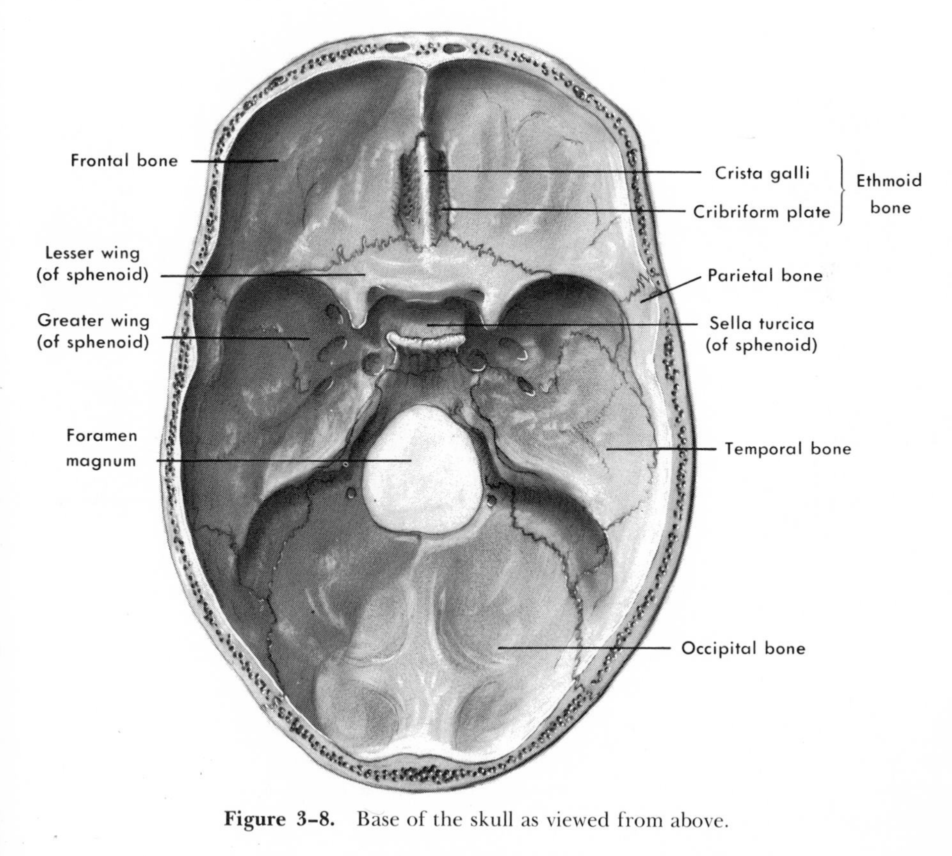

Figure 3–8. Base of the skull as viewed from above.

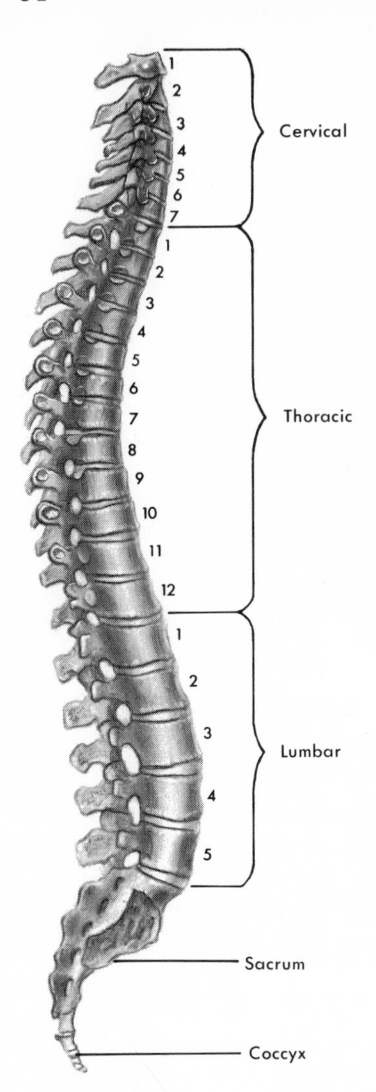

Figure 3–9. The vertebral column.

slight variations occurring in the different regions. With the exception of the first cervical vertebra, or *atlas*, all have a body composed of spongy bone, a *vertebral canal* or foramen for passage of the spinal cord, superior and inferior *articular processes, transverse processes,* and a *dorsal spine* (Fig. 3–10).

The vertebral foramen is formed by the *neural arch*, which is composed of two *laminae* and two *pedicles*. If during fetal life some of these parts fail to fuse, incomplete neural arches will result; and at birth the baby may have part of its spinal cord protruding out of the vertebral canal. Such a condition is called spina bifida.

Cervical Vertebrae. The seven cervical vertebrae are the smallest in the series and are located in the neck region. They may be recognized by the presence of a large, oval foramen in each transverse process and a short, forked dorsal spine.

The first and second cervical vertebrae are modified from this basic plan (Fig. 3–11). The atlas, or first cervical vertebra, which articulates with the occipital bone of the skull, has no body. Instead, the *dens* of the *axis*, or second cervical vertebra, represents the incorporation of the body of the atlas. The dens rises perpendicularly from the upper surface of the axis and forms a pivot around which the atlas, carrying the head, rotates. In addition, the atlas has no dorsal spine; it consists only of anterior and posterior arches and two lateral masses.

The seventh cervical vertebra is distinguished by a long, sharp dorsal spine

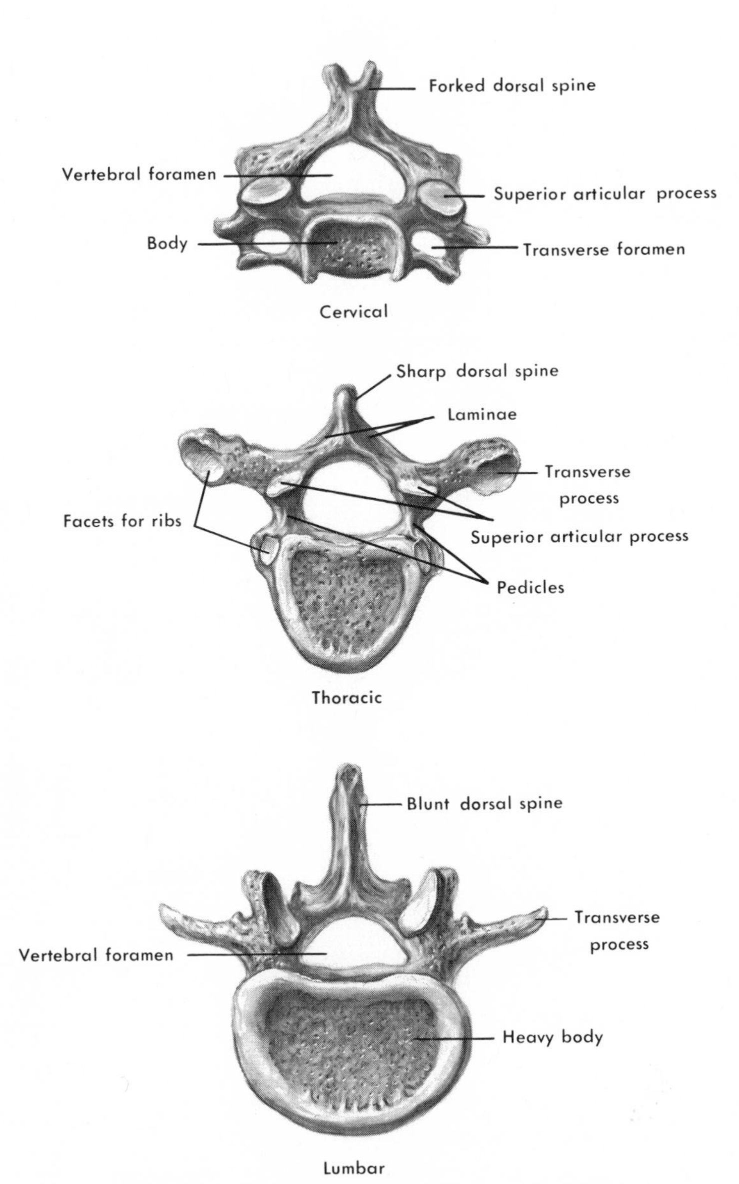

Figure 3–10. Typical vertebrae as viewed from above.

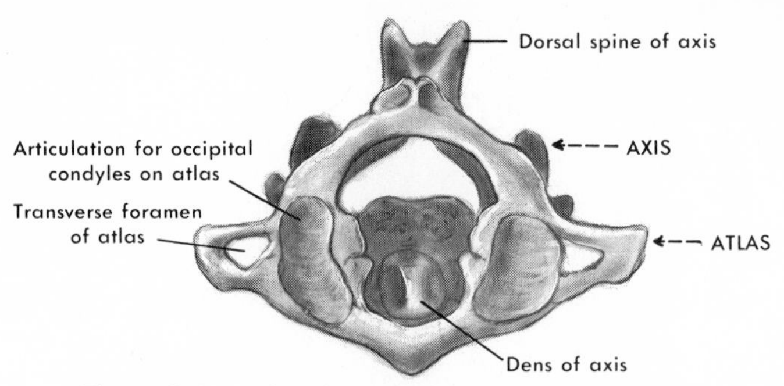

Figure 3–11. The atlas and axis as viewed from above.

and is referred to as *vertebra prominens.* By bending the head forward, one can feel this sharp spine at the base of the neck.

THORACIC VERTEBRAE. The 12 thoracic vertebrae show the least variation from the basic pattern. The bodies are intermediate in size between the smaller cervical and the larger lumbar. The dorsal spine is long and sharp and slopes downward. A characteristic feature is the presence of facets on the body and transverse processes for articulation with each of the ribs (Fig. 3–12).

LUMBAR VERTEBRAE. The five lumbar vertebrae have the largest and heaviest bodies in the series because the body weight supported by the vertebrae

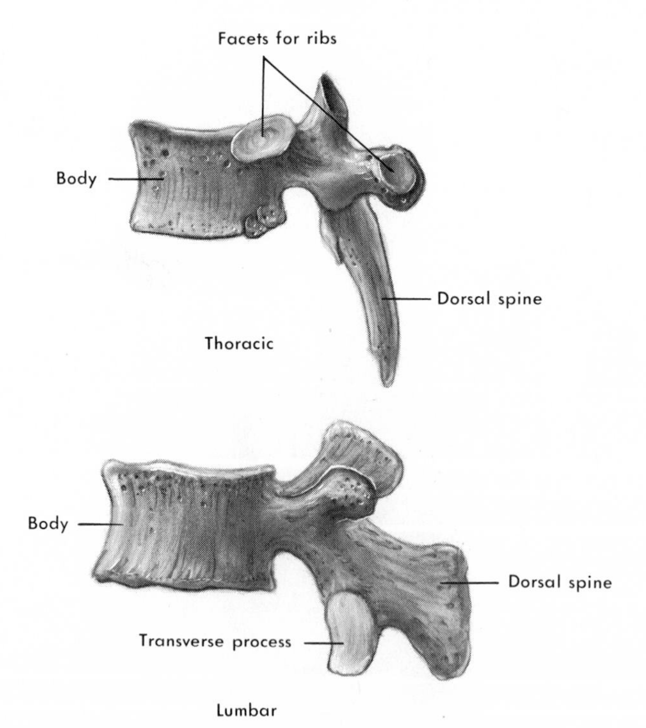

Figure 3–12. Thoracic and lumbar vertebrae – side view.

increases toward the lower end of the column. The dorsal spine is thick and blunt; the transverse processes are long and slender (Fig. 3–12).

SACRUM. The sacrum consists of the five fused sacral vertebrae (Fig. 3–13). It is triangular in shape and concave on its anterior surface. Horizontal ridges mark the separation of the original bones. Foramina at the ends of the ridges allow for the passage of nerves and blood vessels. The two lateral surfaces are smooth for articulation with the two iliac bones of the pelvic girdle.

COCCYX. The coccyx is usually a single bone, formed by the fusion of five separate bones. It has no pedicles, laminae, dorsal processes, or foramina.

The vertebral column as a whole has four curves in the adult: The cervical and lumbar curves are convex anteriorly; the thoracic and sacral curves are concave anteriorly. At birth only the thoracic and sacral curves are present. With the raising of the head and progression to an erect posture, the secondary cervical and lumbar curves develop.

Sternum and Ribs (25 Bones). The *ribs* are curved, narrow strips of bone. There are 12 pairs, each pair being attached posteriorly to a thoracic vertebra. Anteriorly, the first seven pairs attach directly to the sternum and are called the true ribs. Ribs 8, 9, and 10 are attached anteriorly by cartilage to the rib above and are termed false ribs. Ribs 11 and 12 are also false ribs and, because their anterior ends are free, are sometimes called "floating" ribs.

The *sternum* or breast bone is dagger-shaped and has three parts. The upper broad portion is called the *manubrium* (handle). The longer, narrow middle portion is the body. The smallest, most distal portion, the *xiphoid process*, usually does not ossify until adult life. The junction of the manubrium and the body marks the *sternal angle*, an important surface landmark of the chest.

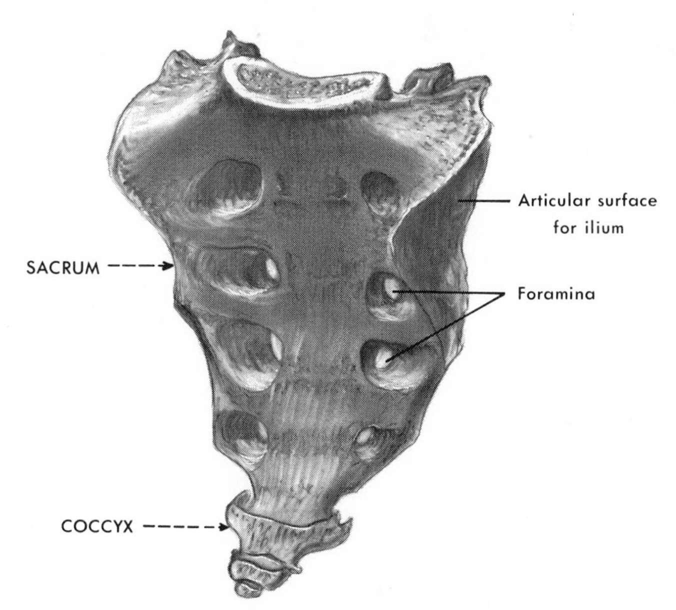

Figure 3–13. Sacrum and coccyx—anterior view.

APPENDICULAR SKELETON (126 BONES)

Upper Extremities (64 Bones). The bones of each upper extremity (Fig. 3–14) consist of the clavicle (collarbone), scapula (shoulder blade), humerus (upper arm bone), the radius and ulna (forearm bones), eight carpal bones (wrist), five metacarpals (the hand), and 14 phalanges (finger bones).

The *clavicle* props the shoulder out from the chest so that the arm can move freely. Its medial end is attached to the sternum, and its lateral end articulates with the acromion of the scapula. The bone has a double curve, giving it the general shape of an **S**. It consists of spongy bone covered by a layer of compact bone and has no marrow cavity. It is the first bone of the skeleton to begin ossifying.

The *scapula* lies against the posterior aspect of the rib cage, overlying ribs 2 to 7. Essentially a flat, triangular-shaped bone, the scapula has certain prominent processes: The *spine* is a large protrusion on the posterior surface; and its continuation is the *acromion*, which forms the point of the shoulder. The *coracoid* (beak) process is a thick, upward projection, which then bends in a horizontal direction. Below the acromion is the shallow *glenoid cavity*, which accepts the head of the humerus.

The *humerus* consists of a shaft and two ends. The rounded head articulates with the glenoid cavity of the scapula to form the shoulder joint. The *anatomical neck* is a slight indentation at the margin of the head. The area where the shaft begins to narrow directly below the head is designated as the *surgical neck*—fractures are more frequent at this point. The inferior end of the bone has articular surfaces for the forearm bones, and the lateral and medial *epicondyles* serve as attachments for muscles of the forearm.

On the thumb side of the forearm is the *radius*. It has two ends and a slender shaft. The *head* is a thick disc with an upper shallow surface for articulation with the humerus. An oval prominence, the *tuberosity*, is an important marking since it is the point of insertion for the tendon of the biceps muscle. The sharp *styloid process* at the distal end forms a margin for the tendons of two muscles to the thumb.

The *ulna* is the bone on the medial or little finger side of the forearm. It is more firmly connected to the humerus in the formation of the elbow joint than is the radius. Since the ulna only indirectly articulates with the wrist and hand, the radius makes the bony contact of the forearm at the wrist joint. A proximal projection of the ulna, the *olecranon*, forms the point of the elbow. The *coronoid process* projects from the front of the ulna, and between it and the olecranon is the *trochlear notch*. On the lateral side of the coronoid process is the *radial notch*, a shallow depression for articulation with the head of the radius. Note that the head of the ulna is at the distal end of the bone. The small, sharp *styloid process* can be palpated (felt with the fingers) along the ulnar border when the hand is pronated (palm down).

Eight small bones make up the *carpals* or wrist bones. They are arranged in two rows of four each and are held together by ligaments.

The five *metacarpal* bones form the skeleton of the hand. They are numbered one through five starting on the thumb side of the hand. The head or digital extremity of each bone has a smooth, rounded surface for articulation with the base of each proximal phalanx. The base or carpal extremity is cuboidal in form and articulates with the carpal bones.

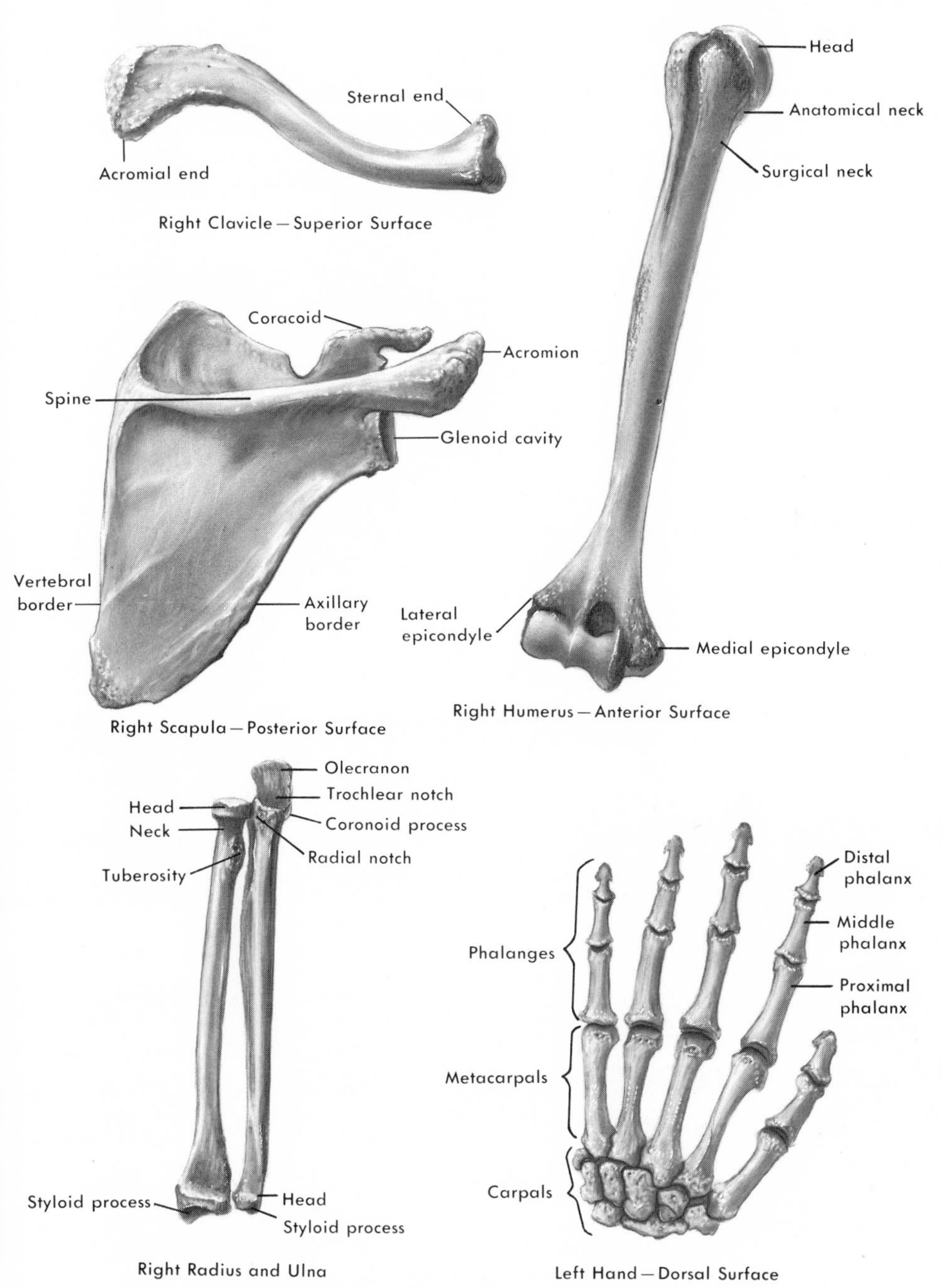

Figure 3–14. Bones of the upper extremity.

There are 14 *phalanges* or digits, three in each of the four fingers and two in the thumb. Each proximal phalanx articulates with the corresponding metacarpal bone of the hand.

Lower Extremities (62 Bones). Each lower extremity consists of an os coxae (hip bone, pelvic bone), the femur (thigh bone), two leg bones (the tibia and fibula), seven tarsal (ankle) bones, five metatarsal (foot) bones, 14 phalanges (toe bones), and one patella (kneecap).

The *os coxae* is a large, irregular bone that has been likened to a propeller with two oppositely bent blades (Fig. 3–15). The two hip bones are joined to each other anteriorly and to the sacrum posteriorly to form the bony pelvis (basin). Each os coxae is composed of three separate bones—the ilium, ischium, and pubis. These come together in the cup-shaped *acetabulum* and by adult life are completely joined.

The rim of the fan-shaped *ilium* is the *iliac crest*, the anterior tip of which forms the prominence of the hip. The *greater sciatic notch* is made into a foramen in the living body by ligaments and transmits blood vessels and nerves from the pelvis to the areas of the buttocks and backs of the thighs.

The *ischium* forms the posterior portion of the hip bone. Its large *tuberosities* receive the weight of the trunk in the sitting position.

The anterior part of the os coxae is formed by the *pubis.* Its main parts are two arms, or *rami.* The superior ramus of one pubic bone meets the superior ramus of the other in the midline to form the *symphysis pubis.* The *obturator foramen* is a large opening formed by the bodies and rami of the ischium and pubis. It transmits nerves and blood vessels to the thighs.

There are important differences between the male pelvis and the female pelvis (Fig. 3–16). Special features of the female pelvis are related to the child-bearing function: The pelvic outlet is enlarged by a wider subpubic angle than that in the male, everted (turned outward) ischial tuberosities, and a shallower symphysis pubis. In addition, the female pelvis has a shallower iliac fossa; a smaller, more triangular obturator foramen; and a wider greater sciatic notch. The lesser or *true pelvis* is shallower and wider in the female. The whole female pelvis is tilted forward, and the bones are thinner and more delicate.

The *femur* is the longest and strongest bone in the body (Fig. 3–17). The shaft is almost cylindrical, but its two ends are irregular. The smooth, round head is received into the acetabulum of the os coxae to form the hip joint. The greater and lesser *trochanters* are prominences on the upper end of the shaft to which muscles are attached. The lower end of the bone is broadened for articulation with the tibia at the knee. The *epicondyles* provide attachment for ligaments and tendons. On the anterior surface of the lower end, located between the two *condyles* (rounded processes seen on the posterior aspect of the femur), is the contact surface for the *patella* or kneecap (Fig. 3–18).

The *tibia,* larger of the two leg bones, articulates with the femur to form the knee joint. It is on the medial side of the leg, and its sharp anterior border, the shin, can be felt through the skin. The *medial malleolus* at the lower end forms the inner or medial prominence of the ankle.

The *fibula* is located on the lateral side of the leg. Its *lateral malleolus* completes the ankle joint on that side. A long, slender bone, it does not function in weight-bearing as does the tibia.

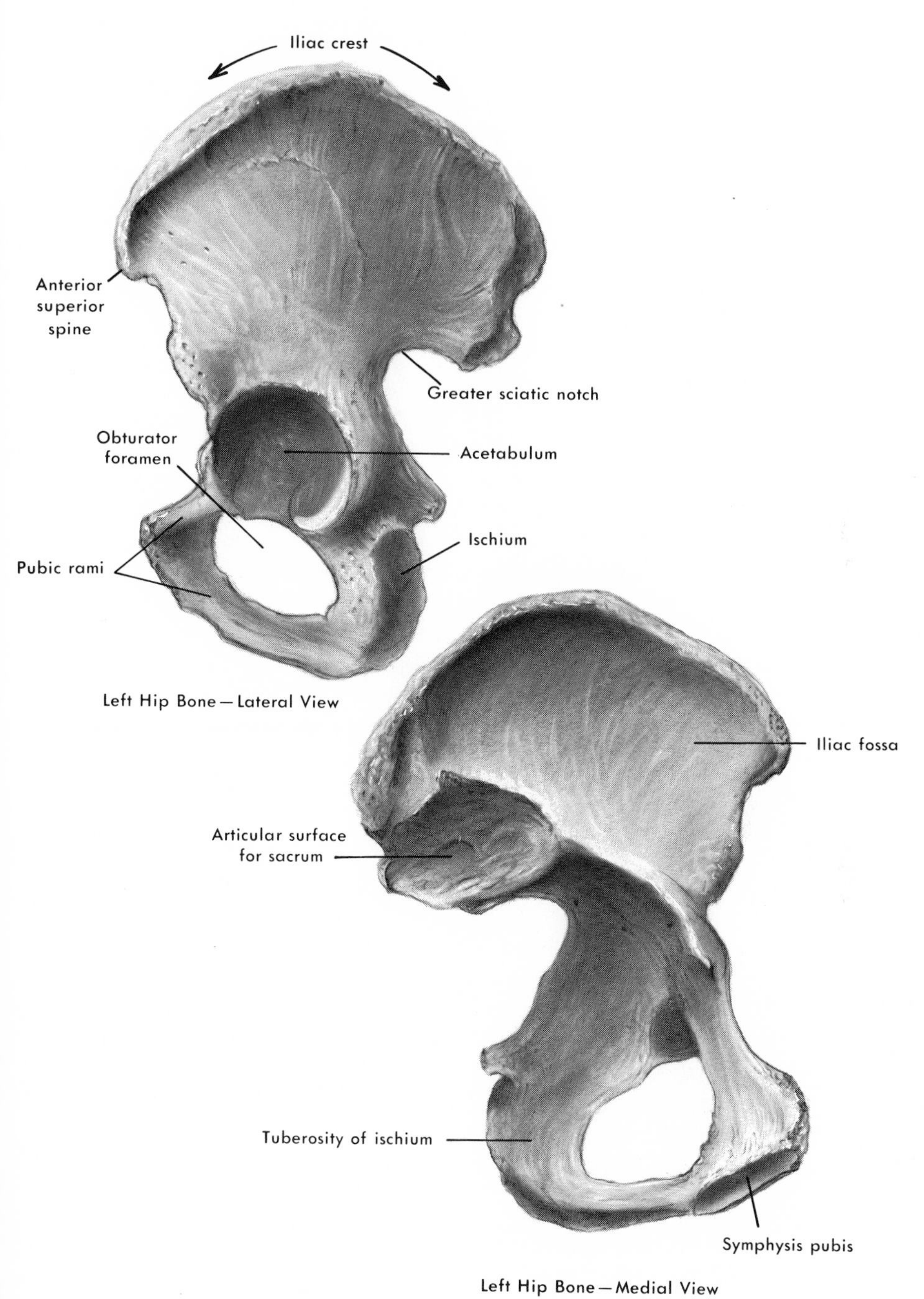

Figure 3–15. The os coxae (hip bone).

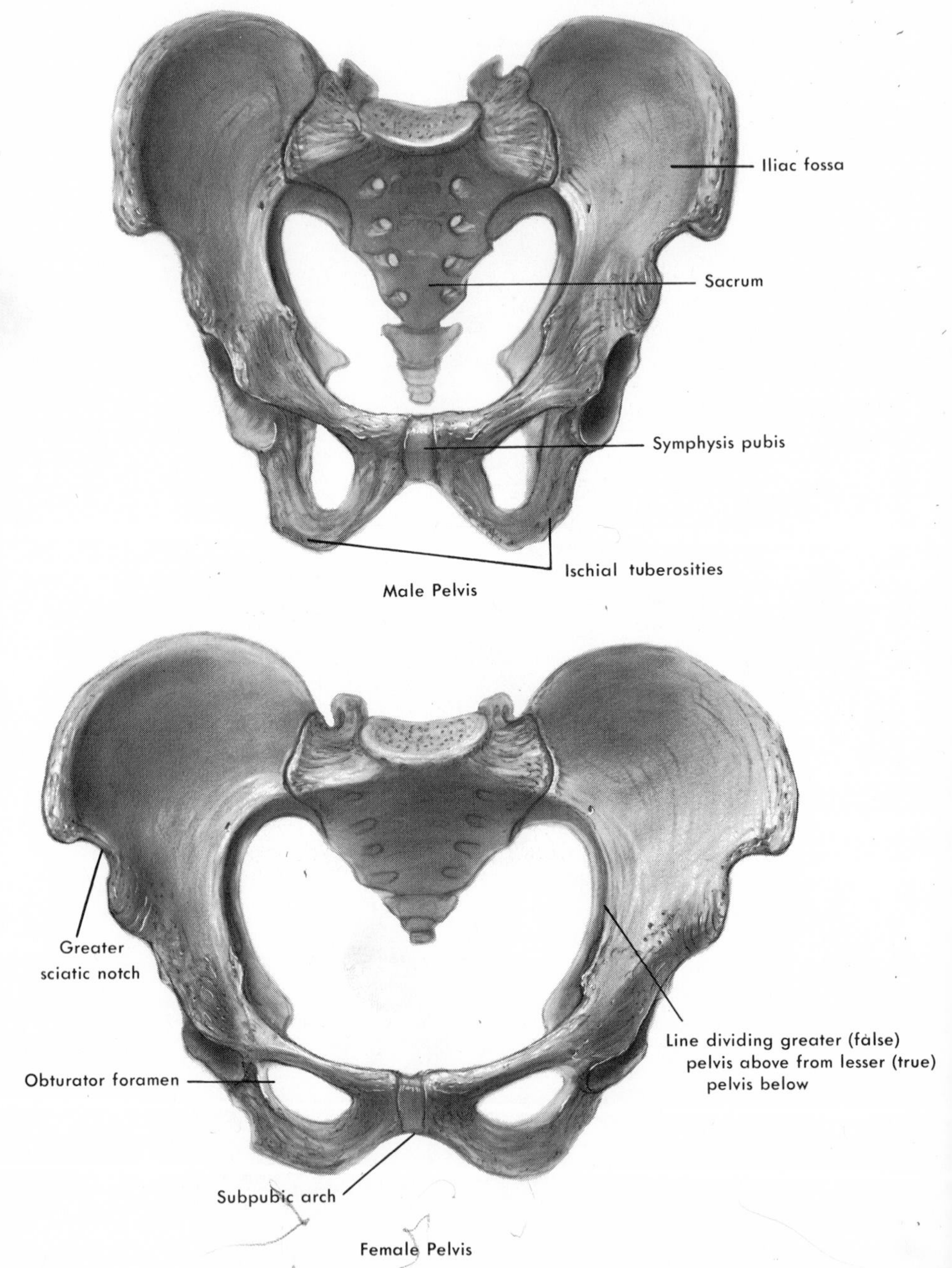

Figure 3–16. The bony pelvis as viewed from the front.

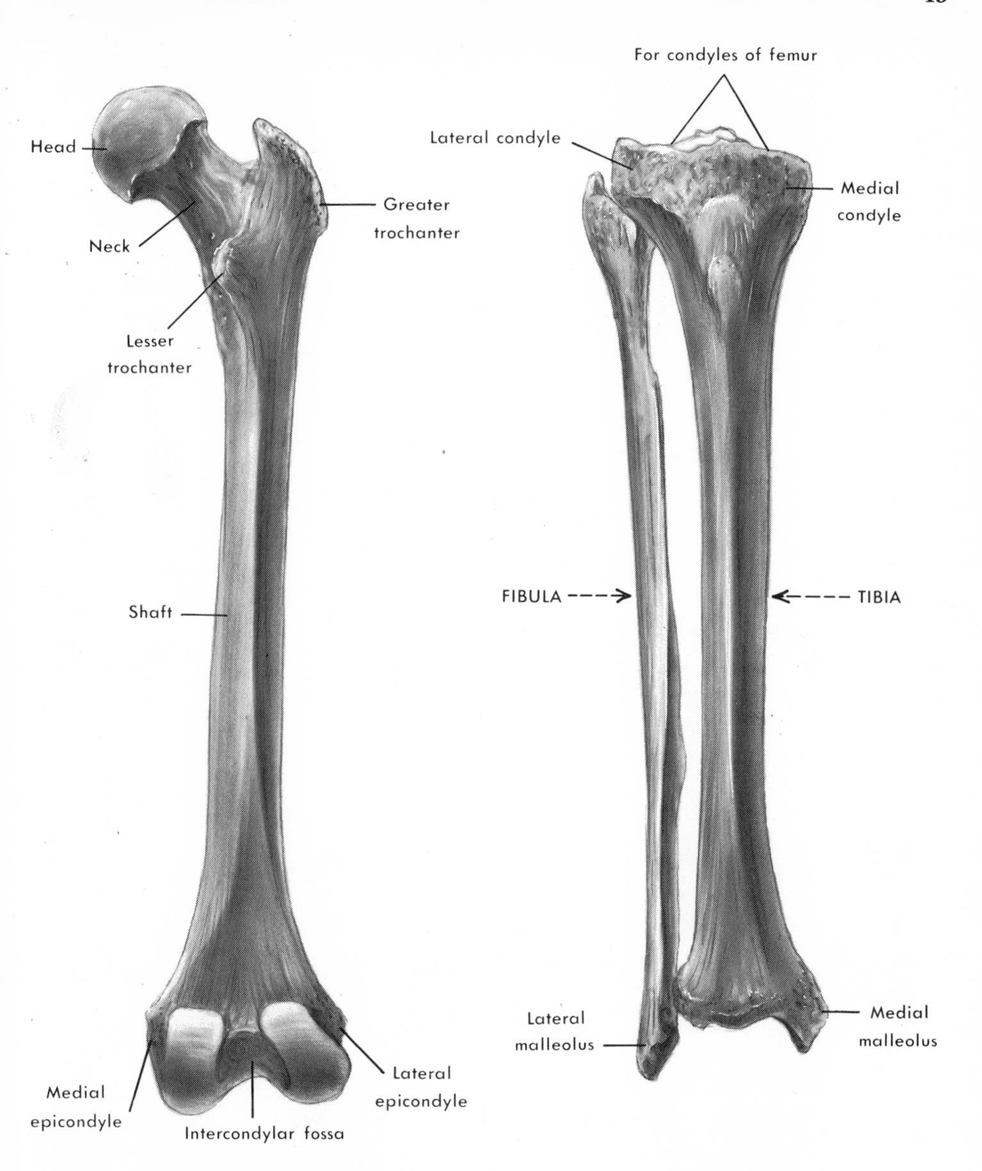

Figure 3–17. Thigh and leg bones.

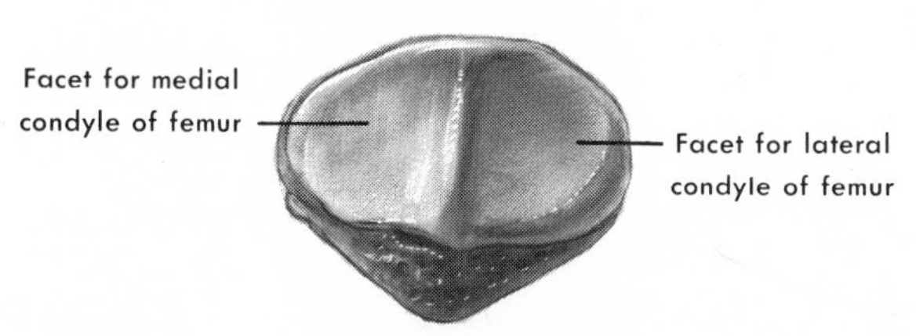

Figure 3–18. The patella (kneecap)—posterior surface.

The bones of the foot fundamentally resemble those of the hand (Fig. 3–19). There are seven *tarsals,* five *metatarsals,* and 14 *phalanges.* The *calcaneus* or heel bone is the largest and strongest bone of the foot. The superior surface of the *talus* articulates with the tibia and so bears the weight transmitted by that bone.

The foot bones are arranged not only for support of the body's weight but also for resilience and shock absorption. In order to perform these functions, it has an arched structure. The *transverse* or metatarsal arch is formed by the bases

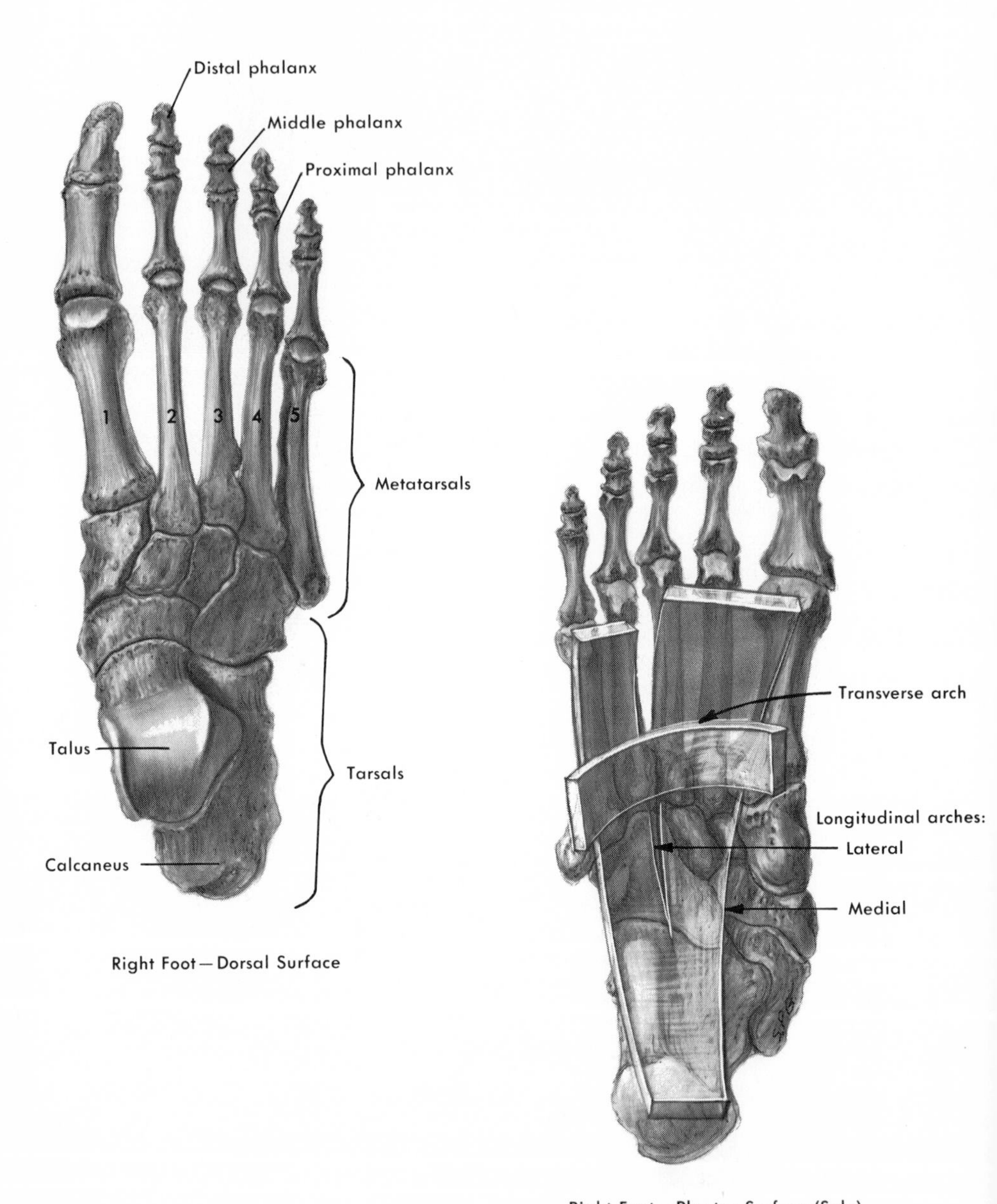

Figure 3–19. The foot.

of the metatarsals and the distal tarsal bones. In the *longitudinal* arches, the lateral arch is low and runs through the calcaneus posteriorly and through the fourth and fifth metatarsals anteriorly. The higher medial arch runs through the calcaneus, the talus, and the first three metatarsals.

ARTICULATIONS (JOINTS)

By definition a *joint* refers to the relationship between two bones at their point(s) of contact. The degree of movement allowed at a joint ranges from no freedom of movement to freely movable and is determined by the structural detail of the joint. There are three main kinds of joints classified according to structure: fibrous joints, cartilaginous joints, and synovial joints (Fig. 3–20).

FIBROUS JOINTS. There are two types of fibrous joints: sutures and syndesmoses. The junction of the flat bones of the skull are examples of sutures. Here the bones are joined by their interlocking saw-toothed edges. During youth there is fibrous tissue in these sutures, but with increasing age the bones progressively fuse together (called *synostosis*), and by old age the sutures become obliterated.

In syndesmosis the bones are joined by a membrane or a ligament of fibrous tissue. Examples of such joints are those between the parallel borders of the radius and ulna and between the fibula and tibia.

CARTILAGINOUS JOINTS. These joints are classified as either a synchondrosis or a symphysis. The synchondrosis is characterized by a cartilage plate separating the shaft (diaphysis) of a bone from the end (epiphysis) of the bone. The cartilage is progressively replaced by bone, a process that allows for the growth of the bone in length. Examples are the "growth" zones in the arm and leg bones.

Hyaline cartilage on the bony surfaces and the union of these surfaces by fibrous tissue characterize the symphysis articulation. This construction allows some movement of the joint, since the pads or discs can be compressed or displaced. Examples are the discs between the vertebral bodies and the joint between the pubic bones.

SYNOVIAL JOINTS. These joints are the most freely movable joints in the body. The ends of the bones have a thin layer of cartilage. A *joint cavity* is present and the entire joint is surrounded by a *capsule* of fibrous connective tissue. The cavity and capsule are lined with *synovial membrane,* which elaborates a clear fluid resembling egg albumin. This fluid lubricates the joint and nourishes the cartilage. Strong ligaments reinforce the joint.

The types of synovial joints are classified according to the degree of movement allowed. From the least movable to the most freely movable they are as follows:

1. *Gliding* or *sliding* joints exist between vertebral processes, the bones of the fingers and hand, and the toe and foot bones. Movement is restricted because of a tight fibrous capsule.

2. *Hinge* joints permit movements of flexion and extension only. Examples are the elbow, knee, and finger joints.

3. *Pivot* joints allow movement around a longitudinal axis and involve a rounded process which rotates in a bony fossa. The joints between the atlas and the axis and between the upper ends of the radius and the ulna are pivot joints.

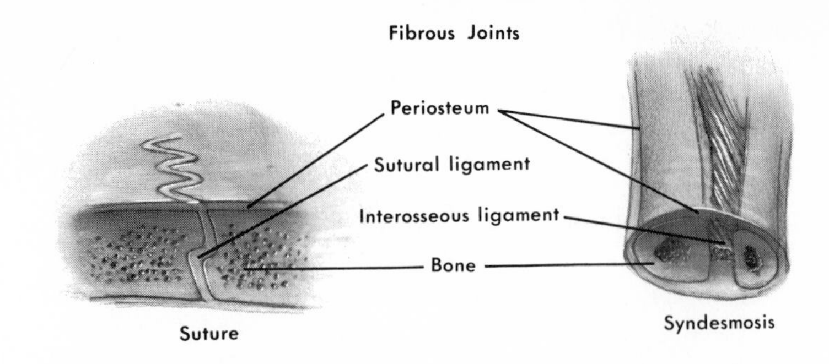

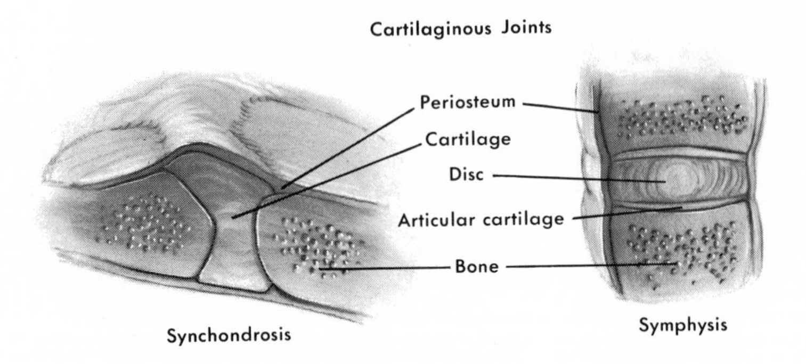

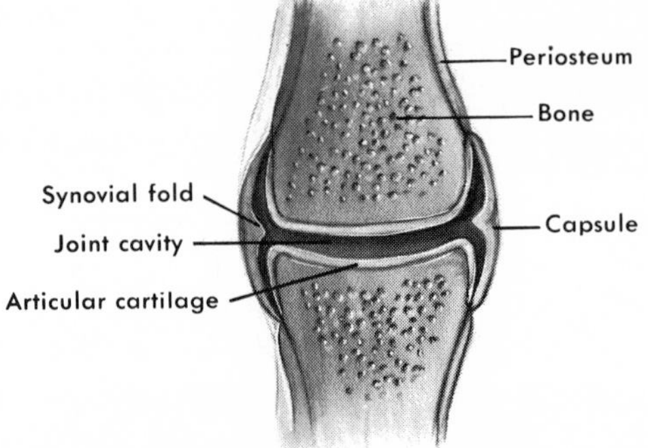

Figure 3–20. Types of articulations.

4. *Ball and socket* joints, the most freely movable in the body, allow all types of movement. The hip and shoulder joints are examples.

TYPES OF JOINT MOVEMENT

A joint movement may be described as flexion, extension, abduction, adduction, circumduction, or rotation (Fig. 3–21). *Flexion* is a movement which decreases the angle between two bones. The opposite of flexion is *extension,* which increases the angle between two bones. *Abduction* is movement away from the midline of the body; *adduction* is movement toward the midline.

Circumduction is movement in which the body part describes a circle; therefore it involves all four of the preceding movements—flexion, extension, abduction, and adduction. *Rotation* involves the turning of a part or an axis without dis-

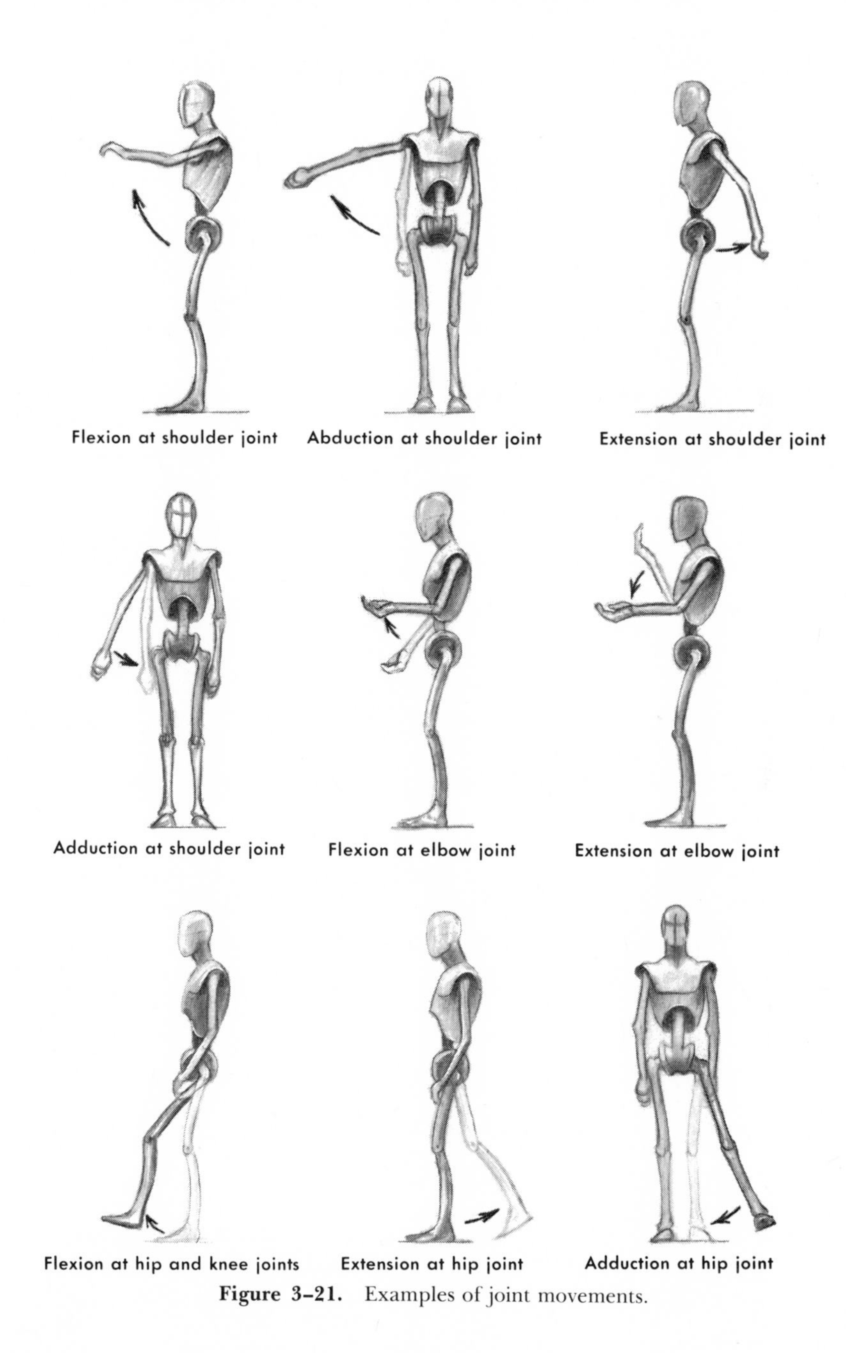

Figure 3–21. Examples of joint movements.

placement, such as turning the head from side to side. Because the head of the radius rotates around the ulna, the forearm and hand may be *supinated* (palm up) or *pronated* (palm down).

The axes of movement through which a joint moves parts of the body may also be used to classify the joint type. In gliding joints, movement is restricted to one plane; hence these are *plane* joints. Pivot and hinge joints allow movement around one axis only, so they may be termed *uniaxial* joints. *Biaxial* joints, such as the carpometacarpal joint of the thumb, permit movement around two axes. The freely movable ball and socket joints, involving movement around several axes, are *polyaxial* joints.

OUTLINE SUMMARY

1. Composition and Function
   a. Complete adult skeleton – 206 bones
   b. Framework of body
   c. Attachment for muscles
   d. Protection of delicate structures
   e. Formation of red blood cells (erythropoiesis)
2. Development and Types of Bones
   a. Formed in cartilage and fibrous membrane
   b. Types – long, short, flat, irregular
   c. Sesamoid bones – formed in tendon
   d. Wormian bones – accessory bones in sutures
3. Bone Markings
   a. Projection – tubercle, tuberosity, condyle, epicondyle, spine, ridge, crest
   b. Depressions – facet, pit, groove, fossa
   c. Hole – foramen
   d. Spaces – sinuses
      1) Four pairs of paranasal sinuses.
      2) One pair of mastoid air sinuses
4. Divisions of the Skeleton
   a. Axial skeleton – 80 bones (descriptions on pages 27–37)
      1) Skull (face, cranium, ear ossicles, hyoid) – 29 bones
      2) Vertebral column – 26 bones
      3) Thorax (sternum and ribs) – 25 bones
   b. Appendicular skeleton – 126 bones (descriptions on pages 38–45)
      1) Two upper extremities (shoulder girdle, arm, forearm, hand) – 64 bones
      2) Two lower extremities (pelvic girdle, thigh, leg, foot) – 62 bones
5. Articulations (Joints) – relationship between two bones
   a. Fibrous joints
      1) Sutures – between skull bones; immovable
      2) Syndesmosis – fibrous tissue; between leg bones and forearm bones
   b. Cartilaginous joints
      1) Synchondrosis – cartilage plate separates shaft and end of bone
      2) Symphysis – surface cartilage; union by fibrous tissue; between vertebral bodies and between pubic bones

   c. Synovial joints – most freely movable
      1) Characteristics – joint cavity, capsule, synovial membrane, reinforcing ligaments
      2) Types (from least to most movement) – gliding or sliding, hinge, pivot, ball and socket
6. Types of Joint Movement
   a. Flexion – decreasing angle between two bones
   b. Extension – increasing angle between two bones
   c. Abduction – moving away from midline
   d. Adduction – moving toward midline
   e. Circumduction – describes circle; includes flexion, extension, abduction, adduction
   f. Rotation – turning on axis without displacement
   g. Supination – turning palm of hand up
   h. Pronation – turning palm of hand down

REVIEW QUESTIONS

1. What is erythropoiesis? What bones of the skeleton participate in this process?
2. What is the difference between immature and mature bone?
3. What are sesamoid bones? Can you name one?
4. What are the soft spots on a baby's skull called? How many are there?
5. Name and locate the bones of the skull. What are sutures? Name and locate them.
6. What is the "key" bone of the cranium? Why is it called this?
7. What is the only movable bone in the skull? With what bones does it articulate?
8. What and where are the paranasal sinuses?
9. Parts of seven skull bones help to form the walls of the bony orbits. Name and locate them.
10. List at least two characteristics by which each of the regional vertebrae (cervical, thoracic, lumbar) can be distinguished from each other.
11. What is a neural arch?
12. Name the two primary curves of the vertebral column which are present at birth.
13. Where is the sternal angle?
14. Name and identify each of the bones of the upper extremity, pointing out the main markings of each bone.
15. What makes up the os coxae? What is the difference between the pelvic bone and the pelvis?
16. Describe some of the differences between the bony pelves of the male and female.
17. Name and identify each of the bones of the lower extremity and give the main markings of each.
18. Describe the arches of the foot.
19. Give at least one example in the body of each of the different kinds of joints.
20. Demonstrate on yourself each of the types of joint movement.

ADDITIONAL READING

Bassett, C. A. L.: Electrical effects in bone. Sci. Am. (Oct.), 1965.
Napier, J.: The evolution of the hand. Sci. Amer. (Dec.), 1962.
Rosse, C., and Clawson, D. K.: Introduction to the Musculoskeletal System. New York, Harper & Row, 1970.

Chapter 4

THE MUSCULAR SYSTEM

MUSCLE TISSUE

In discussing the properties of living material (Chapter 1), it was stated that the property of contractility, the ability to reduce in length upon stimulation, is most highly developed in muscle cells. This property, together with that of elasticity, enables these cells to shorten (contract) and then return to their original length (relax). Muscle tissue is responsible for all movement in the body.

There are three kinds of muscle tissue: skeletal, smooth, and cardiac (Fig. 4–1). Although all three types have features in common, each has additional characteristics which suit it for its special work in the body.

Skeletal Muscle. Sometimes referred to as *striated* or *voluntary* muscle, skeletal muscle is the largest category of muscle tissue in the body. The multinucleated fibers show alternating light and dark bands called *striations*. Striations result from the periodic spacing of the protein molecules that compose the fibers. Each fiber is considered one cell, and these cells vary in length from 1 to 80 mm. (0.04 to 3 inches). Several fibers are bound together in a bundle by connective tissue; this bundle is called a *fasciculus*. Many fasciculi are bound together into a larger bundle by fascia, and this bundle constitutes a skeletal muscle. Because these muscles are supplied by that part of the nervous system which is under the direct control of the will, they are said to be *voluntary*.

The skeletal muscles constitute about 40 per cent of the body by weight and are considered as one of the systems of the body. This chapter, then, will be a discussion of the skeletal muscle system.

Smooth Muscle. This type of muscle is also referred to as *nonstriated* or *visceral* muscle. It forms the muscle layers of the walls of hollow organs (vis-

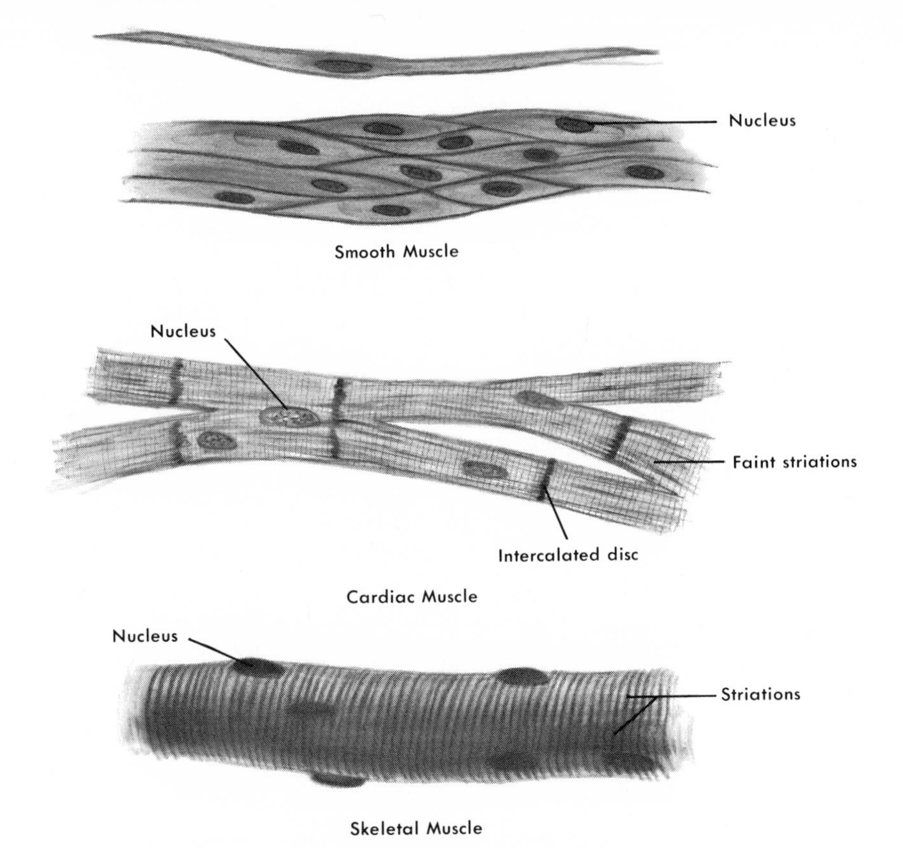

Figure 4–1. Kinds of muscle tissue.

cera) and of blood vessels. The cells, each containing only one nucleus, are spindle-shaped or *fusiform*. They are held closely together in flat sheets or layers by elastic and collagenous fibers. These cells do not show striations. Individual fibers vary in length from 40 to 100 microns (1 micron = 1/25,000 inch).

Smooth muscle contractions are slower and less powerful than those of skeletal muscle, but they are more sustained. Because its action is not under the direct control of the will, smooth muscle is said to be *involuntary*.

Cardiac Muscle. This tissue forms the bulk of the heart wall. Its fibers are about the same size and shape as those of skeletal muscle. Since the fibers appear to run together, it was thought that cardiac muscle formed a *syncytium* or continuous mass. It has been determined, however, that although the fibers do divide and interconnect, each is a separate cell. The cells are joined at their ends by specializations called *intercalated* discs. It is because of these specialized junctions that excitation spreads rapidly from cell to cell, and thus heart muscle does behave as though it were a true syncytium.

Faint cross-striations are present, and the nuclei are centrally located in the fibers. Cardiac muscle also is involuntary, but since it exhibits automatic rhythmic contraction, it is clear that its nerve supply serves to modify but not initiate contraction.

SKELETAL MUSCLES IN GENERAL

SIZE AND SHAPE

Muscles vary in size, shape, and arrangement of fibers. Those which allow considerable range of movement are long and narrow, whereas short, broad muscles are found where strength of movement over a shorter range is needed (Fig. 4–2).

In the simplest arrangement, a muscle may be composed of parallel fibers collected into a bundle and usually tapered at both ends. Such an arrangement is called a *fusiform* muscle. A more powerful type is the *pennate* muscle in which the fibers are arranged obliquely in a serial fashion to the point of insertion. The arrangement of fibers in a *bipennate* muscle may be likened to that of the insertion of the barbs of a feather into the central shaft. In a *circumpennate* muscle, fibers approach a central insertion from all directions.

ATTACHMENTS

Muscles may be attached to other muscles or to skin by connective tissue. Examples of these muscles are some of those which control facial expression. Many muscles are attached to the bones they move by a narrow strip of dense connective tissue called a *tendon*. Others are attached to each other or to bone by a broad, sheet-like piece of dense connective tissue called an *aponeurosis*. Still others may attach to cartilage, such as those in the voice box in the throat.

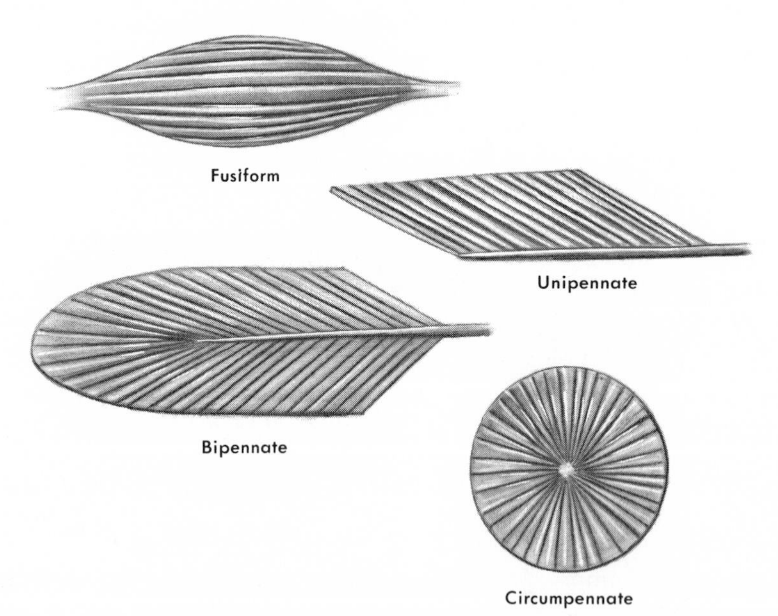

Figure 4–2. Architecture of skeletal muscle.

PARTS

It is important to remember that a muscle can produce movement only by pulling on a body part, never by pushing. The end of a muscle that remains more stationary during the action of pulling is called the *origin*. The opposite end, where the movement of the body part takes place, is the *insertion*. Although muscles can only pull, some of them can move either of their two ends, thus reversing origin and insertion designations.

ACTIONS

Muscles usually move a body part by pulling across a joint, as in bending the knee or elbow. (Review the different types of joint movements described in Chapter 3.)

Muscles usually do not act alone but in groups to effect a particular movement. Certain muscles in the group are responsible primarily for the action; these are called the *prime movers*. Other muscles, called *synergists*, assist the prime movers, often by stabilizing the joint involved. When one group of muscles is contracting to perform a particular action (such as flexion of the elbow), the muscle group which opposes this action (namely, the extensors of the elbow) are called the *antagonists*, and these must be relaxed.

This interaction of various muscles allows for smooth coördinated movements of body parts. By use of the example of extending the fingers, the result of this integrated action can be illustrated. The prime mover is a digital extensor muscle located in the forearm and attached to the dorsal aspects of the finger bones by tendons. Small muscles in the hand are synergists, which help in completely straightening the ends of the fingers. In addition, by helping to keep the wrist immobilized, three flexor muscles prevent it from being bent backward by the extensors. The main antagonists are the digital flexor muscles in the forearm, which must be relaxed.

NAMING OF MUSCLES

The name of a muscle often gives a clue to its action. For example, the flexor digitorum flexes the digits or fingers. For others, the name gives the location of the muscle, as in the case of the tibialis anterior, a muscle on the front part of the tibia. Some muscle names give both pieces of information: The flexor carpi ulnaris is a muscle which flexes the carpals and is located on the ulnar side of the forearm.

Muscles may be named according to the number of *heads of origin*. Examples are the biceps, a two-headed muscle, and the triceps, a three-headed muscle. The shape of a muscle may be indicated by a geometric name. An example is the trapezius, which is shaped like a trapezoid. Perhaps most informative are muscle names that give places of attachment in the body. One such muscle is the sternocleidomastoid, which is attached to the sternum, the clavicle, and the mastoid process of the temporal bone.

REPRESENTATIVE SKELETAL MUSCLES

It is beyond the scope of this book to describe in detail all of the over 400 muscles of the human body. Instead, to illustrate some of the general points about skeletal muscles which have been discussed, a representative list of 12 muscles, together with their location and principal actions, is given in Table 4–1. These muscles should be located in the body by studying Figures 4–3 through 4–9.

In the extremities, note that the bulk of a muscle is not located directly over the part of the body being moved by it. In order to achieve maximum range and leverage, most of the muscle mass will lie proximal to the joint over which it acts. For example, the biceps brachii, which flexes the supinated *forearm*, lies on the anterior surface of the *arm* (Fig. 4–10). Notice also that some muscles—such as the sartorius, quadriceps femoris, and hamstrings—span *two* joints, the hip and the knee.

MUSCLE PHYSIOLOGY

SOURCE OF STIMULUS FOR CONTRACTION

In order to contract, or shorten, skeletal muscle must be stimulated. In the body this stimulation is provided by motor nerve fibers coming from the volun-

(Text continues on page 62.)

Table 4-1. TWELVE REPRESENTATIVE SKELETAL MUSCLES

Name	*Location*	*Part(s) Moved*	*Principal Action(s)*
1. Trapezius	Back and neck	Shoulder	Squaring shoulders by adducting scapulae
2. Latissimus dorsi	Back	Upper arm	Extends humerus; draws arm down and back, rotating medially
3. Deltoid	Caps shoulder point	Upper arm	Main abductor of humerus
4. Pectoralis major	Chest	Upper arm	Adducts and flexes humerus
5. Biceps brachii	Upper arm	Forearm	Flexion and supination of forearm
6. Triceps brachii	Upper arm	Forearm	Extension of forearm
7. Gluteus maximus	Buttock	Thigh	Extension of thigh
8. Sartorius	Thigh	Thigh and leg	Flexion, abduction, lateral rotation of thigh; flexion of knee
9. Quadriceps femoris (rectus femoris and three vastus muscles)	Thigh	Thigh and leg	Extension of knee (all); flexion of thigh (rectus femoris)
10. Hamstrings (biceps femoris, semitendinosus, semimembranosus)	Thigh	Thigh and leg	Flexion of knee; extension of thigh
11. Tibialis anterior	Leg	Foot	Dorsiflexion of foot
12. Triceps surae (gastrocnemius and soleus)	Leg	Foot	Plantarflexion of foot (raises heel in walking)

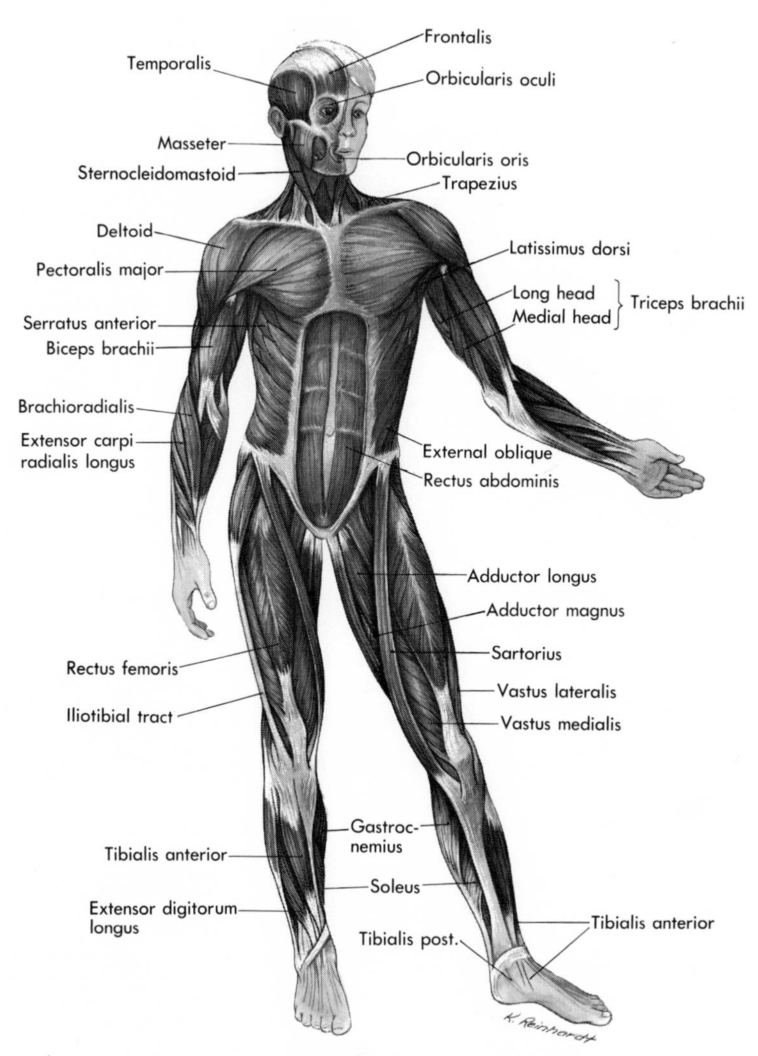

Figure 4–3. Muscles of the body – anterior view.

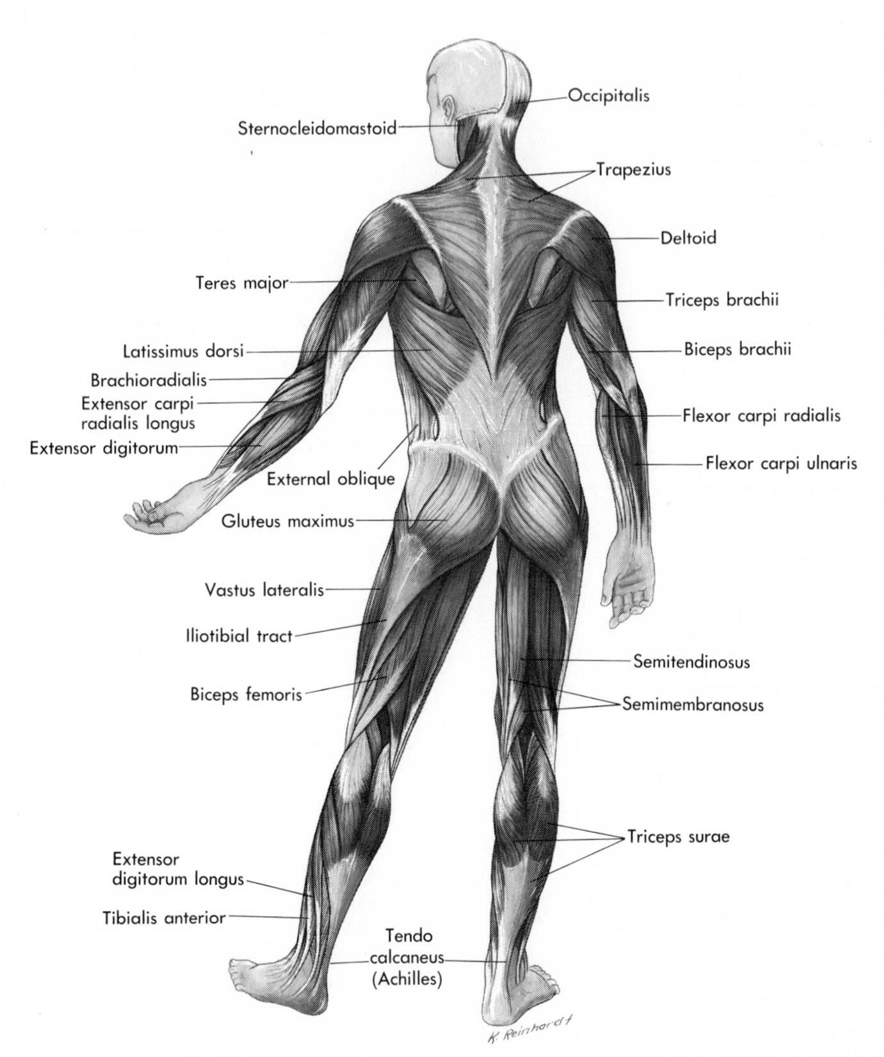

Figure 4-4. Muscles of the body – posterior view.

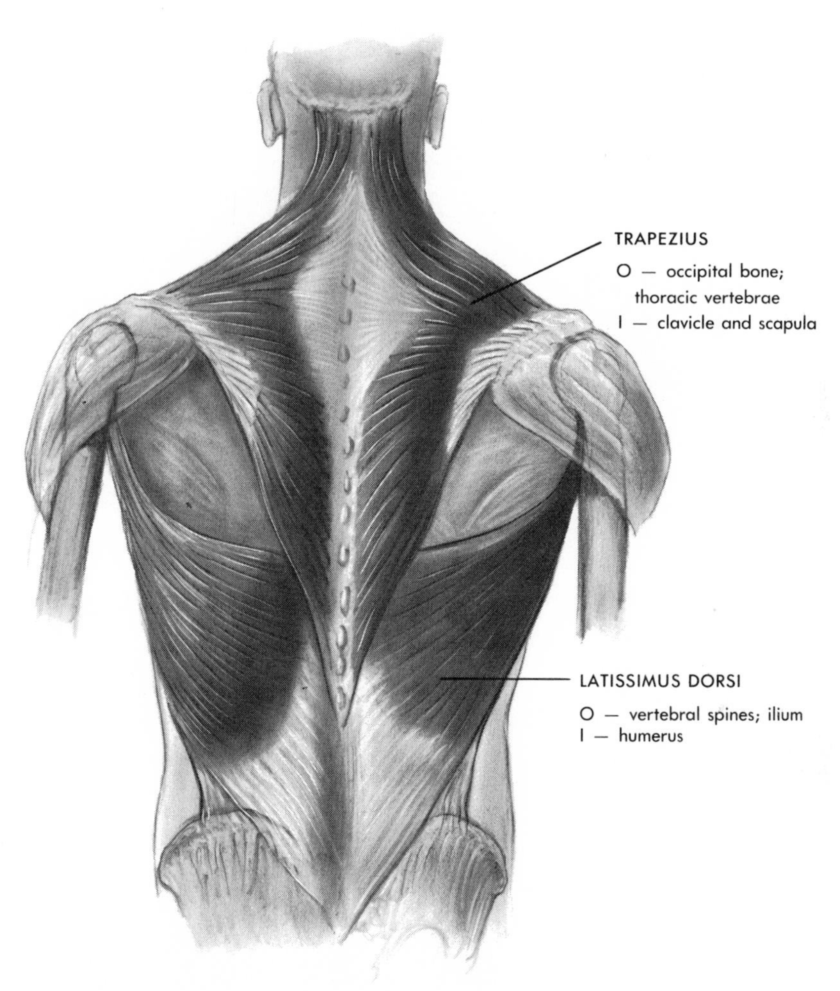

Figure 4–5. Trapezius and latissimus dorsi. O, origin; I, insertion.

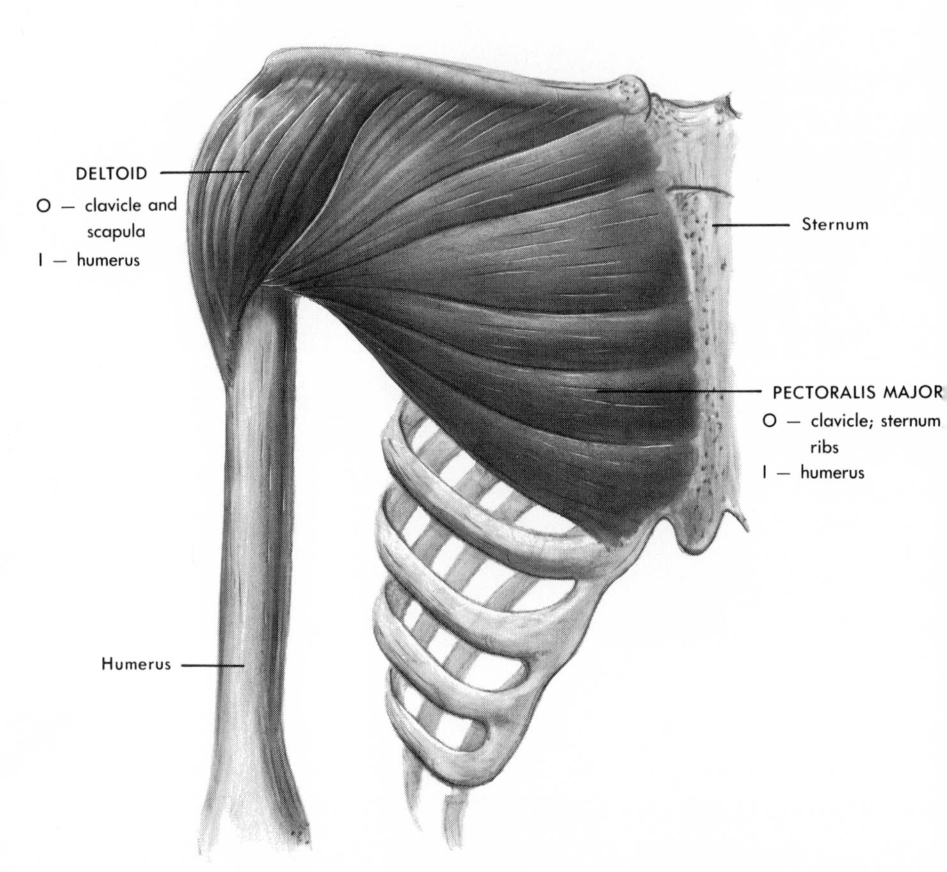

Figure 4–6. Deltoid and pectoralis major—anterior view.

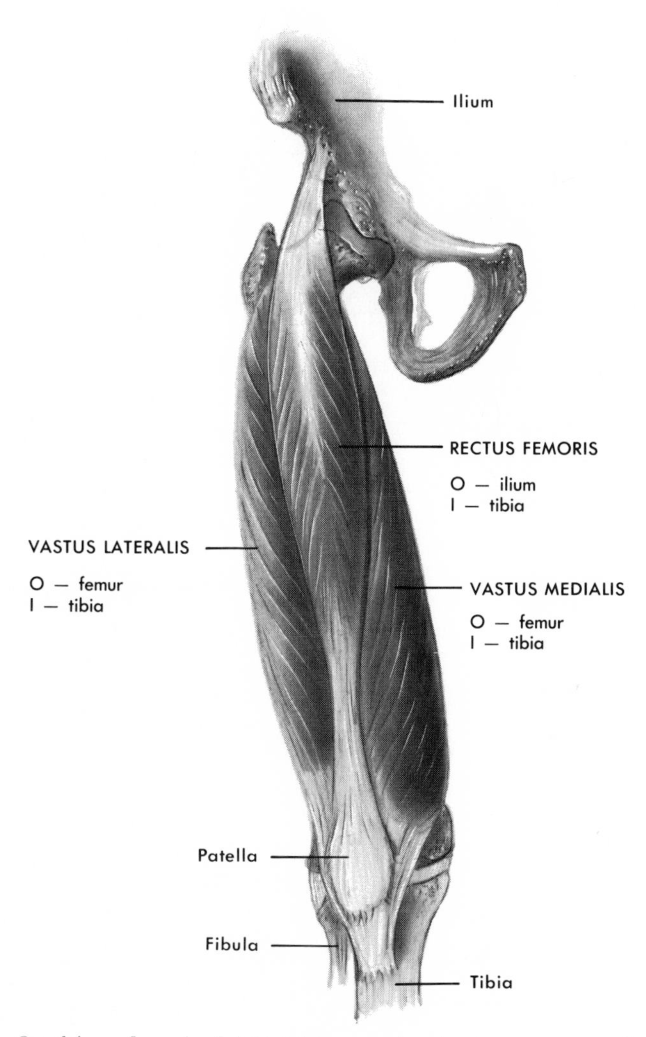

Figure 4–7. Quadriceps femoris of right thigh – anterior view (vastus intermedius not shown).

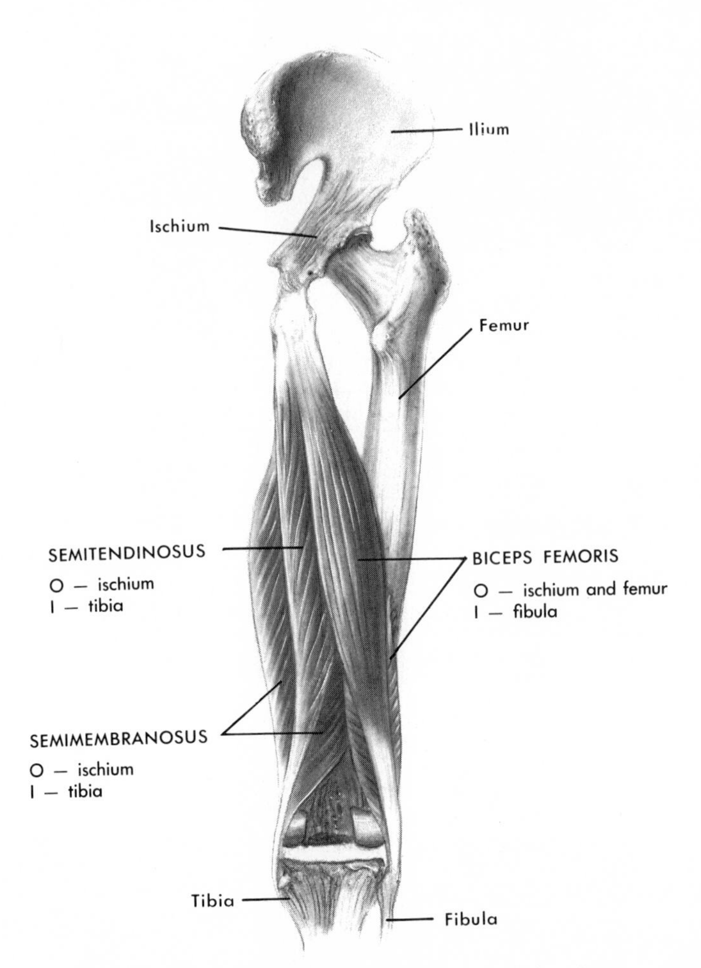

Figure 4–8. Hamstrings of right thigh—posterior view.

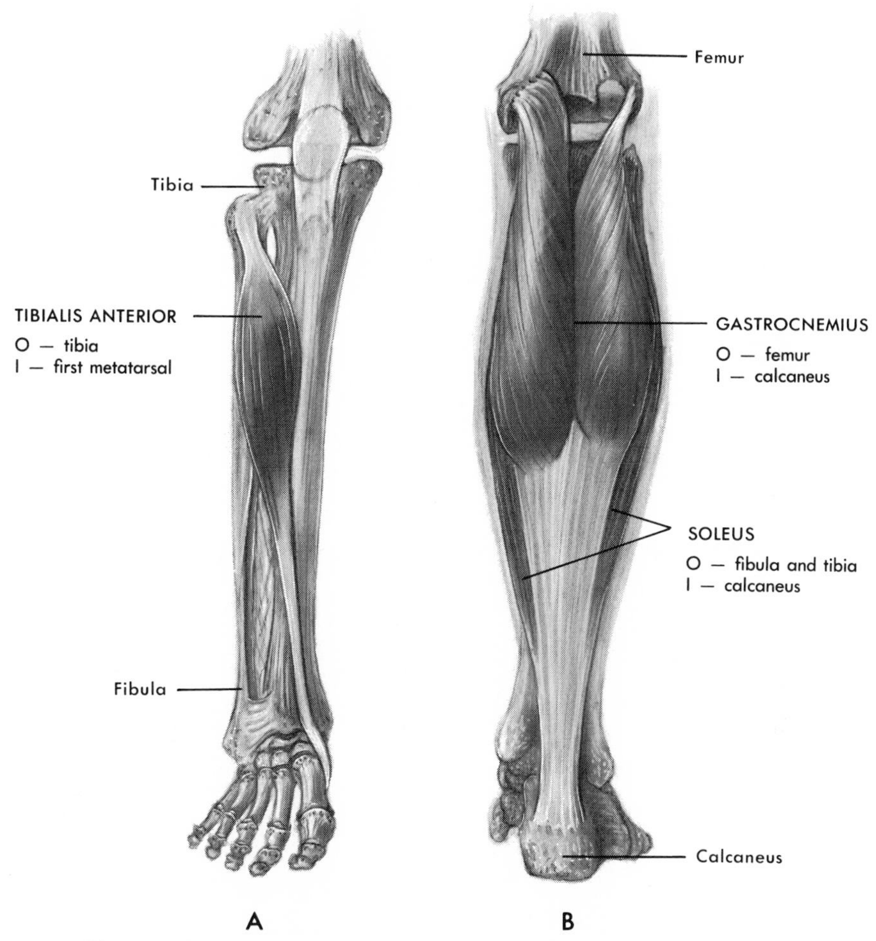

Figure 4–9. Muscles of the right leg. *A*, Anterior view; *B*, posterior view.

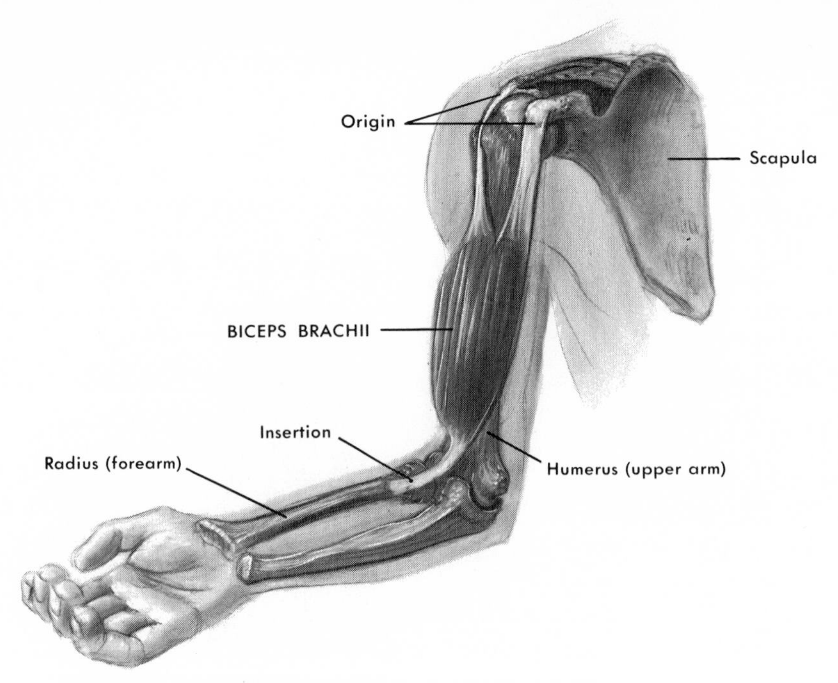

Figure 4–10. Biceps brachii muscle.

tary part of the central nervous system. A motor nerve fiber plus the muscle fibers it innervates is called a *motor unit.* One nerve fiber may supply from one to 200 muscle fibers, depending upon the type of work the muscle is called upon to perform—the more delicately coordinated and highly controlled the movement, the fewer the muscle fibers in each motor unit.

The point where the motor nerve fiber enters the muscle fiber (the neuromuscular junction) is called the *end plate.* A nerve impulse, or action potential, travels as a wave of excitation down the nerve fiber and is transmitted to the muscle, spreading over the fibers and causing them to contract.

MECHANICAL CHANGES IN MUSCLE

All-or-None Law. If a stimulus is strong enough to cause a skeletal muscle fiber to contract at all (a threshold stimulus), the fiber will contract to its fullest extent. This maximal contraction is what is meant by saying that individual skeletal muscle fibers obey the all-or-none law. The whole muscle, however, does not follow this law, since it will not contract maximally unless *all* of its motor units are stimulated by a threshold stimulus at the same time.

Simple Muscle Twitch. Several characteristics of muscle contraction can be studied by observing an electrically stimulated, isolated muscle preparation in the laboratory. When such a muscle is given a single stimulus of maximum strength, it produces a response called a simple muscle twitch. This response can be recorded graphically and shows that there are three phases or time periods involved (Fig. 4–11). The *latent* period is the time lag between the time the stimu-

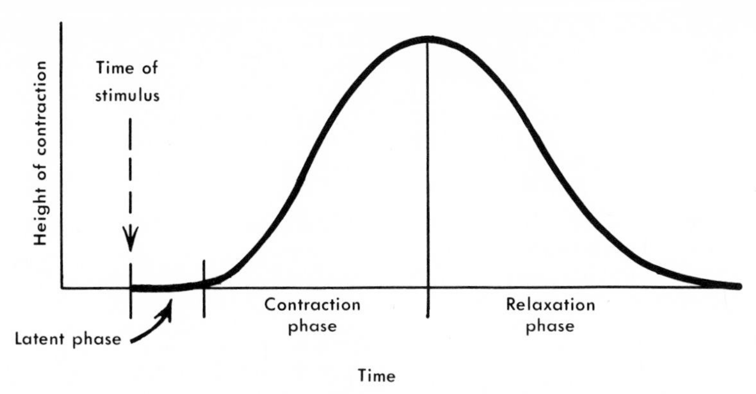

Figure 4–11. Tracing of a simple muscle twitch (isotonic contraction).

lus is applied to the muscle and the time it begins its *contraction* or shortening phase. During the *relaxation* period the muscle returns to its original length.

When the muscle actually shortens but the tension remains the same, the contraction is said to be *isotonic*. If the muscle is firmly attached at one end and the other end is fastened to a very strong torsion spring, it will not shorten when stimulated, but there will be a great increase in tension. This is called *isometric* contraction. In the body, muscles contract in both ways: In standing, a person tenses or isometrically contracts the quadriceps to keep the knee joint tight and the leg stiff. In lifting an object, however, the biceps muscle shortens, or contracts isotonically.

Summation and Tetanus. If several stimuli are sent rapidly into a muscle, they will be added together to produce a strong contraction. This addition of individual muscle twitches is called *summation* (Fig. 4–12). Summation occurs in two ways: (1) by increasing the *frequency* of nerve impulses coming down the nerve fiber, and (2) by increasing the *number* of motor units responding. This phenomenon does not violate the all-or-none law, as one might think. The law applies only when conditions of stimulation are kept constant, and in the case of summation, each succeeding stimulus is applied during the contraction phase of the previous stimulus. This means that the muscle fiber's length is less than when the

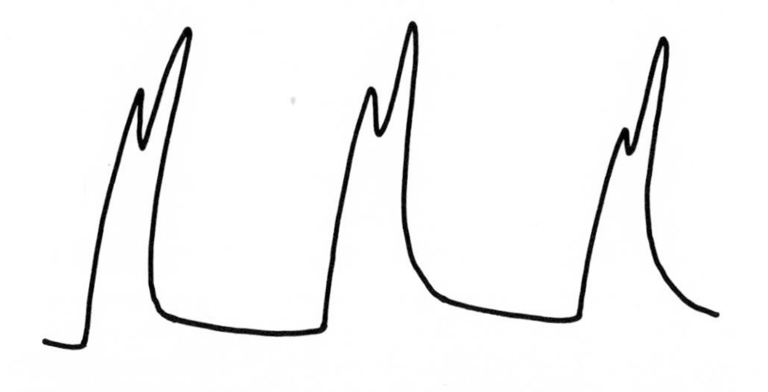

Figure 4–12. Summation tracing. Note that the second contraction of each pair is greater than the first.

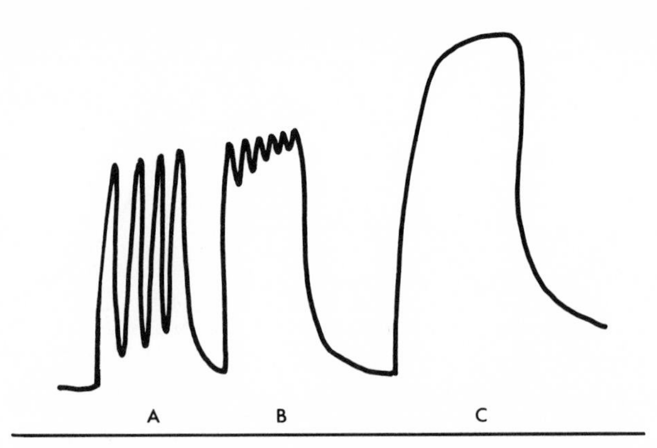

Figure 4–13. Tetanus tracings. *A* and *B*, Incomplete tetanus; *C*, complete tetanus.

muscle is in the relaxation phase; thus the conditions under which the law applies have been changed.

If the muscle is repeatedly stimulated at progressively more rapid rates, a frequency is finally reached at which the successive contractions fuse together. This fused response to repeated stimuli is called *tetanus* (Fig. 4–13). The smooth sustained contraction enables muscles to work to their greatest capacity and so is very useful in the performance of many tasks.

Graded Response. By varying the frequency and number of impulses, one can get a *graded response* from a muscle. An individual muscle's ability to grade its response depends on the number of its motor units and the number of individual muscle fibers per unit. Generally speaking, muscles with the largest number of motor units and the smallest number of fibers per unit will produce the greatest variety of responses. For example, muscles moving the eyeballs and fingers have very few fibers per motor unit and so can produce their fine gradations of response.

Muscle Tone. Even when skeletal muscles are completely relaxed, a varying amount of firmness or tension remains. This state of continued partial contraction in all healthy muscles is called *muscle tone.* It is produced by the simultaneous action of different motor units scattered throughout the muscle. These motor unit groups contract in relays, thus "spelling" each other from time to time.

CHEMICAL CHANGES IN MUSCLE

Muscles are machines that convert chemical energy into mechanical energy. Part of the energy involved in muscle contraction results in movement and part is given off as heat. (The movement of skeletal muscles is a very important method of heat production in the body.) The efficiency factor of muscles—that is, the percentage of energy actually resulting in work or movement— is about 25 per cent. The factor compares very favorably with the efficiency of most man-made machines.

Source of Energy for Contraction. Carbohydrate, stored in the muscle in a complex form called *glycogen* (animal starch), is the source of energy for muscle contraction. When energy is needed, glycogen is broken down to the simple sugar *glucose*, which can be burned by the muscle to produce energy.

A trigger mechanism is needed, however, to allow the transfer of this energy to the protein molecules which make up the muscle fiber. This mechanism involves two other substances found in muscle, phosphocreatine and adenosine triphosphate (ATP). The latter is capable of storing large amounts of energy, some of which combines with calcium ions to form the trigger complex. This complex causes the muscle tissue to become irritable so that it will respond to stimuli.

Much of the energy stored in ATP is used for muscle contraction and is released by an enzyme (a protein in muscle called *myosin*). In this process, a phosphate is split from the ATP, forming another compound called adenosine diphosphate (ADP). As ATP is used up, it can be replenished very quickly by phosphocreatine, which restores phosphate to ADP.

Therefore, although energy for muscle contraction is ultimately derived from the breakdown of glycogen, it is accepted that the immediate source of energy is ATP. The role played by glycogen is that of releasing energy during its breakdown (called *glycolysis*), which is used for the formation of phosphocreatine. The importance of ATP cannot be overemphasized: It is the primary storehouse of chemical energy in the body. In addition to muscle contraction, its energy is used for the absorption of carbohydrate in the intestine, the formation of ammonia by the kidney, the formation of urea by the liver, and the calcification of hard tissues.

Oxygen Debt and Fatigue. In strenuous muscular exercise, the chemical steps already described take place without the presence of oxygen; that is, they are said to be *anaerobic*. This occurs because during prolonged or strenuous exercise a person is not able to breathe in enough oxygen to satisfy the requirements of the muscles. This inability results in an oxygen debt being accumulated in the body. For example, a man running the 100 yard dash requires over 6 quarts of oxygen; yet during the 10 seconds or so that it takes him to run this distance, he can absorb only about one-sixth that amount.

This oxygen debt is paid back during the rest or recovery period following the exercise. While recovering, the runner will breathe rapidly and deeply. The oxygen thus taken in is used to remove a substance called *lactic acid* that accumulates in a muscle during anaerobic contraction.

As lactic acid piles up and the supply of energy-producing materials becomes depleted, the irritability of the muscle gradually becomes depressed. This loss of irritability leads to a condition called fatigue, in which the muscle contractions become weaker and weaker. The oxygen taken in during the recovery period is required for the removal of lactic acid, part of which is broken down to carbon dioxide and water. Energy is released during this breakdown and is used to change the rest of the lactic acid to glycogen, thus replenishing the energy-producing compounds necessary for contraction. Figure 4–14 summarizes the main chemical steps occurring during muscle contraction and recovery.

SPECIAL FEATURES OF SKELETAL MUSCLE

Hypertrophy. Forceful and continued muscular activity causes the muscle fibers to increase in size. This growth is called muscle hypertrophy and is

1. Reaction most closely connected with *contraction*:

$$\text{ATP} \xrightarrow{\text{myosin}} \text{ADP + phosphoric acid + ENERGY (used for contraction)}$$

2. ATP replenished by:

$$\text{ADP + phosphocreatine} \longrightarrow \text{ATP + creatine}$$

3. Phosphocreatine replenished by:

$$\text{Glycogen} \longrightarrow \text{lactic acid + ENERGY}$$

$$\text{Phosphate + creatine + ENERGY} \longrightarrow \text{phosphocreatine}$$

4. During muscle *recovery* glycogen replenished by:

$$\text{1/5 lactic acid + oxygen} \longrightarrow \text{carbon dioxide + water + ENERGY}$$

$$\text{4/5 lactic acid + ENERGY} \longrightarrow \text{glycogen}$$

Figure 4–14. Chemical steps occurring during muscle contraction and recovery.

commonly seen in trained athletes. In addition to an increase in fiber size, the muscle gains more of the energy-producing compounds involved in muscle contraction, that is, glycogen, ATP, and phosphocreatine. It may be noted here that, within limits, hypertrophy is a normal response made by all healthy muscle tissue to usage. In some types of heart disease, however, cardiac hypertrophy is a sign that the heart is attempting to compensate for an increased work load caused by the disease process.

ATROPHY. The opposite of muscular hypertrophy is muscular atrophy. This decrease in fiber size and in the amount of nutrients present occurs under various conditions. When a muscle is not used at all, as in the case of a body part's being immobilized for a period of time, it undergoes disuse atrophy.

If the motor nerve to a skeletal muscle is cut (denervation), atrophy begins immediately. Unless passive stretching is applied daily to such a muscle, the fibers tend to shorten in length. Thus, even if the nerve supply should be reëstablished, the fibers become so shortened they will be of little use.

RIGOR MORTIS. Several hours after death the skeletal muscles of the body contract to a shortened length. This state of contracture, called rigor mortis (rigidity of death), is thought to be caused by the loss of energy materials from the muscles. A temporary state resembling rigor mortis occurs when skeletal muscles are excessively fatigued. Rigor mortis may be delayed somewhat by placing the body under refrigeration. In a few days, however, this rigidity disappears because by this time the muscle proteins begin to break down.

OUTLINE SUMMARY

1. Muscle Tissue
   a. Skeletal muscle (striated; voluntary)
      1) 40 per cent of body weight
      2) Fibers bound into fasciculi
      3) Fasciculi bound into muscle

- b. Smooth muscle (nonstriated; visceral)
  - 1) Walls of hollow organs and blood vessels
  - 2) Arranged in flat sheets or layers
- c. Cardiac muscle (branching)
  - 1) Bulk of heart wall
  - 2) Faint striations; intercalated discs mark cell boundaries

2. Skeletal Muscles in General
   - a. Size, shape, and arrangement of fibers
     - 1) Fusiform – parallel fibers
     - 2) Pennate and bipennate – fibers oblique to the insertion
     - 3) Long and narrow – for greater range of movement
     - 4) Short and broad – for strength of movement
   - b. Attachments
     - 1) To skin or to other muscles
     - 2) To bone by tendon
     - 3) To bone or to each other by aponeurosis
     - 4) To cartilage
   - c. Parts and actions
     - 1) Origin – less movable end
     - 2) Insertion – more movable end
     - 3) Prime mover – mainly responsible for action
     - 4) Synergists – assist prime movers
     - 5) Antagonists – opposite action to prime mover
   - d. Naming of muscles
     - 1) By action – flexor digitorum; adductor longus
     - 2) By location – tibialis anterior; pectoralis major
     - 3) By heads of origin – biceps brachii; triceps; quadriceps
     - 4) By shape – trapezius; deltoid
     - 5) By attachments – sternocleidomastoid
3. Representative Skeletal Muscles (see Table 4–1, p. 54)
4. Muscle Physiology
   - a. Source of stimulus for contraction – motor unit
   - b. Mechanical changes in muscle
     - 1) All-or-none law – obeyed by muscle fiber but not whole muscle
     - 2) Simple muscle twitch – latent, contraction, relaxation phases
     - 3) Summation – series of rapid stimuli
     - 4) Tetanus – fused contractions
     - 5) Graded response – by varying frequency and number of impulses; responses depend on number of motor units and number of muscle fibers per unit
     - 6) Muscle tone – continued partial contraction in healthy muscle
   - c. Chemical changes in muscle
     - 1) Efficiency of muscles as "engines" is about 25 per cent; heat is also produced
     - 2) Energy sources for contraction – glycogen, ATP, phosphocreatine
     - 3) Oxygen debt – incurred during anaerobic contraction; paid back during muscle recovery
     - 4) Fatigue – accumulation of lactic acid and depletion of energy-producing substances
5. Special Features of Skeletal Muscle
   - a. Hypertrophy – increase in size of muscle fibers and in amounts of nutrients

b. Atrophy—decrease in size of fibers and in amounts of nutrients
c. Rigor mortis—state of muscle contracture after death

REVIEW QUESTIONS

1. Why is smooth muscle also referred to as visceral muscle?
2. What is the difference in shape between a smooth muscle fiber and a skeletal muscle fiber?
3. What are striations in a muscle fiber? What causes them?
4. How does a fusiform muscle differ from a pennate muscle?
5. What is a tendon? An aponeurosis?
6. Describe in general how muscles act. What is a synergist? An antagonist?
7. For each of the muscles listed in Table 4–1 on page 00 tell on what basis each is named (shape, location, etc.).
8. What is a motor unit? What is the end-plate?
9. What is meant by the statement that "muscle fibers follow the all-or-none law"?
10. What does the graph of a simple muscle twitch show?
11. What is summation? What is tetanus?
12. Explain the difference between isometric contraction and isotonic contraction.
13. How can one get a graded response from a muscle?
14. What is muscle tone? What causes it?
15. What three chemical compounds are most important in the mechanism of muscle contraction?
16. What is meant by oxygen debt? Why does it occur? How is it repaid?
17. What is the most important chemical compound that is found in a fatigued muscle? How does the body handle this substance?
18. What is the difference between muscle hypertrophy and muscle atrophy?

ADDITIONAL READING

Chapman, C. B., and Mitchell, J. H.: The physiology of exercise. Sci. Amer. (May), 1965.
Hoyle, G.: How is muscle turned on and off? Sci. Amer. (April), 1970.
Huxley, H. E.: The mechanism of muscular contraction. Sci. Amer. (Dec.), 1965.
Rosse, C., and Clawson, D. K.: Introduction to the Musculoskeletal System. New York, Harper & Row, 1970.

Chapter 5

THE NERVOUS SYSTEM

GENERAL FUNCTION OF THE NERVOUS SYSTEM

It has been pointed out that although the human body is composed of a vast number of individual units, these units must function together as a whole. The nervous system is one of the most important factors in the integration or working together of the various individual body parts. It performs this task by acting as the body's "communication" system; that is, it receives information from stimuli outside the body (the external environment) and from stimuli inside the body (the internal environment) and sends messages to the various organs, governing their functions. This reception of information and the ensuing responses may be voluntary, involuntary, or a combination of the two.

In addition to the general coördination of body activities, the nervous system builds up a background of experience (memory) by recording and relating certain stimuli and responses. This process is really what we call *learning*. To use a simple example—one learns by using sight, touch, and taste that a round, rough-skinned fruit of a particular color and flavor is an orange. This information becomes stored in the "memory bank" for use in identification of that object in later instances.

Because of the close functional relationship of the various components of the nervous system, it is difficult to separate one part from another in an anatomical description. Indeed, some authors prefer to discuss the system totally in terms of function, dividing it into voluntary and involuntary parts. This method, however, has its pitfalls and leads to misinterpretations, some of which will be discussed later.

NEURONS

Structure and Function. The basic structural and functional unit of the nervous system is the *neuron* or nerve cell. Although varying in size and shape, all neurons function basically in the same way; that is, each performs as an active irritable and conducting unit of the nervous system.

Nerve cells may be differentiated into (1) *sensory* or *afferent neurons,* which transmit impulses toward the central nervous system,* (2) *motor* or *efferent neurons,* which transmit impulses away from the CNS, and (3) *connector neurons,* which conduct impulses between the other two.

Cell Body and Processes. The cell body of the neuron contains very fine filaments called neurofibrils, which extend into the processes and play some part in nerve conduction. When subjected to a special dye, the cell body is seen to contain tiny, round structures called Nissl bodies, which disappear if the nerve cell is injured. Processes which extend from the cell body are classified into two groups: *axons* and *dendrites.* A unipolar nerve cell has one process, a bipolar cell has two, and a multipolar cell has many processes. Regardless of the total number of processes, there is only *one* axon per cell, and it conducts nerve impulses away from the cell body. However, a cell may have more than one dendrite, and these extensions of the cell body conduct impulses toward it (Fig. 5–1).

Fiber Coverings. Axons of peripheral nerves are surrounded by a protective cellular sheath, the *Schwann cell sheath,* sometimes referred to as the *neurilemma. Myelin* is a fat-like material that covers many nerve fibers; it lies under the Schwann cell sheath. The Schwann cells are responsible for the formation of the myelin covering of peripheral nerves. At intervals, the myelin is inter-

*Hereafter, central nervous system will be referred to as CNS.

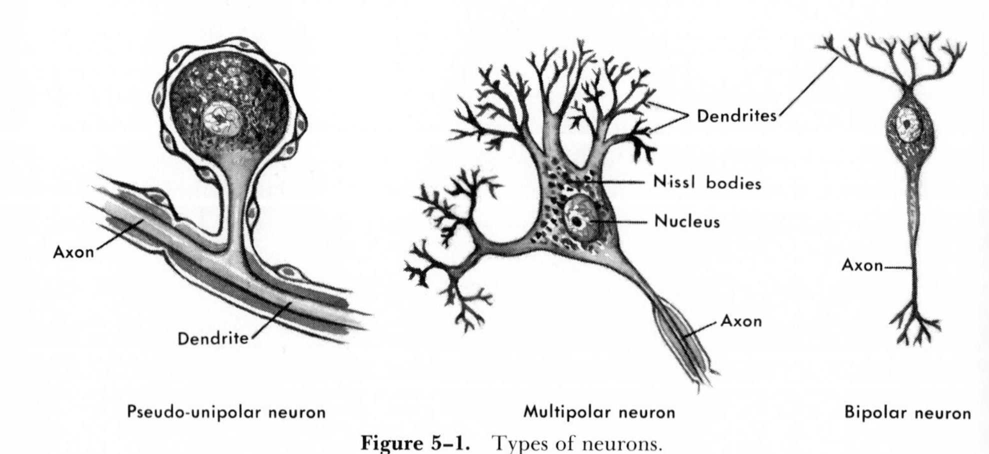

Figure 5–1. Types of neurons.

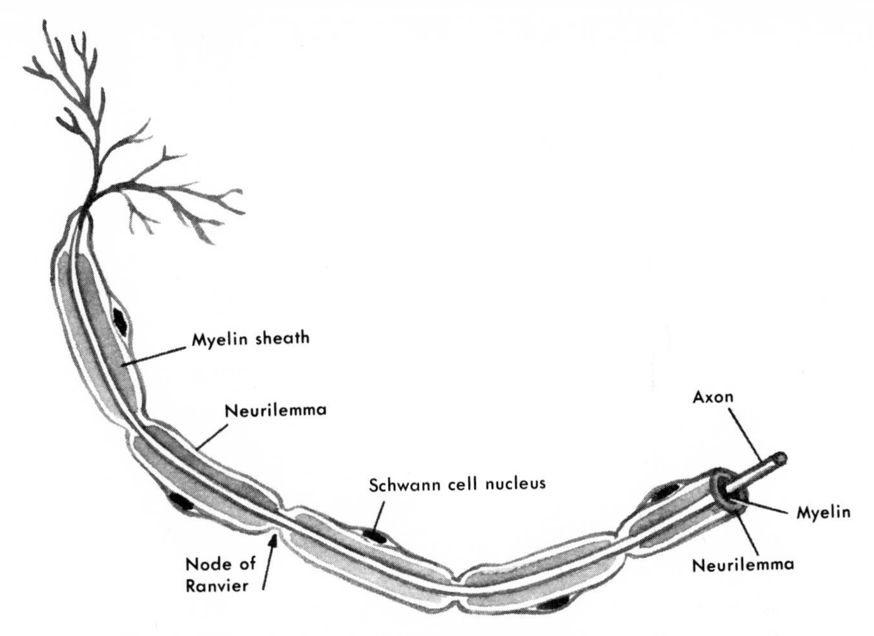

Figure 5–2. Nerve fiber surrounded by myelin and neurilemma.

rupted by constrictions called *nodes of Ranvier,* at which points myelin is absent. Nerve fibers inside the CNS (brain and spinal cord) lack a Schwann cell sheath, although most of them are myelinated. Here the myelin is formed by a type of non-neuronal or supporting cell, the oligodendrocyte (Fig. 5–2).

Myelin is thought to influence the speed of conduction of nerve impulses since the more heavily myelinated fibers conduct impulses more rapidly. Myelin gives nerves their white, shiny appearance. The Schwann cell sheath plays an important role in the regeneration of peripheral nerve fibers. After transection of a fiber, it and its myelin covering degenerate. The Schwann cells, however, multiply; thus neurilemmal tubes are formed which serve as a guide for the nerve fibers which will begin to sprout from the cut ends during regeneration.

The types of nerve fibers can be classified according to their coverings, as follows:

1. Unmyelinated* without neurilemma: some nerves inside the brain and spinal cord
2. Myelinated with neurilemma: cranial and spinal nerves
3. Unmyelinated with neurilemma: nerves of the autonomic nervous system
4. Myelinated without neurilemma: some nerves inside the brain and spinal cord

The Synapse. Since it is vital to maintain the continuity of a nerve impulse, usually by a chain of at least two neurons, individual nerve cells must be related to each other. The place where two neurons come into such a functional rela-

*Actually, all nerve fibers have some myelin, so "unmyelinated" is only a relative term.

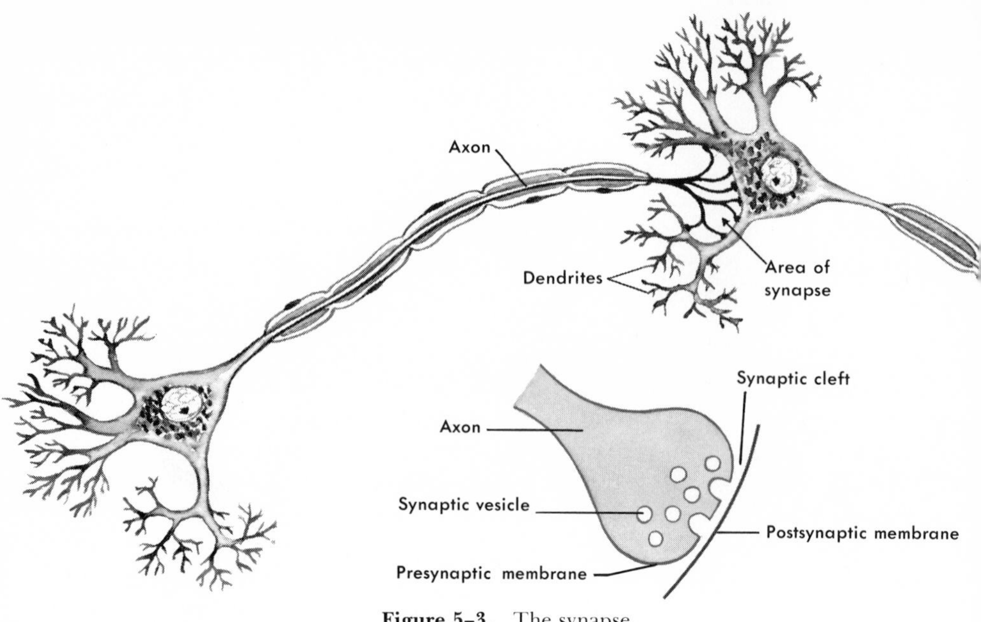

Figure 5–3. The synapse.

tionship is called a *synapse*. The axon of one neuron ends in close proximity to the dendrites or cell body of another neuron, but there is no proof that the substances composing the neurons actually merge or fuse with each other. Indeed, under high magnification with the electron microscope, a very small gap (150 to 200 angstroms) is seen between the presynaptic and postsynaptic membranes. This is called the *synaptic cleft,* across which the nerve impulse must pass (Inset, Fig. 5–3). Evidence suggests that the mechanism of synaptic transmission is chemical. A chemical transmitter substance, stored in vesicles in the tip of the axon, is released into the synaptic cleft when an impulse passes down the axon. This chemical then acts upon the postsynaptic membrane in some manner so that the impulse is propagated.

Nerve impulses are continued across a synapse in one direction only, that is, from the axon of one cell to the cell body or dendrites of another. This principle is called the *law of dynamic polarity.* The synapse is also the point where a nerve impulse may be blocked, as for example in the use of anesthetics.

Receptors and Effectors. In various areas of the body, such as the skin, there are nerve endings which are specialized for receiving certain stimuli; these are called *receptors.* They are usually divided into three classes:

1. *Exteroceptors,* which receive such sensations as touch, pain, temperature, vision, and hearing from sources *outside* the body.
2. *Interoceptors,* which receive visceral sensations such as hunger, thirst, and visceral pain from sources arising *inside* the body.
3. *Proprioceptors,* located in muscles, tendons, and joints, which receive sensations of position, movement, deep pressure, and balance.

Effectors are structures which carry out the motor activities of the body. These are either *somatic effectors,* located in the skeletal muscles, or *visceral effectors,* located in smooth muscle, cardiac (heart) muscle, or secretory glands.

THE NERVE IMPULSE

FUNCTIONAL BASIS

The physiological property of irritability is very highly developed in the neuron. This enables it to respond very rapidly to stimuli by initiating and conducting electrical impulses or *action potentials*. Thus the neuron can be converted from a resting or inactive cell to one which conducts an electrical impulse in just a few thousandths of a second, that is, in milliseconds. The electrical difference or *electrical potential* between that possessed by the fluid outside the neuron (interstitial) and that possessed by the fluid inside the neuron (intracellular) determines whether the cell is in a resting or in an active state. In other words, the cell itself acts as a small generator, producing power by burning (oxidizing) glucose and trapping the energy thus liberated. The power generated is measured in millivolts (thousandths of a volt). Changes in the potential difference across the cell membrane are determined mainly by the concentration differences of two chemical ions, sodium and potassium, on either side of the membrane.

The Resting Potential. In Figure 5–4, it can be seen that in the resting cell, the electrical difference or electrical potential is such that the inside of the fiber is negative with respect to the fluid outside the cell. Sodium ions are in higher concentration on the outside, while potassium ions are in higher concentration inside the cell. Because of these concentration differences, sodium ions tend to diffuse into the neuron, while potassium ions tend to diffuse out into the interstitial fluid. However, the cell returns sodium to the interstitial fluid by means of a metabolic "pump" and recovers potassium by a similar mechanism. These mechanisms aid in maintaining the concentration differences which result in a *polarized* membrane, that is, one that is negative inside and positive outside. This difference in electrical charge across the cell membrane is called the *resting potential* and is determined on the basis of the differences in potassium ion concentration between the inside and outside of the cell.

The Action Potential. When the resting nerve cell, as described above, is stimulated, a significant change occurs. The permeability of the nerve mem-

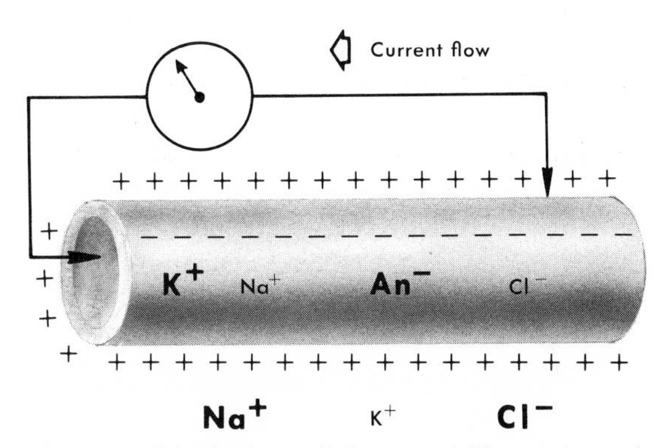

Figure 5–4. Resting potential. The intracellular potential is negative to the extracellular potential. Potassium (K^+) ions and protein (An^-) are in high concentration inside the fiber, and sodium (Na^+) and chloride (Cl^-) ions are in high concentration outside the fiber.

brane is altered so that there is an immediate inrush of sodium ions, followed by an outrush of potassium ions. These changes in ion permeability cause the membrane to become *depolarized;* that is, at the point of stimulation, the potential across the cell membrane is reversed from that in the resting neuron, becoming positive inside and negative outside. (See Fig. 5–4.) The whole process takes less than 1 msec. to occur, and the action potential thus initiated spreads rapidly along the unmyelinated nerve fiber in both directions from the point of stimulus. In the case of myelinated nerves, the action potential "jumps" or "hops" from one node of Ranvier to the next. This greatly increases the speed of conduction, since the axis cylinder is not covered by myelin at the nodes and thus is exposed to the extracellular fluid. Certain chemical agents can block the passage of the action potential at the nodes.

The energy for the propagation of the action potential does not come from the stimulus but from the ionic concentration differences between sodium and potassium. The depolarization of the membrane, and hence the action potential, is determined by the permeability of the cell membrane to sodium ions. After the action potential passes, the original ionic concentrations on both sides of the membrane are restored, and the nerve assumes its resting potential. It is thought that after the initial inrush of sodium ions, the "gates" which opened to permit this inflow are closed, preventing further sodium inflow. To recover its resting potential, then, the nerve fiber must restore its original concentrations of sodium and potassium, which it does between action potentials.

FIBER SIZE AND CONDUCTION VELOCITY

All nerve fibers do not conduct action potentials at the same rate of speed. There are three main types of fibers according to size: A fibers, the largest; B fibers, the next largest; and C fibers, the smallest. A and B fibers are myelinated, while C fibers are unmyelinated. A fibers conduct at the rate of 5 to 100 m. per second, B fibers from 3 to 14 m. per second, and C fibers less than 2 m. per second. Velocity is greatest of all in large myelinated fibers because the internodal distances are greater.

NEUROGLIA

The only other cells composing the nervous system besides neurons are the neuroglial cells. These are non-nervous cells; that is, they take no part in the transmission of nerve impulses. Instead, they have a supportive function and appear to assist in the nourishment of neurons. Although not directly participating in nervous activity, neuroglial cells are important constituents of the nervous system. They undergo changes in certain diseases of the nervous system, and they may be directly involved in such conditions as tumor formation.

TERMINOLOGY

It might be well at this point to define some of the terms which are frequently used when discussing the nervous system. *Nerve fibers* are axons and dendrites. A *nerve* is a bundle of nerve fibers and small blood vessels bound together by connective tissue and located outside the CNS. A *tract* is a bundle of nerve fibers lying within the CNS. A group of nerve cell bodies with similar functions located inside the CNS is called a *nucleus* or *center,* whereas a cluster of nerve cell bodies outside the CNS is referred to as a *ganglion.*

The *gray matter* of the nervous system comprises those portions which are composed primarily of nerve cell bodies. Myelinated nerve fiber tracts make up the *white matter* of the CNS.

THE CENTRAL NERVOUS SYSTEM

GENERAL CONSIDERATIONS

The CNS is composed of the *brain* and *spinal cord,* which are often referred to as the *neuraxis* (Fig. 5–5). In a sense, they comprise the "heart" of the nervous system, because sensations mediated by receptors enter them and impulses going out to effectors originate in them.

Since the "seat" of consciousness and the will is located in the brain, the brain is considered to be the "voluntary" part of the nervous system. Much of the CNS, however, functions below the level of consciousness. In addition, the activities of the so-called "involuntary" portion of the system (to be discussed later) often exert a strong influence over the actions controlled primarily by the voluntary part.

COVERINGS AND SPACES

Meninges. The brain, a soft and easily damaged organ, is enclosed in the rigid bony cranium. Additional protection is afforded by a three-layered covering of tissue called the *meninges* (Fig. 5–6).

The outermost layer, composed of two parts, is called the *dura mater* ("hard mother"). The more superficial part is the periosteal dura, which is actually the periosteum or inner covering of the cranial bones. Underneath it is the meningeal dura, which lies directly over the brain. Both parts are composed of tough, white, fibrous connective tissue and are fused together except at certain intervals where they separate to enclose blood spaces called *venous sinuses.*

Under the dura lies a very delicate layer, the *arachnoid,* so named because it resembles a spider's web. Like the dura, from which it is separated by the subdural space, it does not dip down into the grooves on the brain's surface. Fingerlike projections of the arachnoid, called *villi,* protrude into the venous sinuses. The space under the arachnoid is called the *subarachnoid space* and contains a clear, watery fluid which will be described presently. Branching, delicate filaments of the arachnoid, called *trabeculae,* bridge the subarachnoid space.

The *pia mater* ("tender mother"), the innermost layer of the meninges, is a delicate, transparent tissue which closely follows the contours of the brain and is adherent to it. Blood vessels run in this layer, and as these extend into the brain, the pia covers them for a distance.

The meninges form a continuous protective covering around the brain and

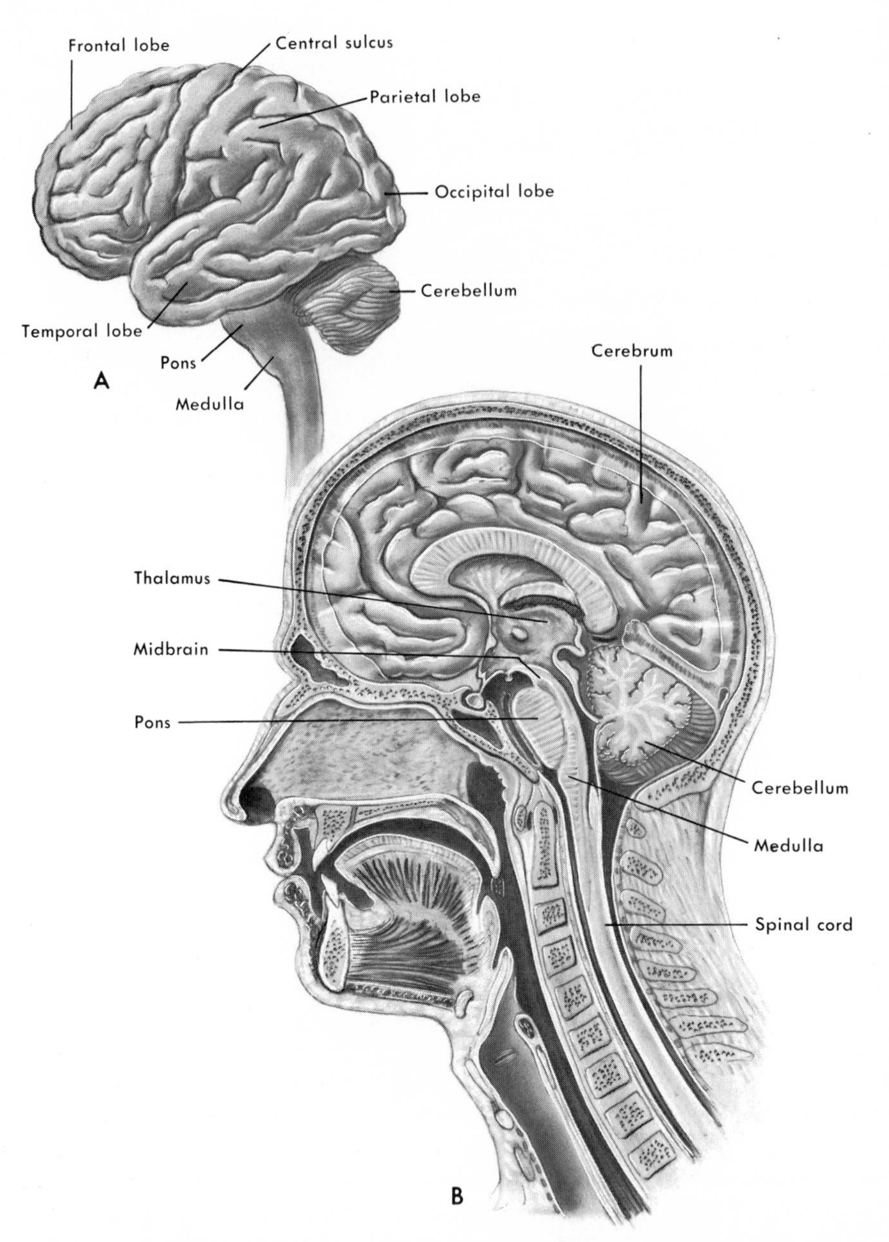

Figure 5–5. *A*, The CNS – brain and spinal cord. *B*, Midsagittal section of the head and neck.

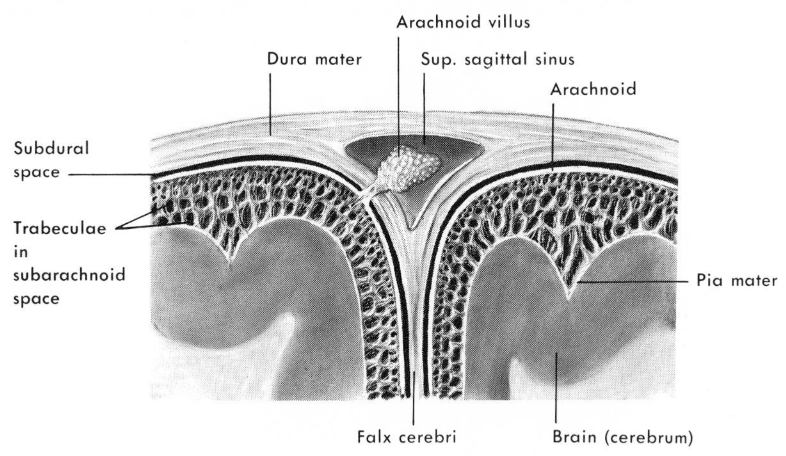

Figure 5-6. Cranial meninges – frontal section through superior sagittal sinus.

around the cord as well. The spinal dura mater, however, is composed of only one part, the meningeal dura, which is separated from the periosteum of the vertebrae.

Ventricles. Four spaces called *ventricles* are found inside the brain (Fig. 5-7). These spaces are to be expected, since the brain develops in the embryo from a tube.

Each of the two *lateral ventricles* occupies one of the halves of the cerebrum, the large upper portion of the brain. They are connected to each other by an *interventricular foramen,* which in turn leads into the *third ventricle.* The third ventricle then opens into the *cerebral aqueduct.* This narrow space opens into the

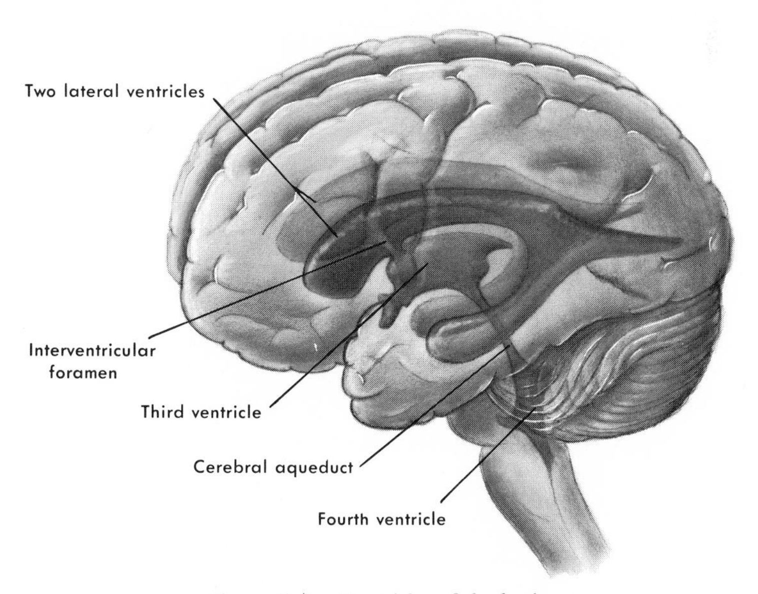

Figure 5-7. Ventricles of the brain.

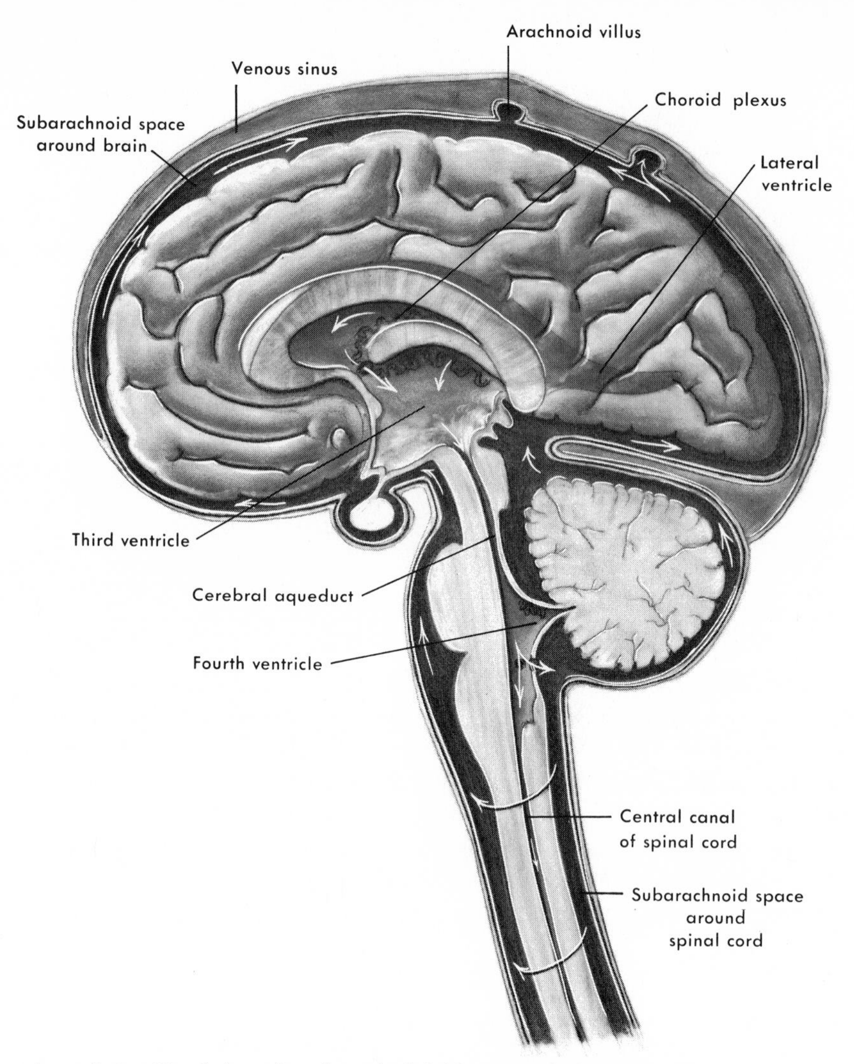

Figure 5–8. Circulation of cerebrospinal fluid. Arrows indicate the direction of flow.

somewhat larger *fourth ventricle,* which has three openings—one *median* and two *lateral foramina.*

These spaces are filled with *cerebrospinal fluid* (CSF) (Fig. 5–8), which circulates through them and, after escaping through the three openings in the fourth ventricle, continues into the subarachnoid space around the brain and cord. The fluid is formed by the *choroid plexuses,* which are tufts of tiny blood capillaries located in the walls of the ventricles. The watery portion of the blood, plus some salts, sugar, and a little protein, filters through the walls of these capillaries and into the spaces.

Because the cerebrospinal fluid is constantly being formed by filtration from the blood, it must also be constantly drained back into the blood. This drainage occurs through the walls of the arachnoid villi, which project into the venous sinuses. Thus, the fluid is returned to the blood from which it originated.

The cerebrospinal fluid acts as a shock absorber for the neuraxis since in the subarachnoid space it serves to suspend the brain and cord in their respective cavities. Normally the volume, pressure, and composition of the cerebrospinal fluid remain constant, but in many diseases of the nervous system (meningitis, for example) some of these factors are altered. Thus, an analysis of the cerebrospinal fluid can be an important diagnostic aid.

THE SPINAL CORD AND SPINAL NERVES

Structure of the Spinal Cord. The spinal cord is a cylindrical mass of nervous tissue about 18 to 20 inches long (40 to 45 cm.) which occupies the upper two thirds of the vertebral canal (Fig. 5–9). It extends from the foramen magnum (where it is continuous above with the brain) down to the level of the second lumbar vertebra. Therefore, the cord does not occupy the entire length of the vertebral canal. The meninges which surround the cord, however, descend to the level of the second or third sacral vertebra. In other words, the subarachnoid

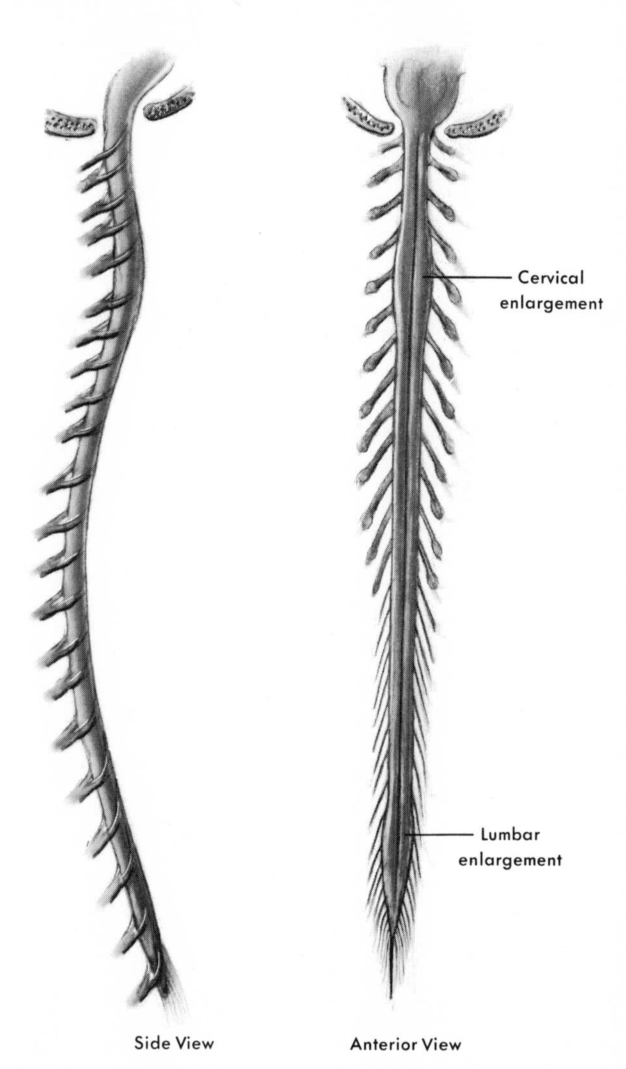

Figure 5–9. The spinal cord.

space extends beyond the length of the spinal cord itself. Thus, when it is desirable to obtain a sample of cerebrospinal fluid for analysis, a hollow needle can be inserted into the subarachnoid space below the level of the second lumbar vertebra without danger of puncturing the spinal cord. At its distal end, the cord is anchored by a thin extension of pia mater, the *filum terminale,* which is attached to the tip of the coccyx.

SPINAL NERVES. When viewed from the side, the spinal cord is seen to have curves which correspond to those of the vertebral column. The spinal cord is composed of 31 segments, or sections, each bearing one pair of spinal nerves. There are eight cervical pairs, 12 thoracic, five lumbar, five sacral, and one coccygeal.

Each spinal nerve passes from the cord through the opening between two vertebrae – the *intervertebral foramen.* Because the cord does not occupy the lower part of the vertebral canal, the lumbar and sacral nerves must travel downward through the canal from their segments of origin until each reaches its proper foramen for exit. Thus there is formed a large bundle composed of the roots of all the spinal nerves below the first lumbar pair. Because of its appearance, it is called the *cauda equina* (horse's tail).

A typical spinal nerve has two roots, a dorsal and a ventral (Fig. 5–10). Each dorsal root has an enlarged area, the *dorsal root ganglion.* These ganglia contain afferent cell bodies. The ventral roots are composed of efferent fibers from cell bodies in the cord. Note that whereas all dorsal roots are *afferent* only, and all ventral roots are *efferent* only, spinal nerves contain both afferent and efferent fibers and so are called mixed nerves. Each spinal nerve will form two main branches, or *rami,* a ventral and a dorsal, each of which will distribute fibers to the muscles and skin of the ventral and dorsal areas of the body respectively.

Internal Structure of the Spinal Cord. Although there are slight differences in form in the different regions, any cross section through the spinal cord shows a fairly typical picture (Fig. 5–11).

Gray matter forms the central portion, and its general shape has been likened to that of the letter **H.** The two dorsal projections are known as the *dorsal horns* or *columns,* and the two ventral projections as the *ventral horns* or *columns.** The cell bodies in the posterior columns are connector cells for afferent fibers entering the cord from their cell bodies in the dorsal root ganglia. Most of the anterior column cells give rise to efferent fibers, which leave the cord via the ventral roots.

The crossbar of the **H,** known as the *gray commissure,* surrounds the small central canal which runs the length of the cord. In the thoracic and lumbar regions the gray matter bulges on either side of the crossbar to form the *lateral gray columns.* These columns contain cell bodies for the sympathetic division of the autonomic nervous system.

White matter, formed by myelinated nerve fibers arranged in ascending and descending bundles, surrounds the gray matter. The *dorsal white columns* are located between the two dorsal projections of the gray matter, and the *ventral white columns* run between the two ventral gray projections. Finally, the *lateral white columns* are located lateral to the gray substance on either side of the **H.**

---

*Since the gray matter forms a continuous column through the cord, it seems more appropriate to designate these regions as "columns" rather than "horns."

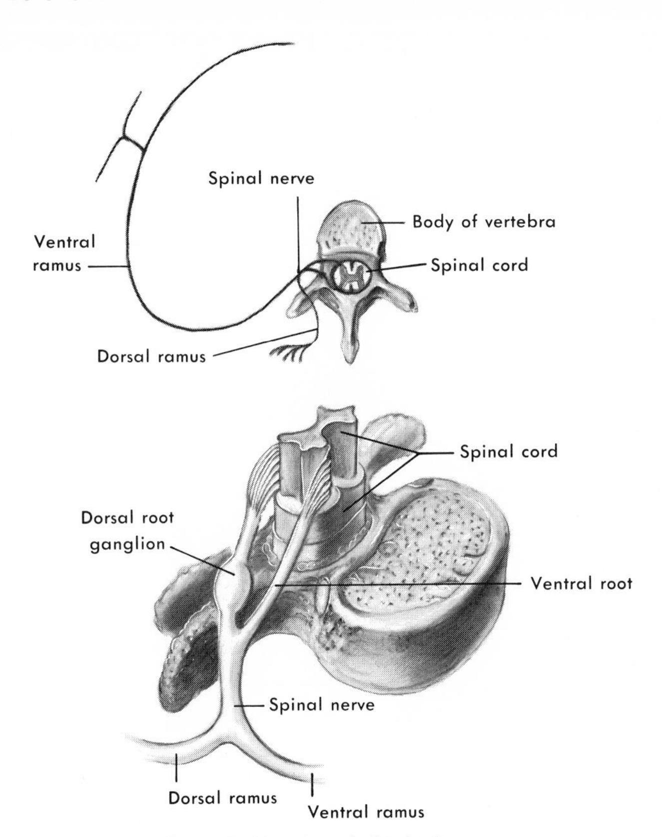

Figure 5-10. A typical spinal nerve.

Functions of the Spinal Cord. The spinal cord has two main functions: It serves (1) as a reflex center and (2) as a pathway for conducting impulses either to or from higher parts of the neuraxis. These two functions will be described separately.

THE SPINAL CORD AS A REFLEX CENTER. A *reflex* may be defined as a stereo-

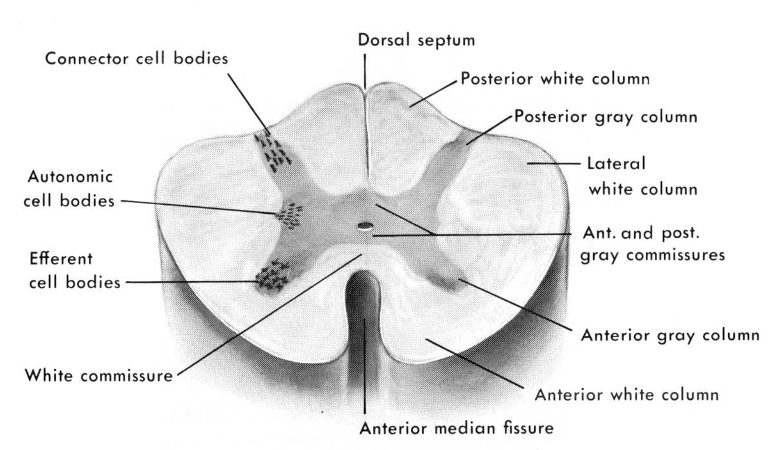

Figure 5-11. Cross section of the spinal cord.

typed, automatic act which is mediated by the nervous system. In lower animals with simpler nervous systems, spinal reflexes constitute a large part of the activity of the system. In man, reflexes become more complicated and involve connections in the brain as well as in the spinal cord. Some physiologists feel that many higher processes of the human nervous system, such as learning, are in reality very complex reflex activities.

In man, involuntary spinal reflexes serve as a sensitive indicator of the condition of this part of the neuraxis. This discussion will be limited to a description of the *stretch* or *myotatic reflex*, which is the simplest spinal reflex seen in man.

The smallest number of components that can mediate such a reflex is two: one afferent and one efferent neuron. An example of such a response in man is the knee jerk reflex. When the tendon of the quadriceps muscle is tapped firmly below the knee, the leg is involuntarily or reflexly extended.

The pathway traveled by the nerve impulses is called the reflex arc. Proprioceptive receptors located in the tendon are stimulated by the blow, which stretches the tendon. Thus initiated, sensory impulses travel via afferent fibers to afferent cell bodies in the dorsal root ganglia of the upper lumbar cord segments of the spinal cord. Axons from these cell bodies enter the spinal cord and synapse with efferent cell bodies in the ventral gray columns. Axons from these motor cells pass out through the ventral roots of the spinal nerves, carrying efferent impulses to motor end-plates in the quadriceps muscle, causing it to contract and extend the leg.

If any of the components of the reflex is not functioning, the arc is interrupted and the response will fail to occur. For example, if the motor cell bodies in the ventral gray columns of the cord segments involved are destroyed by a disease such as poliomyelitis, the reflex will be absent.

If the reflex arc is confined to one segment of the spinal cord, it is said to be *intrasegmental,* as the one shown in Figure 5–12. In some cases the afferent fibers

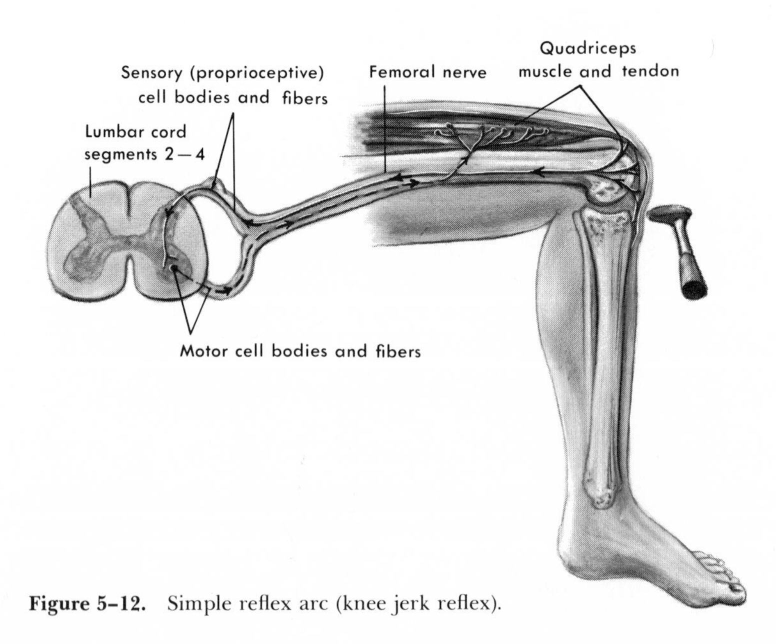

Figure 5–12. Simple reflex arc (knee jerk reflex).

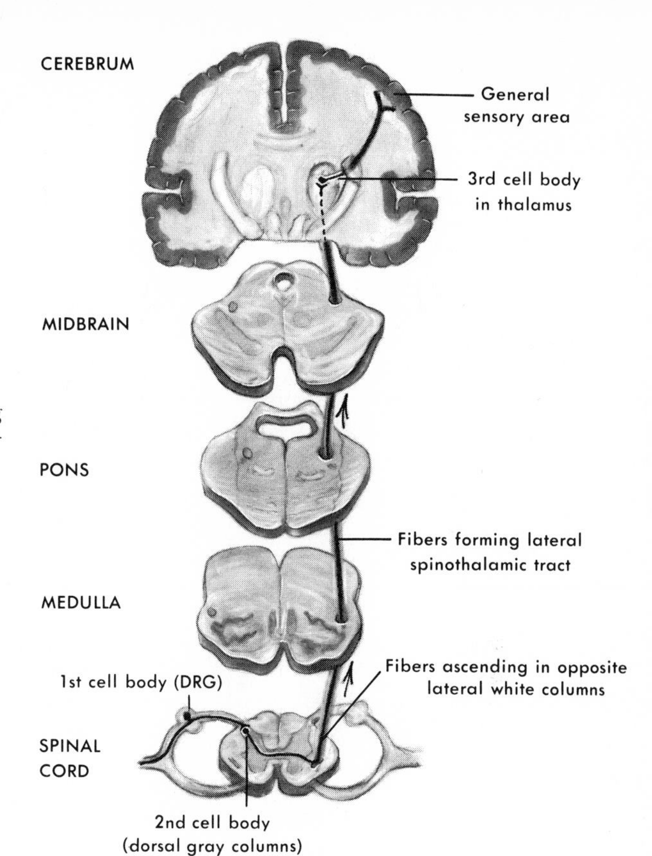

Figure 5–13. Ascending pathway for pain and temperature.

entering the spinal cord may ascend or descend through more than one segment to make connections with several efferent cell bodies in these segments. In this case, the reflex is called *intersegmental.*

The Spinal Cord as a Conduction Pathway. The white matter of the spinal cord is composed of bundles of fibers which conduct impulses either up the cord in ascending tracts or down the cord in descending tracts.

Ascending pathways carry sensory impulses for pain, temperature, touch, and joint position sense up the cord and into various regions of the brain, which interpret them accordingly. To illustrate the mechanism and the structures involved, the pain-temperature pathway will be described. In the spinal cord, the pathway carrying these impulses is called the *lateral spinothalamic tract* (Fig. 5–13).

If a person plunges his hand into water, the temperature of the water is immediately conveyed to his brain. In fact, if the water is very cold or very hot, the sensation may be recorded as one of pain rather than temperature. This information is received by nerve endings in the skin and sent to the conscious center of the brain by a chain of three neurons:

1. Afferent cell bodies in the *dorsal root ganglia.* Temperature (or pain) receptors in the skin of the hand receive the stimulus, which is then conveyed by nerve impulses traveling up the nerves in the arm and into the dorsal root ganglia. Axons from these cell bodies enter the cord where they synapse with–

2. Cell bodies in the *dorsal gray columns* on the *same side*. Axons from these cell bodies then cross over to the opposite side of the cord and ascend in the lateral white columns. They pass through the lower parts of the brain until they reach the *thalamus* (to be described later). Here they synapse with—

3. Cell bodies in the *thalamus*. Axons from these cell bodies proceed upward to the highest part of the brain, the cerebral cortex, where the center for interpretation of the stimulus is located.

Note that the name of the pathway describes its route; *lateral,* because it travels in the lateral white columns; *spino,* because it travels up the spinal cord; and *thalamic,* because this is the location of the last cell body in the chain.

Thus it is that all sensory stimuli received by the body are conducted to the brain. Each particular sensation (pain-temperature, touch, etc.) is conducted through its own special pathway. There are two points to remember about ascending conduction pathways: (1) Fibers carrying such impulses cross over from the side of entrance to the opposite side before they reach the thalamus (this may occur immediately upon entrance, as in the case of the pain-temperature fibers, or farther on up the cord, or even in lower parts of the brain), and (2) all afferent fibers carrying general sensations to the cerebral cortex are relayed through the thalamus.

Descending pathways in the cord conduct efferent or motor impulses to the skeletal muscles. The main voluntary motor pathway is the *cortico-spinal tract;* it begins in the cerebral cortex (*cortico-*) and ends in the spinal cord (*spinal*). These motor impulses are sent from the brain to the skeletal muscles by a chain of two neurons, an upper motor neuron and a lower motor neuron (Fig. 5–14):

1. Large cell bodies in the *cerebral cortex*. Axons from these neurons pass down through the brain and into the cord where they synapse with—

2. Motor cell bodies in the *ventral gray columns*. Axons from these cell bodies pass out through the ventral roots and are distributed by the branches of the spinal nerves to the motor end-plates in the skeletal muscles.

Most of the upper motor neuron fibers (about 80 per cent) cross over to the opposite side from their origin in the cortex. This crossing over or *decussation* takes place in the lowest part of the brain, the *medulla*. These crossed fibers travel down the cord in the *lateral corticospinal* tracts located in the lateral white columns. Those fibers which did not cross in the medulla continue down the same side of the cord in the *ventral corticospinal tracts* located in the ventral white columns. Most of these fibers will cross over later in the cord just before they synapse with the lower motor neurons in the ventral gray columns.

Two points to remember about descending or motor pathways are that (1) motor fibers originating from cell bodies in the cerebral cortex cross over before synapsing with the lower motor cells in the cord, and (2) all upper motor neuron fibers, regardless of where they originate in the brain, synapse with the motor cell bodies in the ventral gray columns of the cord.

THE BRAIN AND CRANIAL NERVES

General Structure of the Brain. The brain, composed of several parts, is the most complex part of the nervous system. It rests on the floor of the cranial cavity and is covered by the cranial bones. There are five main subdivisions of

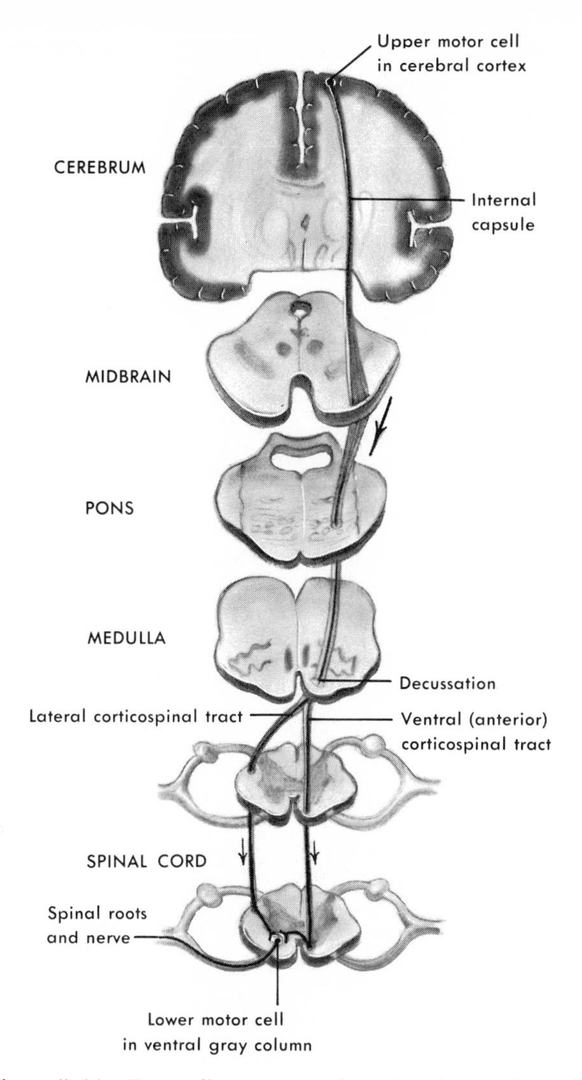

Figure 5-14. Descending motor pathway (corticospinal tracts).

he brain, the names of which describe their development in the embryo from he neural tube. These subdivisions, together with the main parts derived from each, are as follows:

SUBDIVISION	MAIN PARTS
Telencephalon	Cerebral cortex
Diencephalon	Thalamus and hypothalamus
Mesencephalon	Cerebral peduncles and corpora quadrigemina
Metencephalon	Cerebellum and pons
Myelencephalon	Medulla

A sagittal section of the brain will show the location of these main parts and their relationship to each other (Fig. 5–15).

THE MEDULLA. It will be recalled that the spinal cord extends up to the foramen magnum in the occipital bone. At this point, just above the origin of the first cervical spinal nerves, the neuraxis increases in size and becomes the *medulla*. Its ventral surface rests on the base of the occipital bone, and its dorsal surface is covered by the cerebellum. It is a little over an inch in length. In the dorsal part of the medulla, the central canal of the spinal cord opens into the fourth ventricle. Thus the medulla forms the floor of the ventricle.

The ventral portion of the medulla contains the crossing of the corticospinal tract fibers; this is called the decussation of the pyramids. The ascending fiber tracts from the cord are located in the dorsal part. Buried in the substance of the medulla are groups of nuclei, which are called vital centers. These nerve cells control such important functions as respiration, heart rate, and constriction of blood vessels. In addition, the cells of origin for the last four of the cranial nerves (9, 10, 11, and 12) are located in the medulla.

THE PONS. The pons forms the ventral part of the metencephalon. It acts as a bridge between the medulla and cerebral peduncles and lies ventral to the cerebellum. The pons consists mostly of broad bands of myelinated fiber tracts, both ascending and descending. In addition, a transverse band of fibers connects with the cerebellum on either side. Cranial nerves 5, 6, 7, and 8 have their cells of origin in the pons.

THE CEREBELLUM. The cerebellum is the dorsal part of the metencephalon and is the second largest part of the brain (Fig. 5–16). It is located under the posterior part of the cerebrum and separated from it by a tent-like fold of dura

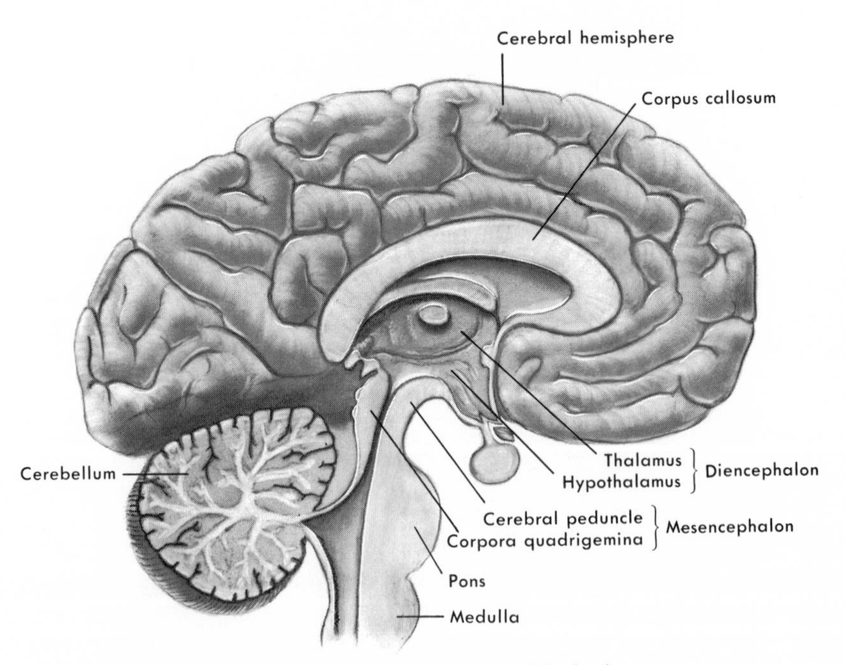

Figure 5–15. Sagittal section of the brain.

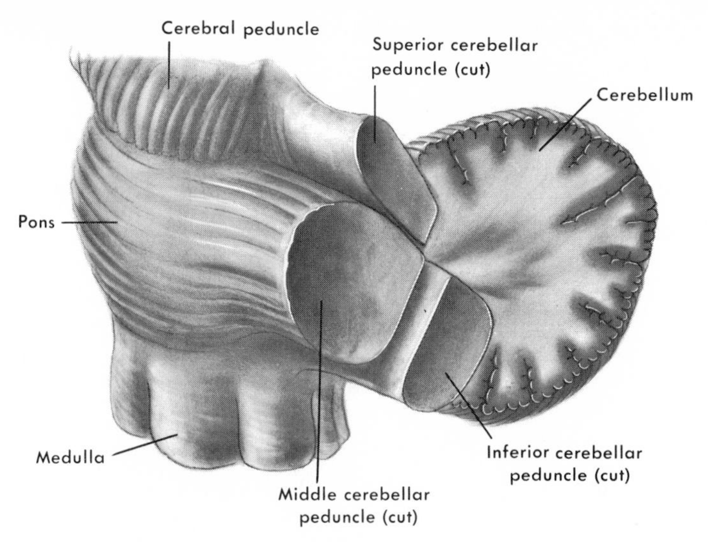

Figure 5–16. Section of the cerebellum, showing cerebellar peduncles.

nater called the *tentorium cerebelli.* The cerebellum is composed of inner white natter and outer gray matter, and it has shallow grooves or *sulci* on its surface. 'urthermore, the cerebellum has two lateral masses or *hemispheres* and a central ›ortion called the *vermis.*

Three pairs of large fiber bands or *peduncles* connect the cerebellum with ›ther parts of the brain. The inferior cerebellar peduncles connect with the nedulla and cord, the middle cerebellar peduncles with the pons, and the supe-ior cerebellar peduncles with the mesencephalon.

The main functions of the cerebellum are concerned with muscle *synergy,* hat is, the fine coördination of prime movers, synergists, and antagonists in keletal muscle activity. Coördination, of course, is necessary for the smooth per-ormance of muscle movements. In addition, since it receives afferent impulses rom the inner ear, the cerebellum plays an important role in the maintenance of alance or equilibrium. Many diseases affecting the cerebellum are characterized ›y uncontrollable tremors and disturbances in walking.

It is of interest to note that in man the hemispheres of the cerebellum are nuch larger in relation to the vermis than they are in lower animals. This dif-erence allows man to have more control over independent limb movements; hat is, he can purposefully move one of his arms or legs independently of the ›ther.

THE MESENCEPHALON. This part of the brain, often referred to as he *midbrain,* lies above the pons and below the cerebrum. The cerebral aque-luct, which connects the third and fourth ventricles, passes lengthwise through t. Forming the ventral part are two rope-like masses of white matter, the *cerebral eduncles.* These large tracts are composed of myelinated fibers which pass hrough the midbrain carrying impulses to or from the cerebrum.

The dorsal portion of the midbrain is composed of four small, rounded levations called the *corpora quadrigemina* ("bodies of four twins"). The two upper elevations are the *superior colliculi* ("hills"), which relay visual impulses. The lower two, the *inferior colliculi*, relay auditory impulses. Cells of origin for cranial nerves 3 and 4 are located in the midbrain.

THE DIENCEPHALON. Sometimes referred to as the interbrain, the diencephalon is located between the cerebrum and the midbrain. The *thalamus* is composed of two lateral masses of gray matter connected by a small intermediate gray mass. It forms the lateral walls of the third ventricle. All afferent impulses going to the cerebrum (except the sense of smell) are relayed through cell bodies in the thalamus.

The *hypothalamus* forms the floor and part of the lateral walls of the third ventricle. The *optic chiasma* (crossing of optic nerve fibers) makes up the anterior part of this area. It has been shown that the hypothalamus is responsible for the control of many body functions and contains centers for hunger, thirst, sleep, and wakefulness. This part of the diencephalon also regulates the release of secretions or *hormones* by the pituitary, one of the endocrine glands. Body temperature is controlled by shivering and sweating to either produce heat or dissipate it. Both these mechanisms are influenced by the hypothalamus.

In addition, the hypothalamus is apparently involved in the expression of emotions; lesions of this area in cats elicit rage and savage behavior. Thus it is evident that this area of the brain is important in the regulation of many of the so-called "vegetative" functions.

THE TELENCEPHALON. The largest part of the human brain is the *cerebrum*, which contains the highest centers of the nervous system (Fig. 5–17). A deep groove or cleft, the *longitudinal fissure*, divides it into right and left *hemispheres*. The hemispheres are joined on their inferior surfaces by a large band of transverse fibers called the *corpus callosum*. The outer surface, or *cortex*, is composed of a layer of gray matter, which is arranged in folds called convolutions or *gyri*. The gyri are separated by grooves called *sulci*.

Because of the folds in the gray matter, only one third of the cortex is exposed on the surface. The gray matter varies in thickness from 4 to 1.5 mm. in different areas of the cortex. Cell bodies composing the cortex number about seven billion.

The inner portion of the cerebrum is composed mostly of white matter, which is formed by *projection fibers* connecting the cerebrum with lower parts of the nervous system, *association fibers* connecting the neurons of one part of the cortex with those in another part in the same hemisphere, and *commissural fibers*, crossing from one hemisphere to the other. (The largest collection of commissural fibers is the corpus callosum.) Scattered areas of gray matter called *basal ganglia* are embedded in the white matter.

There are five regions or *lobes* making up each hemisphere. Four of them are named according to the cranial bones under which they lie; these are the *frontal, parietal, temporal,* and *occipital* lobes. The fifth lobe, the *insula* ("island"), lies under the lower parts of the frontal and parietal areas and thus cannot be seen on the surface of the cerebrum.

In addition to the longitudinal fissure, there are two other deep grooves on the surface of the cerebrum. The *central sulcus* separates the frontal and parietal lobes, and the *lateral fissure* lies between the frontal and temporal lobes.

Main Functions of the Cerebral Cortex. Cerebral functions may be placed

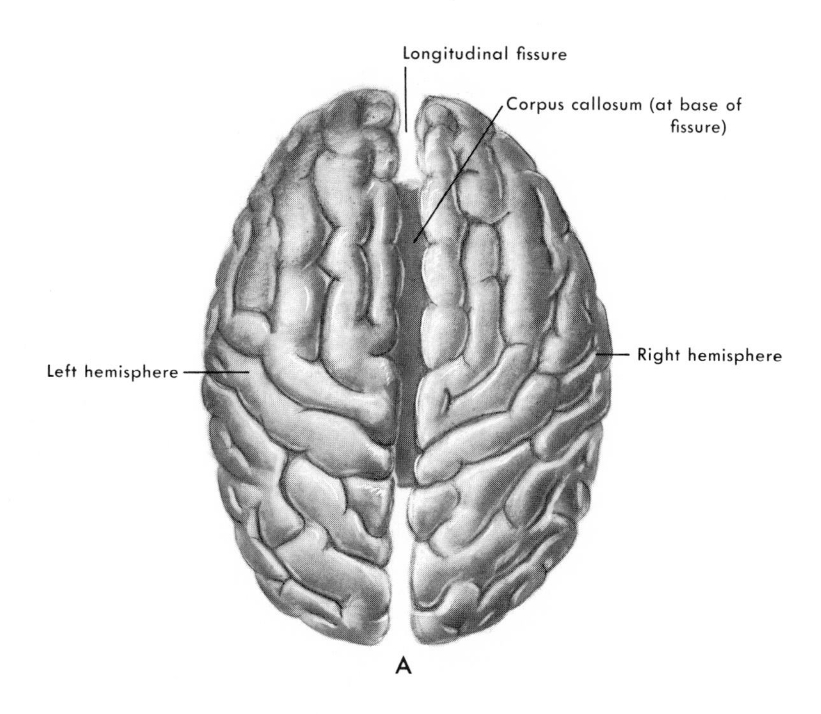

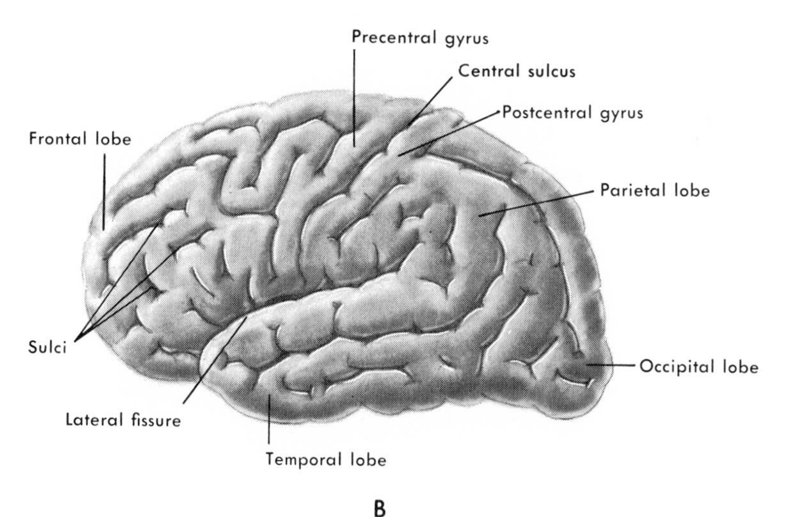

Figure 5–17. The cerebrum. *A*, From above; *B*, from the side.

in one of three main categories: sensory mechanisms, voluntary muscle movements, and association functions. Each will be described briefly.

SENSORY MECHANISMS. All conscious body sensation (pain, temperature, etc.) are sent via their respective tracts to the *general sensory area* in the cerebral cortex. This area, located in the postcentral gyrus, is a fold of gray matter in the parietal lobe that lies directly behind the central sulcus.

Each of the special senses (hearing, vision, smell, taste) has its own reception area. The *auditory area* is buried in the temporal lobe, in the floor of the lateral fissure. The *visual area* is located in the posterior portion of the occipital lobe, specifically in the gray matter which forms the walls of the *calcarine* fissure. Knowledge concerning the *olfactory* center, in which smell and taste are closely related, is somewhat scanty. It is believed to be located on the medial surface of the temporal lobe cortex.

VOLUNTARY MUSCLE MOVEMENTS. The *primary motor area* of the cortex (Fig. 5–18) occupies the precentral gyrus, the fold of gray matter lying immediately in front of the central sulcus in each hemisphere. Each section of this area is specific for a certain region of the body (Fig. 5–19); that is, in this gyrus there is a definite area whose cells send impulses to the toes, another area for impulses to the foot, still another to the leg, and so on for the entire body. The amount of gray matter in the gyrus allotted to a specific body part is not proportional to the amount of muscle in that part, but rather to the elaborateness of movement that is entailed. Thus the face, mouth, and hand have relatively larger representations than the arm, leg, or trunk.

Large motor neurons in this area give rise to fibers which form the *corticospinal* or *pyramidal tracts.* (See Fig. 5–14.) These tracts, described with the spinal cord, pass uninterrupted from the motor areas in both hemispheres down through the lower parts of the brain and into the cord.

Although the pyramidal tracts are the main motor pathways, other parts of the cerebral cortex contain motor cells that give rise to fibers that also pass to the brain stem and spinal cord. Because such fibers do not pass through the pyramids in the medulla, they are called *extrapyramidal* pathways. As they descend through the neuraxis, these fibers make one or more synapses with cell bodies in various parts of the brain below the cortex, such as the basal ganglia, midbrain, and medulla. After the final synapse, fibers proceed down into the cord to synapse with the motor cells in the ventral gray columns. It is probable that the extrapyramidal pathways are concerned with either facilitating or inhibiting the motor impulses sent out from the primary motor area. Major signs of damage to the areas of the cortex containing extrapyramidal cells are spasticity and exaggerated reflexes.

ASSOCIATION FUNCTIONS. Such functions as learning, memory, judgment, and emotional states are also handled by the cerebral cortex. Although they cannot be sharply localized in the cortex as the sensory and motor functions, experiments have indicated the general areas involved. For example, memory of past experiences appears to be centered in an area of the temporal lobe. Because people who have had the connections to their frontal lobes severed (frontal lobotomy) often develop distinct changes in personality, these areas probably are concerned with such emotions as rage and fear.

The Cranial Nerves. The brain gives rise to 12 pairs of cranial nerves, which, unlike the spinal nerves, are designated by names as well as numbers (Fig. 5–20). Three of these pairs are purely sensory (S), whereas the other nine, containing both sensory and motor fibers, are mixed nerves (M). The following table gives the principal features of each cranial nerve – number, name, type (S or M), and distribution:

Table 5-1. THE CRANIAL NERVES

Number	*Name*	*Type*	*Distribution* *Sensory*	*Motor*
1.	Olfactory	S	Nasal mucous membrane (sense of smell)	None
2.	Optic	S	Retina of eyes (sense of sight)	None
3.	Oculomotor	M	Proprioceptives from four eyeball muscles	To four eyeball muscles
4.	Trochlear	M	Proprioceptives from one eyeball muscle	To one eyeball muscle (superior oblique)
5.	Trigeminal	M	Exteroceptive from skin and mucosa of head and teeth Proprioceptives from chewing muscles	To chewing muscles
6.	Abducens	M	Proprioceptives from one eyeball muscle	To one eyeball muscle (lateral rectus)
7.	Facial	M	Taste from anterior two thirds of tongue Interoceptives for deep sensations of face Proprioceptives from muscles of facial expression	To superficial muscles of face and part of scalp To salivary glands (sublingual and submandibular)
8.	Acoustic	S	Exteroceptives from inner ear (sense of hearing) Proprioceptives from inner ear (sense of balance)	None
9.	Glosso-pharyngeal	M	Taste from posterior one third of tongue General sensation from tongue and pharynx	To parotid salivary gland To stylopharyngeus
10.	Vagus	M	Sensation from pharynx, larynx, trachea, skin of ear, thoracic and abdominal organs Taste from epiglottic area	To muscles of pharynx, larynx, thoracic and abdominal organs
11.	Accessory	M	Proprioceptives from sterno-cleidomastoid and trapezius	To trapezius and sterno-cleidomastoid muscles
12.	Hypoglossal	M	Proprioceptives from tongue muscles	To tongue muscles

THE RETICULAR FORMATION

Structure. Located in the core of the brainstem (medulla, pons, midbrain) is a meshwork of nerve cells and fibers designated as the reticular formation or RF. Although they present a diffuse appearance, these neurons are organized into groups, each having certain functions. In addition, these neurons make many connections with other nerve fibers from both higher and lower levels of

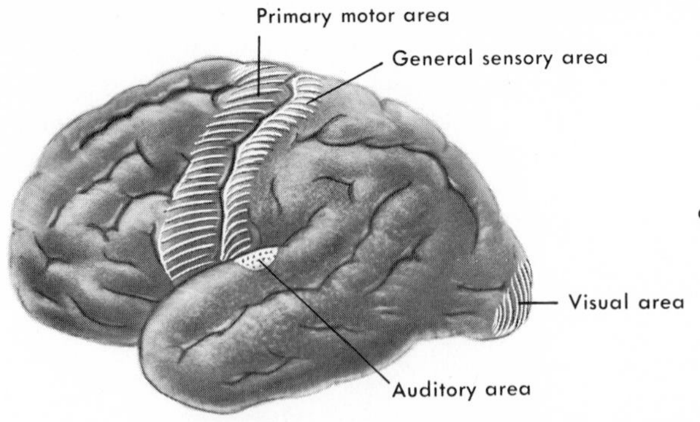

Figure 5–18. Localization of cerebral functions.

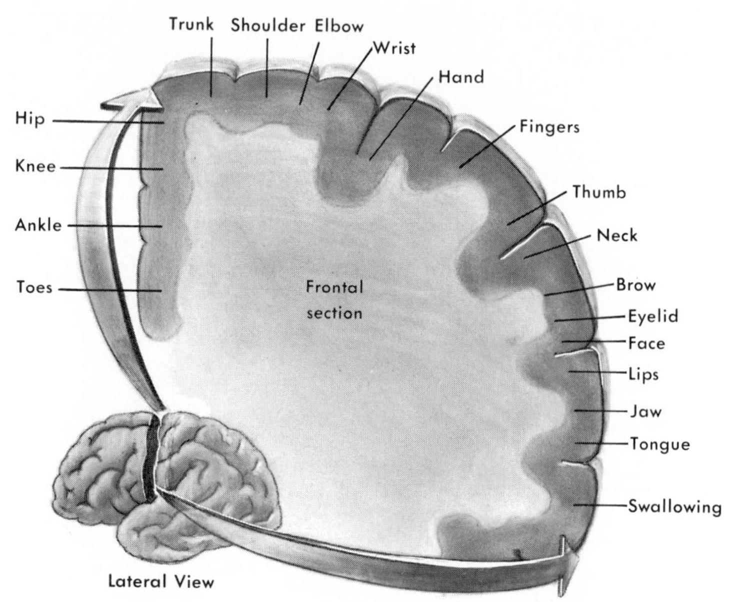

Figure 5–19. Frontal section through precentral gyrus, showing motor representation of body parts.

the CNS. It has been estimated that each RF neuron has about 30,000 such connections.

The RF system receives impulses from ascending spinal cord tracts, the cranial nerves, the cerebrum, and the basal ganglia, cerebellum, and hypothalamus. In addition, the RF has extensive projections to many of these same areas (Fig. 5–21).

Function. At present, it is thought that the RF plays two primary roles, one sensory in nature and the other, motor. Via its ascending connections, the RF is responsible for the "arousal response." This is shown by the fact that destruction of an animal's RF causes unconsciousness from which the animal cannot be aroused. On the other hand, electrical stimulation of the RF results in widespread activity of the cerebral cortex. It thus appears that the RF serves to alert or arouse the cortex in a general, nonspecific manner.

The descending connections of the RF are motor in nature, apparently influencing motor neurons in the spinal cord. Stimulation of its more medial portions results in the inhibition of spinal reflexes, while stimulation of the lateral RF causes the facilitation or enhancement of these reflexes.

From an evolutionary point of view, the reticular formation is a very old part of the CNS; in lower forms such as fish, the RF and the spinal cord are responsible for the control of all behavior patterns. The reticular system is very sensitive to general anesthetics. Indeed, it may be that one effect of general anesthetics is the inhibition of certain responses in the RF, thus leading to a general depression of the cerebral cortex.

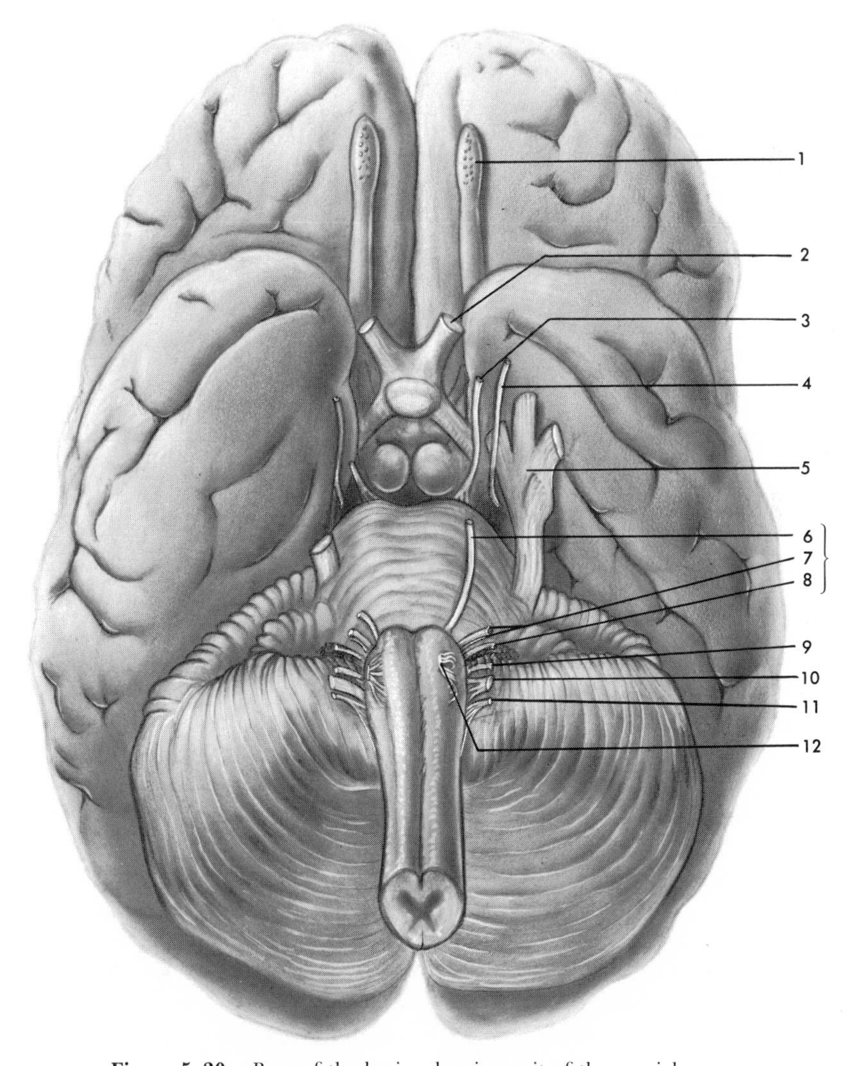

Figure 5–20. Base of the brain, showing exit of the cranial nerves.

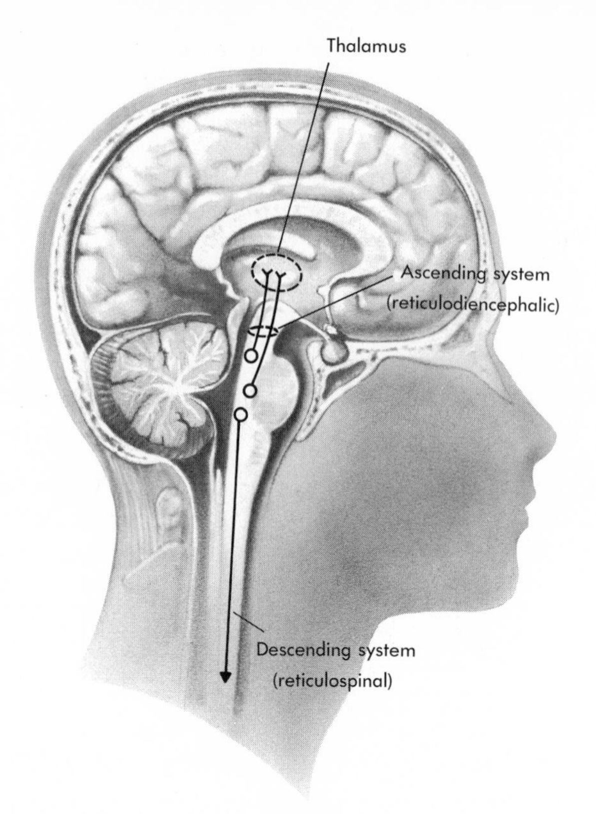

Figure 5–21. Ascending and descending reticular systems.

THE AUTONOMIC NERVOUS SYSTEM

GENERAL FEATURES

The autonomic portion of the nervous system is very important in the regulation of the internal environment of the body; that is, by increasing or decreasing its activity, it assists the body in adjusting to changing physiological conditions. More will be said about this function later.

There are two parts or divisions of this system, the *sympathetic* and the *parasympathetic*. With a few exceptions, all the organs receiving autonomic nerve fibers are innervated by both divisions. The two divisions are antagonistic to each other: If, for example, sympathetic stimulation of an organ speeds up or accelerates its action, then parasympathetic stimulation of that organ will result in slowing down or retarding its activity. Normally, the target organs of this system are constantly bombarded equally by both divisions. However, in certain changes, such as a stress situation, some organs will receive more impulses from one division than from the other.

The autonomic system is a motor system innervating the viscera, or internal organs; hence it is often referred to as the *visceral motor system* to distinguish it from the voluntary motor system, which supplies the skeletal muscles. Its target organs are: (1) glands that form secretions, such as the liver, salivary glands, and sweat glands; (2) those organs that contain smooth muscle, such as the stomach, intestine, uterus, urinary bladder, blood vessels, and iris of the eye; and (3) the cardiac muscle of the heart.

There are two main anatomical differences between the autonomic nervous system and the voluntary motor system: First, the autonomic system employs a chain of *two* neurons to convey impulses from the neuraxis to its target organs. Second, the fibers of these two neurons synapse in *autonomic ganglia* which lie *outside* the brain and cord. It will be remembered that all synapses in the voluntary motor system take place inside the central nervous system. Figure 5–22 shows part of the autonomic nervous system and the organs innervated.

THE SYMPATHETIC DIVISION

The first or *preganglionic* cell bodies of the sympathetic division are located in the lateral gray columns of all the thoracic segments (T1 to T12) and lumbar segments (L1 to L4) of the spinal cord. (Thus, this division is also known as the "thoracolumbar" division.) Preganglionic fibers from these cell bodies pass out of the cord, along with the voluntary motor fibers, via the ventral spinal roots of these segments. Each of the spinal nerves of these segments is attached by two branches (rami communicantes) to a sympathetic ganglion. These ganglia are connected to each other and so form two chains on either side of the spinal cord,

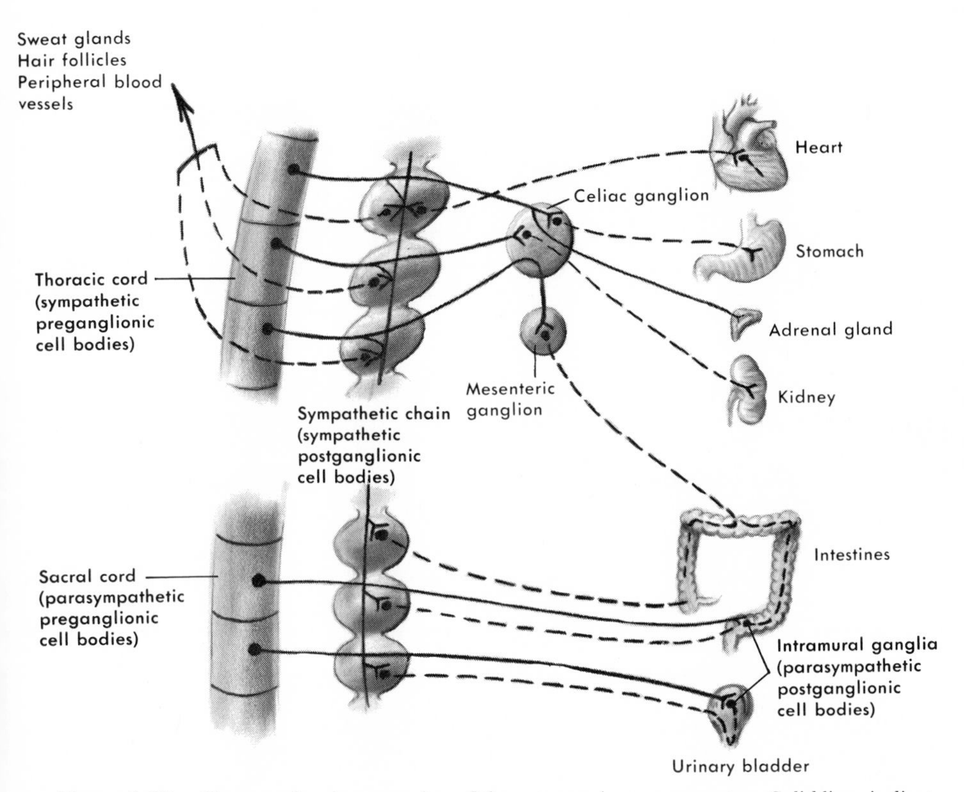

Figure 5–22. Diagram showing a portion of the autonomic nervous system. Solid lines indicate preganglionic fibers; dotted lines indicate postganglionic fibers.

extending from the cervical region to the tip of the coccyx. Here the fiber will either (1) synapse with cell bodies in these ganglia, or (2) simply run through them without synapsing and pass to one of four large ganglia that lie on large blood vessels in the abdomen and synapse there. In either case, *postganglionic* fibers leave the chain ganglia or outlying ganglia and pass to the target organs.

It should be noted that while the sympathetic chain ganglia extend beyond the thoracic and lumbar regions of the cord in both directions, in the cervical and sacral regions, each ganglion is attached to the spinal nerve by only one branch or ramus communicans. Since there are no preganglionic sympathetic cell bodies in these cord segments, there are obviously no preganglionic fibers to form white rami communicantes. The gray rami, however, are present since they consist of axons from the postganglionic cell bodies located in the chain ganglia.

Preganglionic fibers to the sweat glands, hair follicles, blood vessels in the skin, and organs in the head, neck, and chest synapse in one of the chain ganglia. Fibers going to the abdominal and pelvic organs synapse in the outlying ganglia in the abdomen. It should be noted that since there are *no* preganglionic cell bodies in the cervical cord segments, fibers destined for organs in the head and neck must originate in thoracic cord segments and then run up the chain, synapsing in the cervical chain ganglia before exiting.

The sympathetic division exerts widespread influence over the body. Because it is greatly concerned with the expenditure of energy, it has been referred to as the body's emergency mechanism. For example, in an emergency condition such as stress, one body response will be a rise in blood pressure. This response occurs partly because of sympathetic stimulation of the heart, which increases its rate, and partly because of stimulation of the small internal blood vessels, which causes them to constrict. In addition, the blood sugar will rise because sympathetic stimulation to the liver causes it to pour glucose into the blood. Furthermore, the tubes to the lungs will dilate to permit an increased intake of air. Arteries supplying contracting skeletal muscles will dilate, bringing more blood to these organs. All these responses prepare the body for "fight or flight," and all are the result of increased activity by the sympathetic division.

THE PARASYMPATHETIC DIVISION

Preganglionic parasympathetic cell bodies are located in the sacral segments of the spinal cord (S2 to S4) and in the nuclei of cranial nerves 3, 7, 9, and 10 in the medulla and midbrain. (This division is also called the craniosacral division.) Fibers from these cell bodies run to outlying ganglia for synapse with postganglionic cell bodies. These ganglia are located either near or actually in the walls of the organs being supplied. Four of these ganglia are located near target organs in the head. Postganglionic fibers from these cell bodies supply the smooth muscle inside the eyeball, the lacrimal (tear-producing) glands, the nasal glands, and the salivary glands.

All other preganglionic parasympathetic fibers synapse in *intramural* ganglia located in the walls of the heart, bronchi, stomach, intestine, urinary bladder, etc. Short postganglionic fibers from these cell bodies supply the smooth muscle, cardiac muscle, and glands of these organs.

In the body's quiet state, the parasympathetic division is more demonstrable than the sympathetic. The former calls forth responses from the stomach and intestine needed for digesting food, glycogen formation by the liver, digestive enzyme secretion by the pancreas, and slowing of the heart rate. These and other parasympathetic responses are concerned with body conservation processes, such as the digestion and storage of materials necessary for the body's well-being.

SUMMARY OF THE AUTONOMIC NERVOUS SYSTEM

Although most structures supplied by this part of the nervous system receive fibers from both divisions, there are some exceptions. Sweat glands, smooth muscle of the hair follicles, blood vessels in the skin, and part of an endocrine gland, the adrenal medulla, have no parasympathetic supply. In general, sympathetic stimulation results in a widespread response by organs throughout the body, whereas parasympathetic responses are more localized and individual.

Three main points to keep in mind about the autonomic nervous system are (1) that the two divisions of the system are each antagonistic in action to the other, (2) that both divisions use two neurons to relay impulses from the CNS to structures innervated,* and (3) that this system supplies all smooth muscle, cardiac muscle, and secretory glands in the body.

Although the autonomic nervous system is often termed "involuntary," in the sense that its actions cannot be controlled by direct use of the will, the voluntary nervous system does exert a very definite influence over autonomic functions. For example, when one is frightened by seeing or hearing something, this information not only may result in a voluntary response such as running away, but also will be accompanied by involuntary changes such as an increased heart rate, dilatation of the coronary arteries and the pupils of the eyes, and increased excretion of sweat. All such changes are responses controlled by the autonomic nervous system.

The following table lists some of the organ responses to stimulation by each of the divisions of the system:

*Only one sympathetic neuron innervates the adrenal medulla because, having developed from the same embryological source as the autonomic ganglia, it serves as a ganglion itself.

Table 5-2. ORGAN RESPONSES TO AUTONOMIC STIMULATION

Organ Supplied	*Sympathetic Response*	*Parasympathetic Response*
Sweat glands	Secretion	No supply
Salivary glands	Secretion—small amount of sticky saliva	Secretion—large amount of watery saliva
Pupil of eye	Dilatation	Constriction
Heart rate	Increases	Decreases
Intestine: Smooth muscle	Inhibition	Contraction
Intestine: Glands	Inhibition	Secretion
Lung tubes	Dilatation	Constriction
Urinary bladder muscle	Inhibition	Contraction
Skin blood vessels	Constriction	No supply

SPECIAL SENSES

THE EYE

Accessory Structures. The eye is really an extension of the brain, and as such it can be considered as part of the central nervous system. For protection, the eyeballs are located in the bony orbits of the skull and are surrounded by fat and connective tissue. Only one sixth of the eye is open to the external environment, the rest being recessed in the orbit.

Added protection is given by the eyelids positioned over the exposed surface of the eyes. The edges of the lids are lined with lashes and contain small glands which secrete an oily substance. *Lacrimal glands,* located in the upper outer corner of each orbit, constantly secrete tears. The tears spread over the exposed surface of the eye, washing out foreign particles and keeping the surface moist.

Structure of the Eyeball (Fig. 5–23)

LAYERS OR "COATS." The eyeball is a hollow sphere composed of three layers of tissue. The outer coat or *sclera,* composed of tough, white, fibrous connective tissue, has a protective function. The anterior portion of the sclera, the *cornea,* is transparent so that light rays can enter the eye. The sclera has no blood vessels in it, but it does contain sensory nerve endings for pain.

The middle coat or *choroid* contains dark brown pigment and is the vascular layer, containing many blood vessels. The *ciliary body* forms a circular area around its anterior portion and consists of smooth muscle from which *suspensory ligaments* extend to hold the *lens* in place. The colored part of the eye, the *iris,* is a circular structure attached to the ciliary body. Its smooth muscle fibers control constriction and dilatation of the *pupil,* which is a hole in its center.

The inner layer, or *retina,* is an incomplete coat having no anterior portion. It contains two types of specialized nerve cells that are sensitive to light rays. The

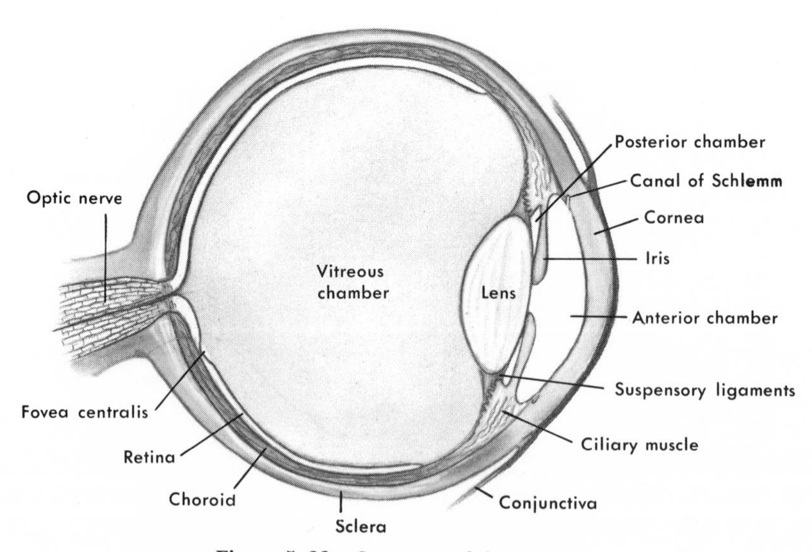

Figure 5–23. Structure of the eyeball.

cones, seven million in number, are responsible for daylight vision; they interpret fine details of form, contrast, and color. Most of them are located in a depression near the posterior pole of the eyeball called the *fovea centralis*, and so this is the area of keenest vision. The *rods*, numbering about 100 million, are located in the more peripheral parts of the retina and are activated only in dim light. These cells, unlike the cones, cannot distinguish fine detail or color. Thus, at night when the rods are functioning, it is difficult to see sharp outlines or distinguish colors.

There are two other groups of nerve cells in the retina. These cells serve to relay the nerve impulses initiated by the rods and cones. The axons of one of these groups form the fibers of the optic nerve, which leaves the posterior region of the eyeball a little to the nasal side of center. The portion of the retina where the optic nerve exits contains neither rods nor cones and is referred to as the *blind spot*.

CAVITIES. The hollow portion of the eyeball is divided into two main areas. The more anterior area is subdivided by the iris into an *anterior chamber*, which is posterior to the cornea and anterior to the iris, and the *posterior chamber*, which lies posterior to the iris and anterior to the lens. Both chambers are filled with clear, watery *aqueous humor*, which is secreted by the ciliary process.

Behind the lens is the large *vitreous chamber*, which is filled with a jelly-like substance called the *vitreous humor*. This fluid has the consistency of raw egg white and serves to maintain intraocular pressure so that the eyeball will not collapse when subjected to outside pressures.

Conduction of Visual Impulses to the Cortex. The field of vision for each eye is divided into halves (Fig. 5–24). Rays of light from objects in the outer or *temporal* half of each field fall on the inner or *nasal* half of each retina. Conversely, rays from objects in the nasal field fall on the temporal half of each retina. Nerve fibers from both eyes travel through the two optic nerves which pass posteriorly out of the bony orbits through the optic canals. A short distance farther on they meet in the midline, and this junction is called the *optic chiasma*. Here the fibers from the nasal halves of both retinas cross over to the opposite side, but the temporal fibers remain uncrossed. Thus, each of the *optic tracts* leaving the chiasma contains temporal retinal fibers from the optic nerve on the same side plus nasal retinal fibers from the opposite optic nerve.

Both optic tracts continue posteriorly on each side until they reach a region of the thalamus called the *lateral geniculate bodies*. Here most of the fibers synapse with cell bodies from which axons continue posteriorly until they reach the visual area in the occipital lobe of the cerebral cortex.

In summary then, light rays from objects in the visual fields focus on the retinas where the rods or cones transform them into nerve impulses. These impulses in turn are carried by the optic nerves to the optic chiasma; then the optic tracts relay them to the thalamus. From the thalamus another set of neurons relays the impulses to the cerebral cortical cells, which "develop" them into "pictures" or images.

Note that the destruction of one optic *nerve* will result in total blindness in the eye on that side. Destruction of one optic *tract*, however, will produce partial blindness in *both* eyes.

Physiology of Vision. As we have seen, the rods and cones of the retina

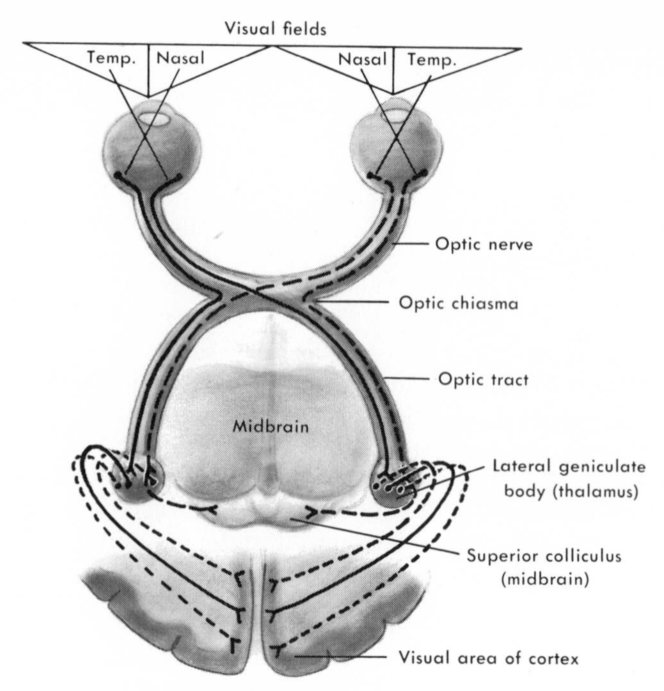

Figure 5-24. Conduction pathway for impulses to visual area of cortex.

are specialized nerve cells that are stimulated only by light rays. In order for vision to occur, rays of light must enter the eye and be brought to a focus on the retina. The nerve impulses so initiated must then be carried to the visual area of the cortex. So that these events can occur properly, various structures in the eyes must perform certain functions.

REFRACTION. As light rays enter the eye, they are bent or *refracted* so that they come to a focal point on the retina (Fig. 5-25). Refraction occurs at the anterior surface of the cornea, the anterior surface of the lens, and the posterior surface of the lens. In the normal eye this optical system is adequate for bringing light rays from objects more than 20 feet away to a sharp focus on the retina.

Near-sightedness, or *myopia*, and far-sightedness, or *hypermetropia*, are errors in refraction caused either by a faulty refraction apparatus in the eye or by abnormal length of the eyeball. In myopia, light rays come to a focus in front of the retina, whereas in hypermetropia the rays focus behind the retina. Specially ground lenses placed in front of the eye bend the light rays properly to correct these errors.

ACCOMMODATION. When one looks at an object closer than 20 feet, light rays from the object require a greater degree of refraction or bending in order to focus on the retina. The *lens* is the part of the eye that performs this task. It does so by changing the focal length, which is the distance from the center of the lens to the point of focus on the retina. This adjustment is called *accommodation*, and it is accomplished by contraction of the ciliary muscle, which releases tension on the suspensory ligaments supporting the elastic lens. As tension is released, the lens "bulges" or becomes more spherical, thus increasing its refractive power.

The near point of vision is the closest point to which an object can be brought to the eye and still remain in focus. In older persons, the lens gradually loses its elasticity and thus some of its refractive power. This causes the near point to recede farther and farther from the eye and results in the far-sightedness of old age, which is called *presbyopia*.

The "work" performed by the eyes is that which is involved in the process of accommodation because of the contraction of the ciliary muscle. Objects more than 20 feet distant send light rays that are almost parallel and thus focus on the retina without muscular effort on the part of the eyes.

Constriction and dilatation of the pupil regulate the amount of light entering the eye. This regulation is accomplished by contraction of either the circular (for constriction) or radial (for dilatation) smooth muscle fibers of the iris. The former is controlled by parasympathetic fibers and the latter by sympathetic.

CONVERGENCE. Man has binocular vision; that is, although he has two eyes, he sees but a single image. To produce this single image the eyeballs must move together in perfect unison, and they do—thanks to the action of the six pairs of finely coördinated *extrinsic eye muscles* (Fig. 5–26). There are four *rectus* and two *oblique* muscles firmly attached to the scleral coat of each eyeball. By means of these muscles, the eyeballs rotate around various axes.

The movement of the two eyeballs inward when the eyes are focusing on a close object is called *convergence*. Convergence allows the light rays from the object to fall on corresponding points of both retinas simultaneously, thus resulting in a single image. Obviously, the nearer the object, the greater the convergence.

VISUAL ACUITY. If a person can see clearly objects of a certain size (such as letters on an eye chart) at a distance of 20 feet, he is considered to have

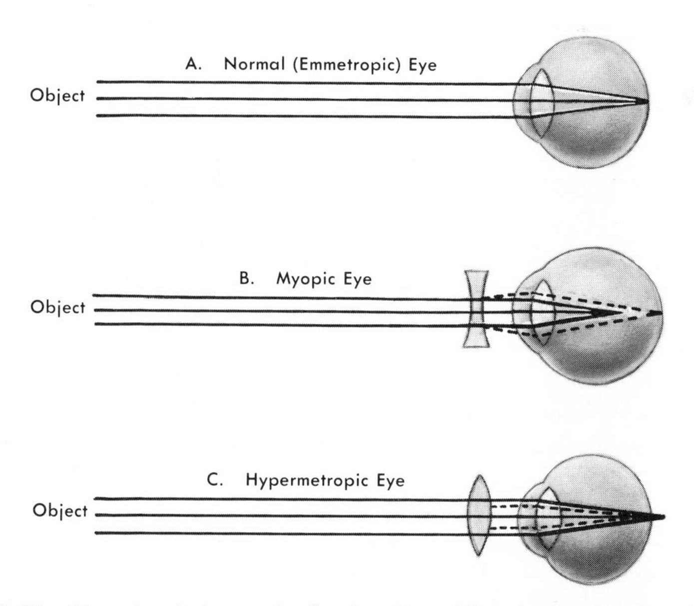

Figure 5–25. Normal and abnormal refraction. Dotted lines in *B* and *C* show how specially ground lenses correct the refractive errors by bringing the light rays to a point of focus on the retina.

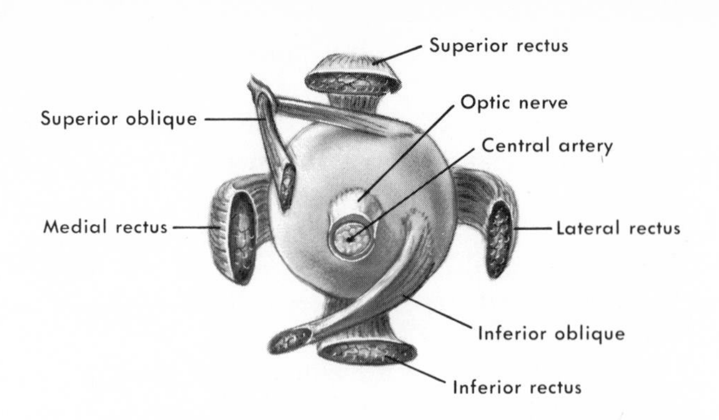

Muscle Involved	*Action*	*Related Actions*
Lateral rectus	Rotates eye outward	When eye is turned out: Superior rectus elevates Inferior rectus depresses
Medial rectus	Rotates eye inward	When eye is turned in: Inferior oblique elevates Superior oblique depresses

Figure 5–26. Extrinsic eye muscles—posterior view of right eyeball (muscles spread out to show insertion).

normal or *emmetropic* vision. Normal vision is usually written as 20/20 vision. If he can see clearly an object at 20 feet that the normal eye sees at 30 feet, he has 20/30 vision, or two thirds of the normal. Thus, visual acuity is the sharpness with which detail is perceived and is often measured by finding the smallest distance by which two lines can be separated without appearing as a single line.

THE EAR

General Features. The ear is another structure which can be considered a part of the central nervous system. It contains the receptors of the eighth cranial nerve (acoustic), and it performs two main functions, hearing and equilibrium. The ear consists of three parts: the external ear, the middle ear, and the internal ear (Fig. 5–27).

The External Ear. The *auricle* or *pinna* is the expanded outer part of the ear which projects from the side of the head. It has a cartilage framework covered with skin. Its function is to collect and direct sound waves into the ear.

The *external acoustic meatus* is a canal 1 inch long that opens from the outside to the tympanic membrane. It has a double curve, so that when one wishes to view the tympanic membrane, the auricle must be pulled up and back to straighten it. The meatus is lined with skin which has fine hairs, oil (sebaceous) glands, and wax (ceruminous) glands.

The Middle Ear (Tympanic Cavity). The middle ear is a hollowed-out area in a portion of the temporal bone. It is connected to the nasopharynx (throat) by the *auditory tube*. When opened by swallowing, this tube helps to equalize the air pressure in the cavity with that outside the body. The entrance to the middle ear is guarded by a fibrous partition called the *tympanic* or *drum membrane.* The roof of the cavity is formed by a thin plate of bone which separates it from the middle cranial fossa. Thin bone also forms the cavity's floor, separating it from an opening in the floor of the skull. The *oval window* is the opening at the other end of the cavity.

Three tiny bones, or *ossicles*, suspended by ligaments from the walls of the cavity, form a chain across the middle ear. These are named the *hammer, anvil,* and *stirrup.* The handle of the hammer is attached to the tympanic membrane, and the foot-plate of the stirrup is connected to the oval window. The anvil lies between them. A small muscle, the *tensor tympani*, extends from the auditory tube to the handle of the hammer, and upon contraction, it tenses the drum membrane. Another tiny muscle, the *stapedius*, is attached to the stirrup, and its contraction causes the foot-plate to tilt.

The three bones are united by true joints and form a lever system which transmits sound waves through the middle ear. This energy is used to rock the foot-plate of the stirrup in the oval window. In this way sound vibrations are transmitted across the middle ear, through the oval window, and into the internal ear.

The Internal Ear. The internal ear houses the essential organs for hearing and equilibrium. It is composed of a series of canals tunneled out of the temporal bone. This bony labryinth in turn is lined by a membranous labryinth, much as a coat is lined. The *vestibule* is the main "room" of the labyrinth and has three *semicircular canals* opening into its posterior wall. The round and oval windows are at either end of the vestibule.

A snail-shaped bony canal, the *cochlea* (Fig. 5–28), winds two and a half times around a pillar of bone and opens into the wall of the vestibule. A slice through the cochlea shows that it is composed of three connected compartments. These

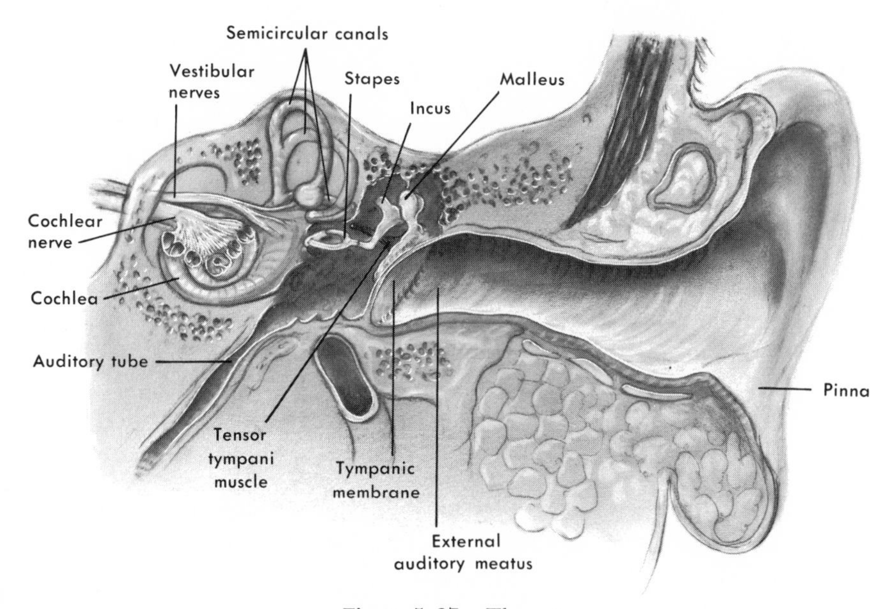

Figure 5–27. The ear.

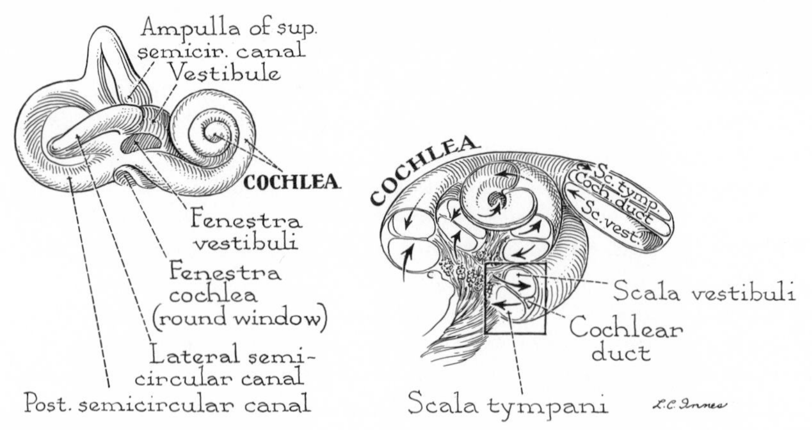

Figure 5–28. The cochlea. (Modified from King, B. G., and Showers, M. J.: Human Anatomy. 5th ed. Philadelphia, W. B. Saunders Co.)

compartments are the upper *scala vestibuli* bounded laterally by the oval window, a middle *cochlear duct*, and a lower *scala tympani*, bounded laterally by the round window.

The organ of hearing is the spiral *organ of Corti* (Fig. 5–29), and it is contained within the cochlea. It has a *basilar membrane*, which forms the floor of the cochlear duct, separating it from the scala tympani. Resting on the membrane

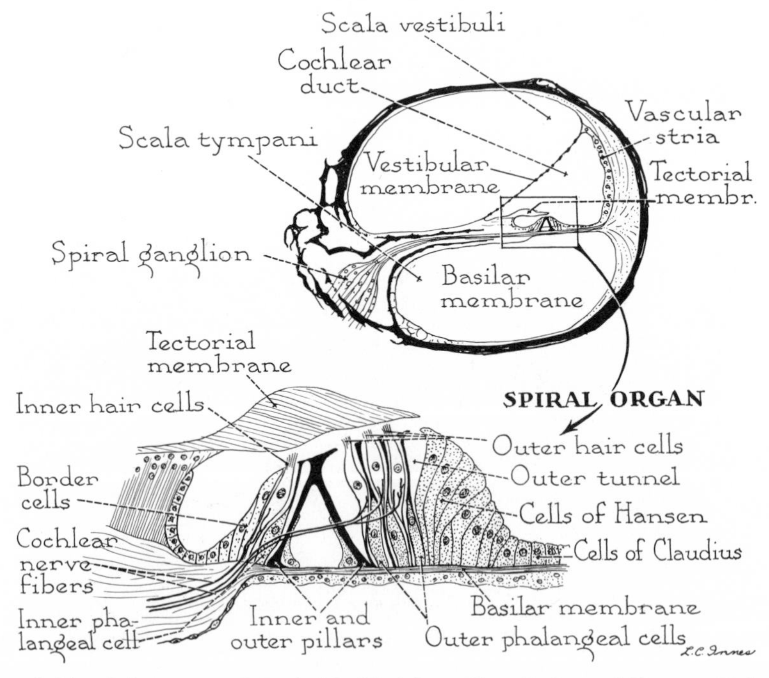

Figure 5–29. Spiral organ of Corti. (Modified from King, B. G., and Showers, M. J.: Human Anatomy, 5th ed. Philadelphia, W. B. Saunders Co.)

are columnar cells with cilia called hair cells which are in contact with fibers of the cochlear nerve.

The membranous labyrinth lining the bony labyrinth is separated from it by a fluid called *perilymph. Endolymph* is a fluid which fills the membranous channels themselves.

Physiology of Hearing. Just as the eye is adapted to receive light rays and transform them into nerve impulses, the ear is adapted to receive sound waves and change them into nerve impulses. Sound waves, channeled into the external acoustic meatus by the auricle, strike the tympanic membrane causing it to vibrate. The mechanical energy of these vibrations is transmitted across the middle ear by the chain of ossicles, finally causing the foot-plate of the stirrup to rock against the oval window. The bulging inward of the oval window causes the perilmyph in the membranous cochlea to be set in motion. The wave begins in the scala vestibuli, proceeds through the cochlear duct, and continues into the scala tympani, finally expending itself by pushing against the round window.

In the cochlea, the sound waves are translated into nerve impulses. The basilar membrane, set in motion by the fluid wave, initiates the action of the hair cells resting upon it. The fibers of the cochlear branch of the acoustic nerve form a network around the hair cells, which, affected by the movements of the basilar membrane, set up nerve impulses in them. It is the basilar membrane which detects the pitch and amplitude of the sound and transmits this information, via the hair cells, to the cochlear nerve fibers.

The auditory pathway by which the nerve impulses reach the cerebral cortex is complex and involves at least four neurons. Impulses pass from the cochlear nuclei to a second set of nuclei in the medulla. After synapse here, fibers run to the inferior colliculi in the midbrain for synapse and on to nuclei in the thalamus. Fibers leaving the thalamus relay the impulses to the auditory region in the temporal lobe of the cerebral cortex.

Equilibrium. The three semicircular canals, placed at right angles to each other, open into the vestibule. They are filled with endolymph, which is set in motion by movements of the head. This motion stimulates nerve endings of the vestibular branch of the acoustic nerve, located in the vestibule. From here nerve impulses travel into the brain so that righting reflexes may be initiated if necessary to maintain balance.

Other receptors in the body aid this vestibular mechanism in the maintenance of equilibrium. Images received by the retinas of the eyes and proprioceptive impulses from the neck muscles are also utilized by the brain for this purpose.

OLFACTION

Importance. Olfaction, or the sense of smell, is mediated by the first cranial nerve. Phylogenetically, it is a very old sensory function which was important even in primitive forms that existed many millions of years ago. Even today, there are certain animals for whom the sense of smell is essential for protection and the finding of food. Although this special sense is not highly developed in man, it is a very complex mechanism both anatomically and physiologically. Olfaction is not restricted simply to the perception of odors; it elicits responses from other neural systems as well. For example, the aroma of a substance can initiate such diverse activities as salivation, gastric secretion, and vomiting. In animals, odors play a role in initiating mating activity as well as rage and fear responses. One interesting aspect of olfaction is that since it is a subjective sense, it cannot be objectively measured or classified as in the case of sound or vision. Since the olfactory sense is not a vital function in man, its loss can hardly be compared to blindness or deafness.

Olfactory Pathway. This pathway will be described briefly. The bipolar receptor cells are located in the upper half of each nasal cavity. Axons from these cells form the olfactory nerves which enter the cranial cavity via the cribriform plate of the ethmoid bone and end in the olfactory bulbs. From cells in the olfactory bulbs, axons pass via olfactory tracts into that part of the brain known as the *rhinencephalon* or "smell brain." Olfaction is the only sense that does not project to the thalamus.

TASTE

Taste is an even more subjective quality than smell since it is an even cruder sensation. Because taste is less sensitive, one usually smells a substance while tasting it. Although four basic taste sensations have been described – sweet, sour, salt, and bitter – many taste sensations do not fit easily into any of these categories.

The taste buds on the tongue (see Chapter 8, p. 165) are supplied with nerve endings which respond to chemicals in the tasted substance; thus a substance must be in solution in order to be tasted. After a person passes the age of about 50, the number of taste buds decreases, as does the sensitivity of taste. Cranial nerves 7, 9, and 10 convey impulses from the taste buds to a nucleus (nucleus solitarius) in the medulla. From here, axons project to nuclei in the thalamus from which fibers travel to the taste center in the parietal lobe of the cerebral cortex.

OUTLINE SUMMARY

1. General Function of the Nervous System
    a. Body's communication system, receiving information and sending messages
    b. Site of memory and learning processes
    c. Parts interrelated in structure and function
2. Neurons – active conducting units
    a. Types according to function – sensory, motor, connector
    b. Cell parts; cell body and processes (axon and dendrites)
    c. Fiber coverings
        1) Neurilemma – aids in fiber regeneration
        2) Myelin – fatty insulating material; speeds impulse conduction
    d. Synapse – functional relationship between neurons
    e. Receptors – nerve endings in body parts receiving stimuli
    f. Effectors – nerve endings in body parts carrying out actions
3. Neuroglia – non-nervous supportive cells of the nervous system
4. The nerve impulse
    a. Resting potential
    b. Action potential
    c. Fiber size and conduction velocity

5. Terminology
    a. Nerve fibers—axons and dendrites
    b. Nerve—bundle of nerve fibers, blood vessels, and connective tissue outside CNS
    c. Tract—bundle of myelinated nerve fibers inside CNS
    d. Nucleus or center—group of nerve cell bodies inside CNS
    e. Ganglion—group of nerve cell bodies outside CNS
    f. Gray matter—composed of nerve cell bodies
    g. White matter—composed of myelinated nerve fibers
6. Central Nervous System: Spinal Cord and Brain
    a. Coverings and spaces
        1) Meninges—three-layered protective covering around brain and cord
        2) Ventricles—space inside brain; cerebrospinal fluid formed here
    b. Spinal cord—located in vertebral canal
        1) Cylindrical shape; 18 to 20 inches long; 31 segments
        2) Extends from foramen magnum to second lumbar vertebra
        3) Gray matter—**H**-shaped; centrally located
        4) White matter—surrounds gray matter; long nerve fiber bundles
    c. Spinal nerves
        1) 31 pairs, one for each cord segment
        2) Each has dorsal and ventral roots
        3) Dorsal root ganglia; contain afferent cell bodies
    d. Functions of spinal cord
        1) As a reflex center; reflex arc is functional unit
        2) As a conduction pathway; ascending and descending tracts to and from brain
    e. Brain—located in cranial cavity
        1) Main parts—medulla, pons, cerebellum, midbrain, diencephalon, cerebral hemispheres
        2) General functions—sensory reception, voluntary movement; association functions
    f. Cranial nerves—12 pairs (see Table 5–1, p. 91)
    g. Reticular formation
7. Autonomic Nervous System: Visceral Motor System
    a. Sympathetic (thoracolumbar) division—widespread activity; "flight or fight" reaction
        1) Preganglionic cell bodies—gray matter of thoracic and lumbar cord
        2) Postganglionic cell bodies—sympathetic chain and abdominal ganglia
    b. Parasympathetic (craniosacral) division—localized activity; body conservation
        1) Preganglionic cell bodies—sacral cord and cranial nerves 3, 7, 9, 10
        2) Postganglionic cell bodies—four head ganglia; intramural ganglia
    c. Functions of two divisions antagonistic (see Table 5–2, p. 97)
    d. Innervates smooth muscle, cardiac muscle, glands
8. The Eye
    a. Structure
        1) Three coats—sclera, choroid, retina
        2) Cavities—anterior (aqueous humor) and posterior (vitreous humor)

b. Conduction pathway to visual cortex (see Fig. 5–22, p. 100)
c. Physiology of vision – refraction, accommodation, convergence, visual acuity

9. The Ear
   a. Three parts
      1) External ear – auricle and external acoustic meatus
      2) Middle ear – tympanic membrane; ossicles
      3) Internal ear – vestibule, semicircular canals, cochlea
   b. Physiology of hearing
      1) Organ of Corti in cochlea
      2) Sound waves conducted through ear; transformed into nerve impulses
   c. Equilibrium – semicircular canals and vestibule
10. Olfaction – sense of smell
    a. Mediated by first cranial nerve
    b. Not as vital to man as to other mammals
    c. Pathway does not project to thalamus
11. Taste
    a. Associated with olfaction
    b. Mediated by cranial nerves 7, 9, and 10 from taste buds

REVIEW QUESTIONS

1. How does a nerve impulse travel through a neuron?
2. Describe the coverings of the cranial and spinal nerves.
3. What is a synapse?
4. Name and describe the different kinds of receptors. What are effectors?
5. What is the main difference between the cranial meninges and the spinal meninges?
6. Where is the cerebrospinal fluid formed? How does it circulate? How is it reabsorbed?
7. Where is a lumbar puncture performed? Why?
8. What is the difference between a spinal nerve and spinal nerve roots?
9. Describe (or draw) the pathway of a spinal reflex arc.
10. If someone stepped on your toe, how would this information travel to the brain? Trace the pathway.
11. To what subdivision of the brain does the medulla belong? What is the decussation of the pyramids?
12. What does the cerebellum do?
13. Name some body functions regulated by the hypothalamus.
14. What and where is the corpus callosum?
15. What tracts originate in the primary motor area? Trace their course.
16. Name the cranial nerves that control the extrinsic eye muscles.
17. What body structures are innervated by the autonomic nervous system?
18. Describe the pathway traveled by sympathetic impulses to the heart, to the stomach, and to the iris of the eye.
19. In general, what functions does the parasympathetic division control?
20. Where in the eye is the area of keenest vision? Why?
21. How would vision be affected by a lesion to the right optic *tract*?
22. How do sound waves travel through the ear?
23. What parts of the ear are involved in maintaining balance?

24. The concentrations of which two chemical ions determine the electrical potential across a nerve cell membrane? Of these, which one determines the resting potential?
25. What structural feature of the myelin sheath is responsible for the high conduction velocity of large, myelinated nerve fibers?
26. From what other parts of the brain does the reticular formation receive input?
27. Describe the probable role played by the reticular formation in the "arousal" response.
28. What is the only sensory modality that is not projected to the thalamus?
29. What other special sense is closely related to the sense of taste?

ADDITIONAL READING

Benzinger, T. H.: The human thermostat. Sci. Amer. (Jan.), 1961.
Gatz, A. J.: Clinical Neuroanatomy. 3rd ed. Philadelphia, F. A. Davis Company, 1966.
Katz, B.: The nerve impulse. Sci. Amer. (Nov.), 1952.
Melzack, R.: The perception of pain. Sci. Amer. (Feb.), 1961.
Noback, C. R.: The Human Nervous System. New York, McGraw-Hill Book Company, Inc., 1967.
von Bekesy, G.: The ear. Sci. Amer. (Aug.), 1957.
Wald, G.: Eye and camera. Sci. Amer. (Aug.), 1950.

Chapter 6

THE CIRCULATORY SYSTEM

INTRODUCTION

The circulatory system is the "transportation" system of the body. It carries food and oxygen to the body cells and waste products away from them. It carries hormones, which help regulate body processes, and antibodies, which protect against infections. In addition, this system aids in the control of body temperature and the maintenance of homeostasis.

The parts of the circulatory system are the blood, the heart, the blood vessels, and the lymphatic system.

BLOOD

Blood is a highly specialized kind of connective tissue. It is composed of formed elements (red corpuscles, white cells, and platelets) and liquid intercellular material, the plasma. It gets its red color from an iron-containing pigment called *hemoglobin*, which carries the oxygen in the blood.

Blood is a slightly sticky, or viscous, fluid because of the red blood corpuscles and the proteins in the plasma. The average amount of blood in a normal adult is about 4 to 5 quarts, depending on the size of the individual.

RED BLOOD CORPUSCLES

The erythrocyte, or red blood corpuscle (RBC), is the only "true" formed cellular element of the blood since it alone performs its functions while inside the intact vessels. By definition, the RBC is not a cell since, in its mature form, it lacks a nucleus which is lost during maturation in the bone marrow. Therefore, the presence of nucleated RBC's in the circulating blood is considered an indication of abnormal RBC formation. Through usage, however, the term "cell" is often

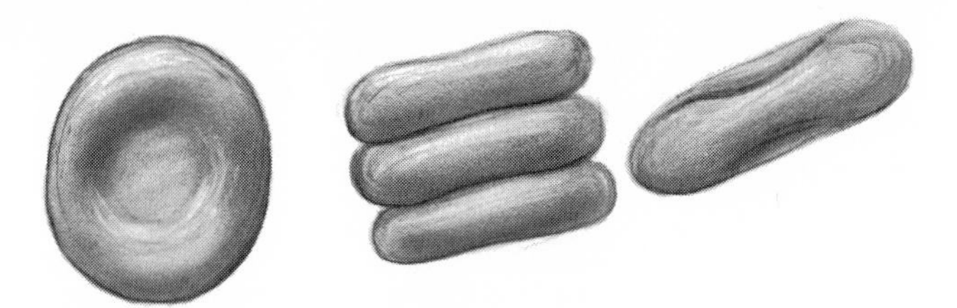

Figure 6–1. Typical red blood corpuscles.

used instead of the more correct "corpuscle." Erythrocytes make up about 45 per cent of the total blood volume; this volume percentage is called the *hematocrit*.

RBC's are round, with both sides being concave, or indented (Fig. 6–1). Because they must squeeze through the tiniest blood vessels, the capillaries, each cell is only 1/3000 of an inch in diameter and has an elastic cell membrane.

The normal adult male has about 4½ to 5 million red corpuscles per cubic millimeter (cu. mm.) of blood or a total of approximately 30 trillion. The red cell count in the adult female is slightly lower, from 4 to 4½ million per cu. mm. An excess of RBC's is called *polycythemia*; an abnormally low count is found in some kinds of *anemia*.

The process of erythrocyte formation is called *erythropoiesis*, and in the adult it takes place in the red bone marrow of the sternum, ribs, scapulae, vertebrae, in the diploë of the cranial bones, and in the proximal ends of the femur and humerus. The average life span of an RBC in the circulating blood is 120 days.

The primary purpose of the erythrocytes is to carry oxygen, which they pick up as they pass through the lung capillaries. The oxygen combines with the hemoglobin (oxyhemoglobin) and is carried to all the body cells. In the body cells the hemoglobin releases its oxygen load, and the blood is returned to the lungs by the veins. Because of its higher oxygen content, arterial blood is a brighter red than venous blood.

WHITE BLOOD CELLS (LEUKOCYTES)

There are five kinds of white blood cells, or leukocytes: neutrophils, eosinophils, basophils, lymphocytes, and monocytes (Fig. 6–2). The first three can be distinguished by the characteristic specific granules in the cytoplasm that have an affinity for certain dyes; hence these cells are called granular leukocytes or granulocytes. Lymphocytes and monocytes are referred to as nongranulocytes, although their cytoplasm may contain some fine, nonspecific granules. The average life span of white cells in the circulating blood is nine days.

Under normal conditions, the granulocytes are formed in the bone marrow by the differentiation of primitive cells, the myeloblasts. Lymphocytes are produced in the lymph nodes, the spleen, the tonsils, and the mucous membranes of the digestive, genitourinary, and respiratory tracts. The origin of monocytes is a matter of debate; possible sources are the lymph nodes, the spleen, and the lymphocytes themselves. As in the case of the red cells, the presence of immature white cells in the circulating blood is abnormal: Their presence there may indicate *leukemia*, a disease of the blood-forming organs.

The normal white cell count in the adult ranges from 5000 to 10,000 per cu. mm. of blood. In addition to the total white count, it is important to know the

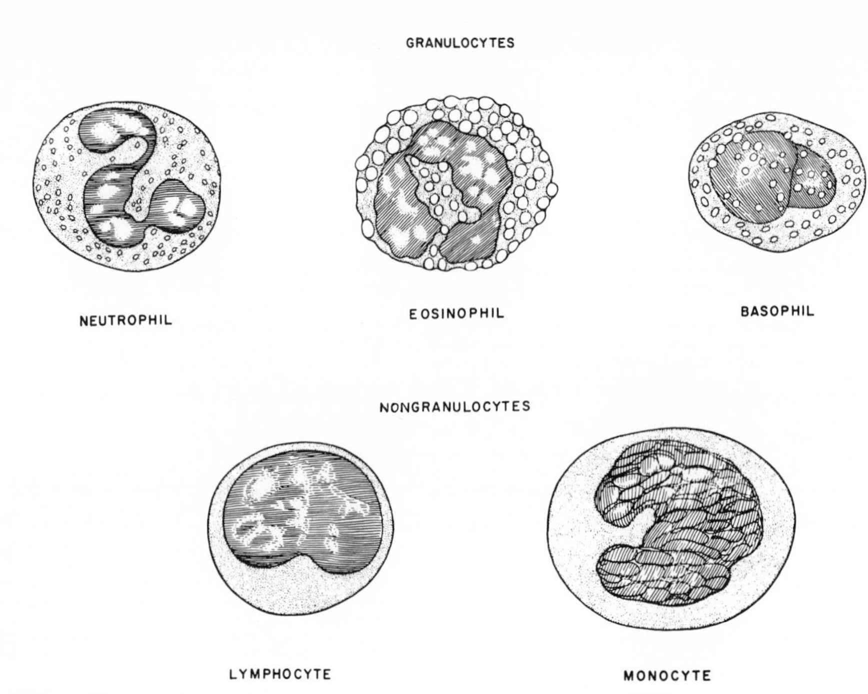

Figure 6–2. Five kinds of white blood cells. Neutrophils contain lilac-colored granules. Eosinophil granules stain orange or red. The cytoplasm of basophils contains granules that stain deep blue or black with basic stains. Lymphocytes have a nucleus that stains a deep purple and a cytoplasm that stains pale blue. The nucleus of a monocyte stains a pale blue; its cytoplasm stains a gray-blue. (Modified from Grollman, S.: The Human Body. New York, The Macmillan Company, 1964.)

number of each of the five kinds of cells. Such a count is called a differential white count.

NEUTROPHILS. These are often referred to as the polymorphonuclear leukocytes, or polys, because the nucleus has from three to five lobes. When the cytoplasm is studied, it shows numerous pale lavender specific granules. Neutrophils are the most numerous of the white cells, composing 50 to 80 per cent of the total count.

EOSINOPHILS. Normally, eosinophils compose 1 to 3 per cent of the total white count. The nucleus has two lobes, and the cytoplasm contains large, round specific granules which stain a bright orange-pink.

BASOPHILS. These are the least numerous of the leukocytes, composing only up to 1 per cent of the total white cell count. Indeed, one may look at a drop of normal blood under the microscope and find not even one basophil. The nucleus is large, dark purple, and blotchy; it may be **S**-shaped or indented on one side. Large, dark purple specific granules in the cytoplasm are characteristic of basophils. They frequently overlie the cell's nucleus.

LYMPHOCYTES. These are the second most numerous of the white cells, composing 20 to 50 per cent of the total. Lymphocytes vary greatly in size. The nucleus is large and usually round but may be slightly indented. The cytoplasm is pale blue and may contain a few very fine granules.

MONOCYTES. Next to basophils, monocytes are found in fewest numbers, composing 2 to 10 per cent of the total count. They are large cells and have a relatively large amount of dull gray-blue cytoplasm in relation to the nucleus.

The granules are usually fine, but may be prominent. The nucleus is kidney-shaped or indented and often has folds or grooves which help identify the cell.

Functions of the White Cells. The polys form a very important part of the body's defense against infections. They are usually the first cells to arrive on the scene in cases of acute inflammation because of their ability to move quickly out of the capillaries and into the tissues. This process is called *diapedesis*. While in the tissues they engulf and destroy bacteria, a process called *phagocytosis*.

Eosinophils also play a part in fighting infections, although they rank behind neutrophils in this respect. They appear on the scene of infection after the acute stage has passed. Their job is to "clean up" dead tissue cells, bacteria, and neutrophils.

The role of basophils is not clearly understood. They are considered by many to be a vascular source of heparin and histamine.

Since lymphocytes possess limited powers of phagocytosis, their part in body defense is based largely on their ability to be transformed into other cells, such as plasma cells and macrophages, which assist eosinophils in the cleaning up process after infections. Lymphocytes are also concerned with antibody formation.

Monocytes, like eosinophils and neutrophils, are active in phagocytosis. They are especially important in chronic infections such as tuberculosis. Although some believe that monocytes produce antibodies, evidence shows that this is probably an indirect function.

Leukocytosis means an increase above the normal number of leukocytes in the circulating blood. This increase may occur as a normal physiological response under certain conditions. Strenuous exercise, onset of labor in the pregnant female, and emotional states—all may be accompanied by leukocytosis. In addition, increased white cell counts are found in certain diseases. (See Table 6–1.)

Table 6-1. SOME CAUSES OF LEUKOCYTOSIS DUE TO DISEASE

Type of Cell Increased	*Causes*
Neutrophil	1. Infections—especially pus-forming bacteria 2. Cancers—leukemia, carcinoma 3. Poisoning from drugs and chemicals 4. Acute hemorrhage
Lymphocyte	1. Infections—infectious mononucleosis, whooping cough, syphilis, tuberculosis 2. Certain leukemias 3. Hyperthyroidism
Monocyte	1. Infections—tuberculosis, typhoid fever, malaria 2. Some leukemias 3. During recovery from infections 4. Polycythemia
Eosinophils	1. Allergic reactions—hay fever, asthma 2. Parasite infections—hookworm, tapeworm trichinosis 3. Skin diseases 4. Carcinoma, Hodgkin's disease

Table 6-2. SOME CAUSES OF NEUTROPENIA

1. Infections—viral (infectious hepatitis); bacterial (typhoid fever); malaria; overwhelming blood infections (septicemia)
2. Bone marrow damage—poisonous drugs and chemicals; drug sensitivity
3. Disorders associated with a large spleen
4. Nutritional deficiencies—vitamin B_{12}; folic acid

A decreased number of circulating leukocytes is called *leukopenia*. Usually the granulocytes are most affected. Table 6–2 lists some of the causes of neutropenia, or reduction in neutrophils.

PLATELETS (THROMBOCYTES)

Platelets are small bits of cytoplasm that have broken off from giant cells in the bone marrow called megakaryocytes. When stained and viewed under the microscope, they look like tiny saucers or plates—hence the name platelets. The normal platelet count is 250,000 to 500,000 per cu. mm. of blood. They live in the circulation about four days. Their chief role is in blood clotting, in which they perform both mechanical and chemical functions. Blood clotting will be described later.

PLASMA AND SERUM

Plasma is the liquid part of the blood, or blood with the cells removed. It consists mostly of water (about 90 per cent) in which small amounts of many important substances are dissolved. Proteins, sugar, fats, vitamins, hormones, inorganic salts (sodium, calcium, magnesium, etc.), gases (oxygen, carbon dioxide), and metabolic waste products (urea, uric acid) are all important in normal body function. When any of these materials is not present in the blood in the correct amount, it is a sign that some organ or system is not working properly. Plasma is often used to give transfusions to persons who have lost a large quantity of blood from severe hemorrhage in order to restore their blood volume quickly.

Serum is that liquid part of the blood that remains after blood clots. In other words, it is plasma minus the fibrin clot.

BLOOD CLOTTING (HEMOSTASIS)

Because everyone has experienced the phenomenon of blood clotting many times, it might seem to be a rather simple process. Actually, the mechanism of *hemostasis* is a very complicated puzzle and has been studied for many years. Only recently have all the pieces been sorted out and fitted together.

Essentially, hemostasis can be considered to involve three mechanisms: clumping (agglutination) of platelets, constriction of blood vessels, and formation of the blood clot. Blood normally does not clot inside intact vessels, but when a vessel is injured, the hemostatic process is triggered.

Platelet agglutination may mechanically plug the injured vessels if they are small enough. In addition, these platelet masses serve as a point where the clot itself begins to form.

Some controversy exists about the role played by vasoconstriction. It is thought by some that damage to a vessel leads to reflex spasm of the vessel, causing its constriction, thus helping to reduce blood loss. This process may be of significance only in larger vessels, since capillaries do not have a muscular coat.

Formation of the clot itself occurs in three stages, each involving the production of a specific chemical substance. Only a general outline of the process will be given here.

In the **first stage**, the interaction of several coagulation factors found in the blood and the tissue juices outside the broken vessel results in the formation of a substance called *thromboplastin*. Blood platelets and calcium are very important in this process.

The **second stage** involves the changing of *prothrombin* to *thrombin*. This change occurs in the presence of thromboplastin and calcium. Prothrombin is a substance formed in the liver, and it requires vitamin K for its production.

The **third stage** is the conversion of *fibrinogen* to *fibrin* in the presence of thrombin. Fibrinogen is a plasma protein and is also made in the liver.

These three steps can be summarized as follows:

STAGE I:

$$\text{Blood factors + tissue factors} \xrightarrow[\text{Ca}^{++}]{\text{platelets}} \text{thromboplastin}$$

STAGE II:

$$\text{Prothrombin} \xrightarrow[\text{Ca}^{++}]{\text{thromboplastin activity}} \text{thrombin}$$

STAGE III:

$$\text{Fibrinogen} \xrightarrow[\text{Ca}^{++}]{\text{thrombin}} \text{fibrin (clot)}$$

There are 11 coagulation factors in all. One of them, calcium, is needed for all three stages. The absence or deficiency of any one of these factors leads to a defect in the clotting mechanism and subsequent prolonged clotting time. For example, hemophilia is an inherited blood disorder affecting predominantly the first stage of coagulation. It is caused by a deficiency of factor VIII, which is also called the antihemophilia factor (AHF). Certain laboratory tests can be performed on a blood sample to determine which factor is missing.

Normal blood plasma has several safeguards which help to maintain the fluidity of the blood in the intact vessels and thus prevent abnormal clotting. Blood contains a small amount of thromboplastin since platelets and other tissues containing this substance are constantly being broken down. This means that some thrombin is also constantly present. However, as long as the thrombin level does not rise above a critical point, the plasma fibrinogen will not be converted to a fibrin clot. Furthermore, any thrombin thus formed is destroyed by the action of plasma *antithrombin* and *heparin*. Heparin is an anticoagulant found in *mast cells* as well as in basophilic leukocytes. Mast cells are present in large numbers throughout the connective tissue in the body and secrete heparin into the body fluids and hence into the blood vessels. In addition, there are fibrinolytic enzymes in the blood which will break down any fibrin clot that might be formed. Perhaps the most important safeguard against intravascular clotting is a structural one—the smooth endothelial lining of the vessels tends to prevent blood elements from adhering to the vessel walls.

ABNORMAL BLOOD CLOTS

A *thrombus* is an abnormal clot which develops in an intact blood vessel. If a thrombus breaks free from its attachment and flows through the vessels, it is then called an *embolus*. Naturally, when an embolus arrives in a vessel whose diameter is too small to allow it to pass, it plugs that vessel, stopping blood flow. If the vessel so plugged is an important one, it could result in grave consequences.

Very briefly, the causes of abnormal clot formation may be included in one of two general categories: (1) roughening of the vessel lining because of some trauma or disease process, and (2) conditions which seriously slow the rate of blood flow. In the latter case, slow flow allows local concentration of thromboplastin to increase to the level necessary for clot formation.

BLOOD TYPES

When a person needs a blood transfusion, his blood type must be determined so that it can be matched with the same type of donor blood. All human blood belongs to one of four basic inherited types: A, B, AB, or O. This classification is based on the presence or absence of two red cell antigens, A and B. An *antigen* is any substance which is foreign to the body. The body's defense is such that it will react to an injected foreign substance by forming *antibodies* to destroy it.

In the case of the blood antigens, if a person with blood type A (he has A antigen in his red cells) is given a transfusion of blood type B, his serum, which has antibodies to these foreign cells, will cause the donor blood cells to clump or *agglutinate*. This reaction can be serious, because the clumped cells may act to plug up the receiver's blood vessels. Therefore, a person with type A blood has antibodies against type B blood. Obviously he does not have antibodies against type A antigens, or he would destroy his own blood cells. Type AB blood has both A and B antigens and therefore will have neither a nor b antibodies.* Persons with type O blood have neither antigen but have antibodies to both. (See Table 6–3.)

Most people (about 47 per cent) in the United States have type O blood. The O antigen is very weak and no antibodies against it develop in the plasma. For this reason, people with type O blood have been called universal donors. It is important to remember that when blood is mismatched, it is the *donor cells* that are clumped by the *receiver's serum*.

---

*Antigens are usually represented with capital letters and antibodies by small letters.

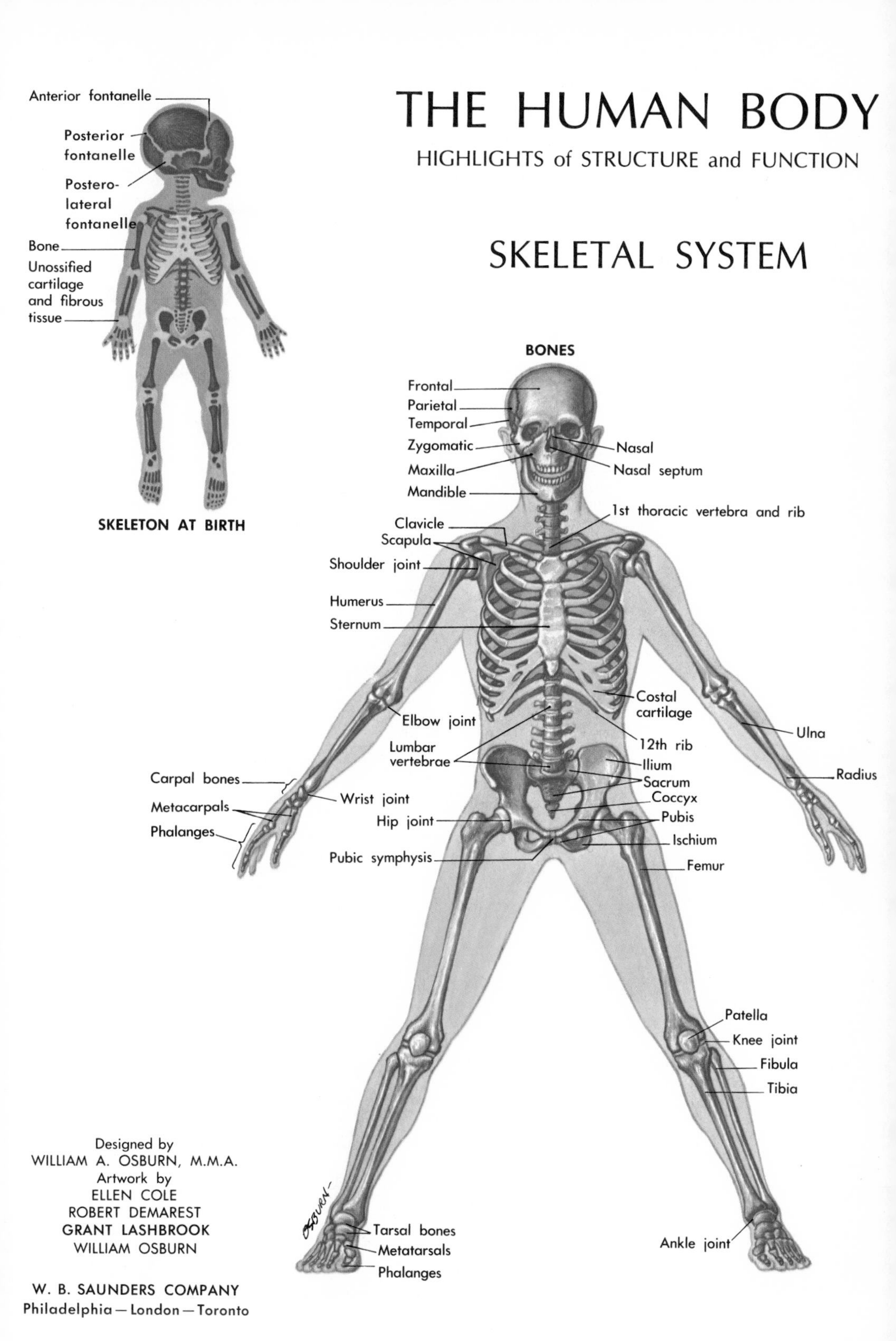

Designed by
WILLIAM A. OSBURN, M.M.A.
Artwork by
ELLEN COLE
ROBERT DEMAREST
GRANT LASHBROOK
WILLIAM OSBURN

W. B. SAUNDERS COMPANY
Philadelphia — London — Toronto

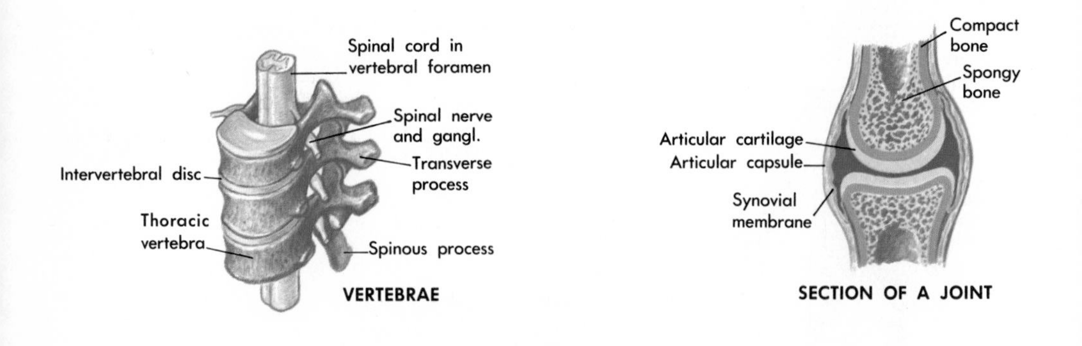
Spinal cord in vertebral foramen
Spinal nerve and gangl.
Transverse process
Intervertebral disc
Thoracic vertebra
Spinous process
VERTEBRAE
Compact bone
Spongy bone
Articular cartilage
Articular capsule
Synovial membrane
SECTION OF A JOINT

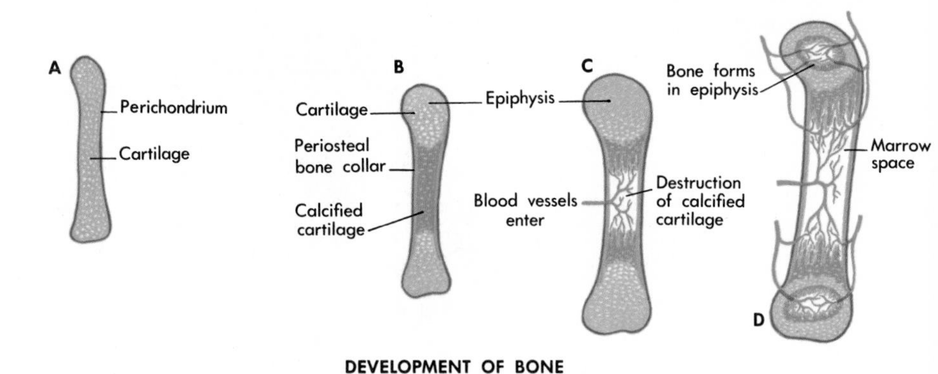
A
Perichondrium
Cartilage
B
Cartilage
Periosteal bone collar
Calcified cartilage
Epiphysis
C
Blood vessels enter
Destruction of calcified cartilage
Bone forms in epiphysis
Marrow space
D
DEVELOPMENT OF BONE

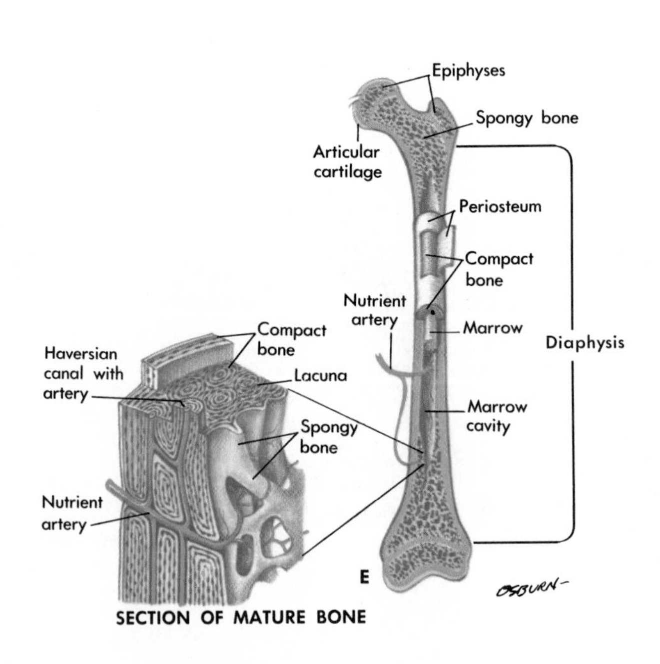
Epiphyses
Spongy bone
Articular cartilage
Periosteum
Compact bone
Nutrient artery
Marrow
Diaphysis
Marrow cavity
Haversian canal with artery
Compact bone
Lacuna
Spongy bone
Nutrient artery
E
OSBURN
SECTION OF MATURE BONE

SKELETAL MUSCLES

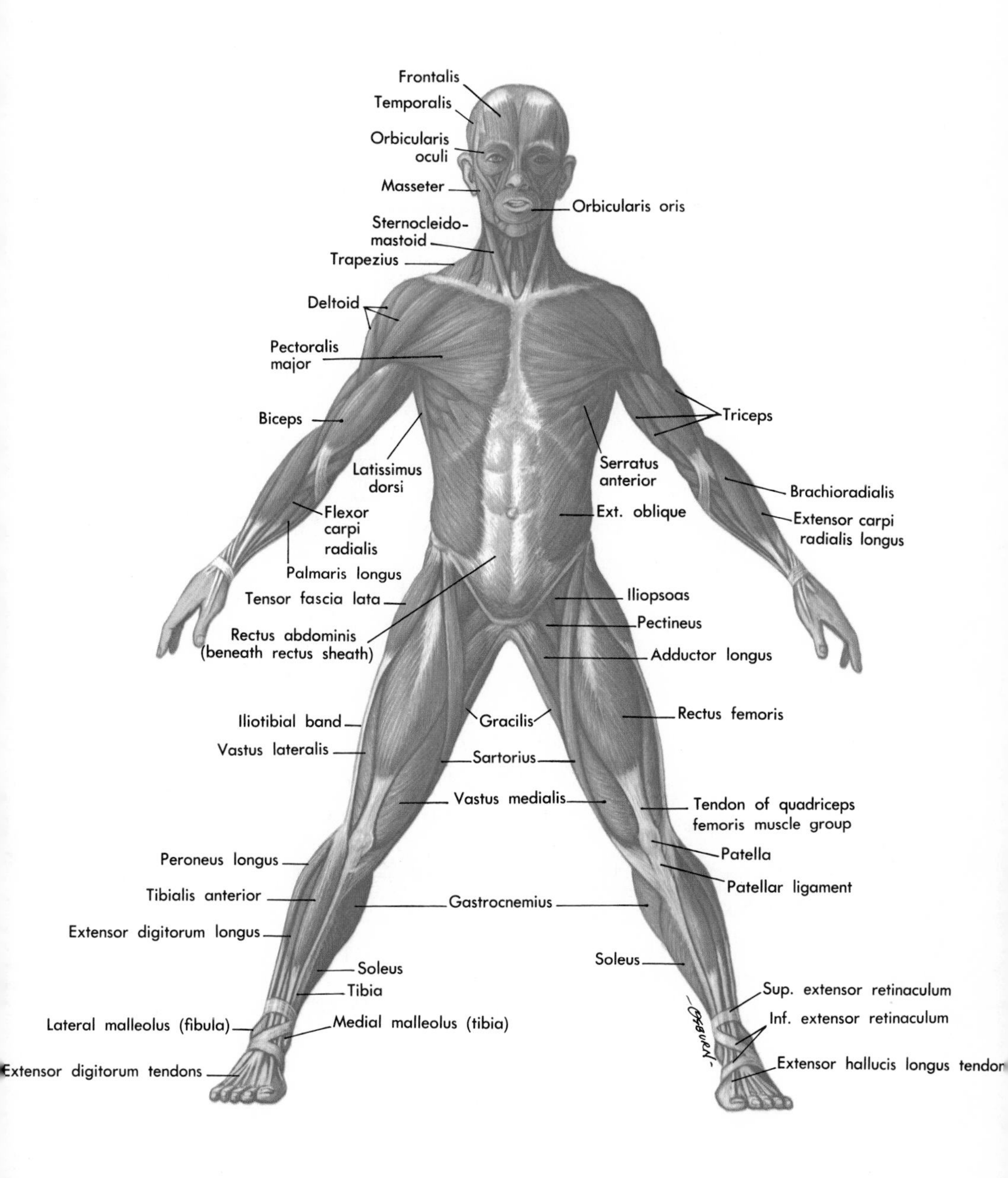

HOW A MUSCLE PRODUCES MOVEMENT

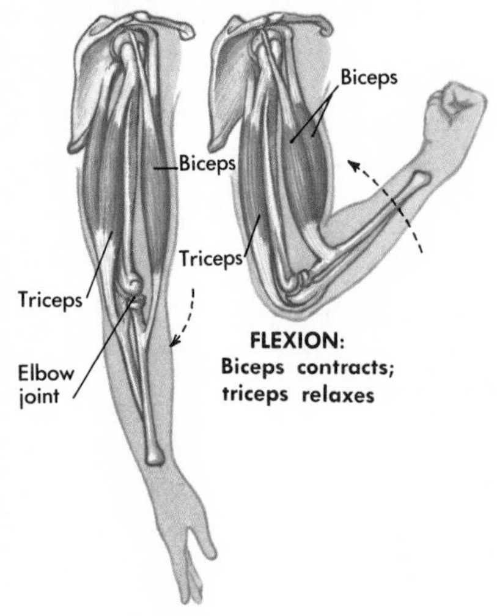

FLEXION:
Biceps contracts;
triceps relaxes

EXTENSION:
Triceps contracts;
biceps relaxes

HOW A MUSCLE ATTACHES TO BONE

Penetrating fibers
Periosteum
Muscle fiber
Int. perimysium
Ext. perimysium
Muscle fasciculus
Tendon

The connective tissue which surrounds the muscle fibers and bundles may (1) form a tendon which fuses with the periosteum, or (2) may fuse directly with the periosteum without forming a tendon.

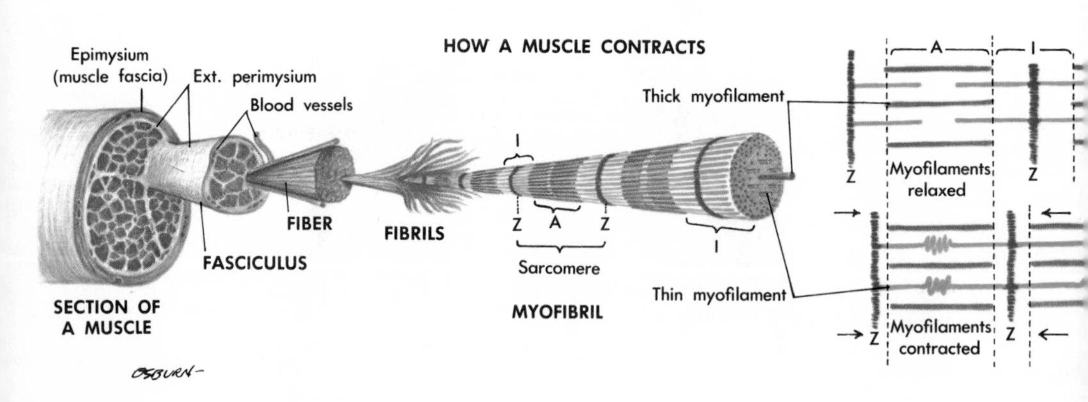

RESPIRATION AND THE HEART

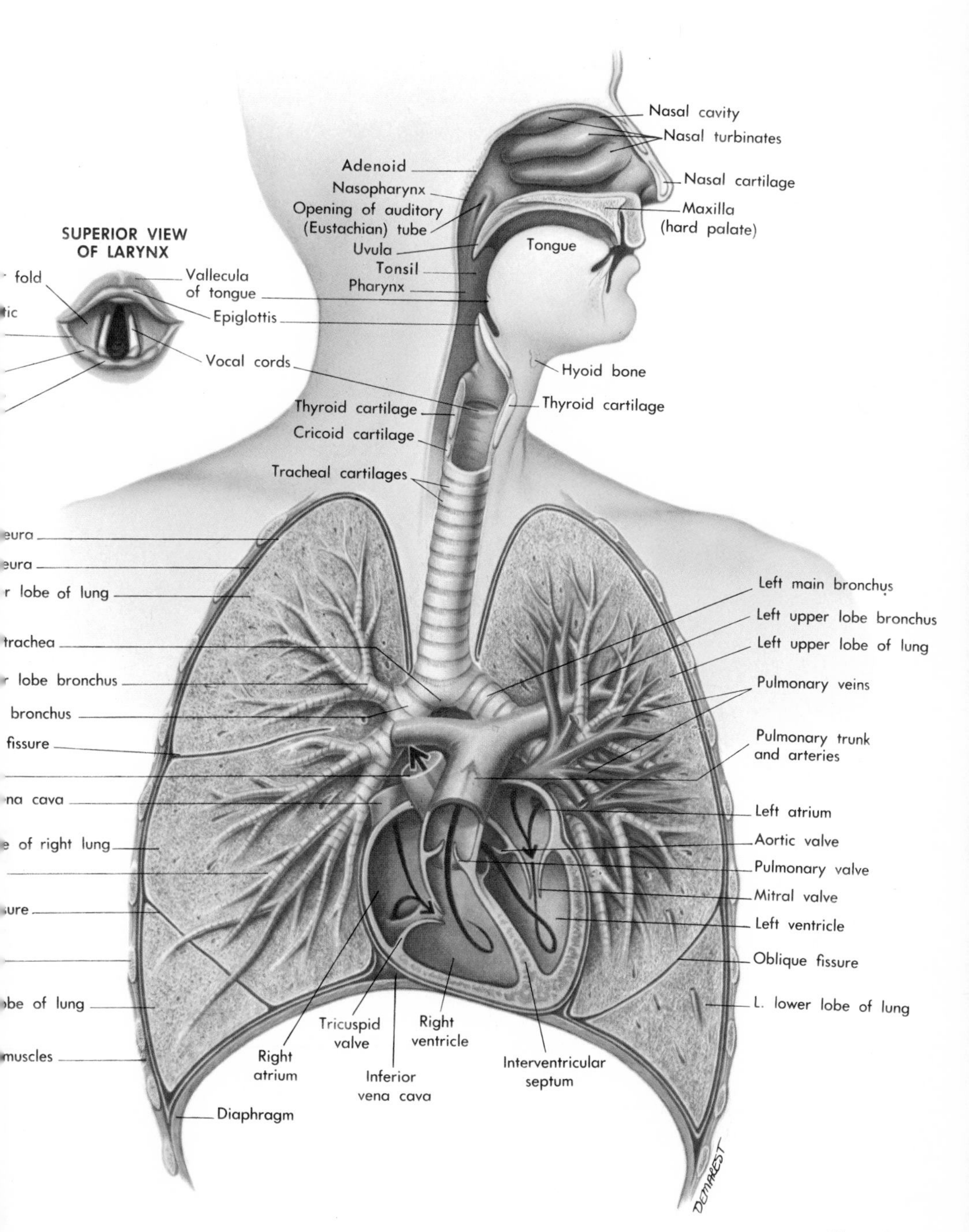

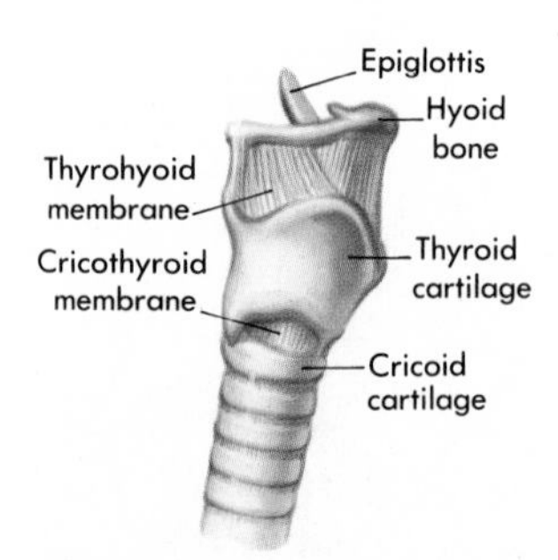

LATERAL VIEW OF THE LARYNX

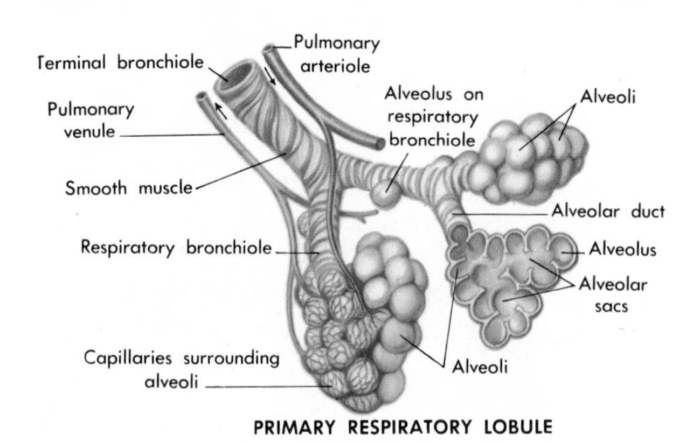

PRIMARY RESPIRATORY LOBULE

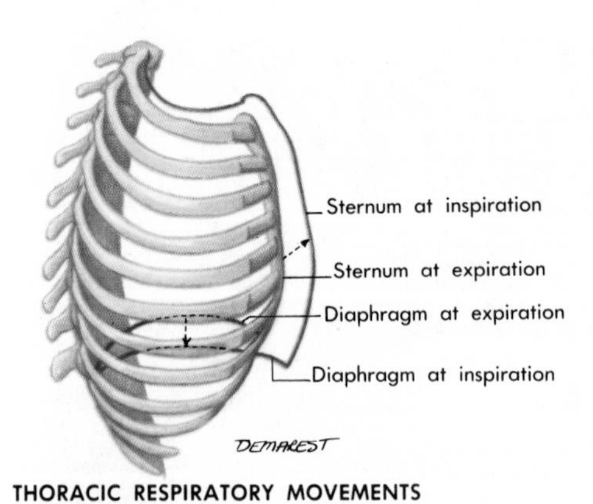

THORACIC RESPIRATORY MOVEMENTS

BLOOD VASCULAR SYSTEM

VEINS

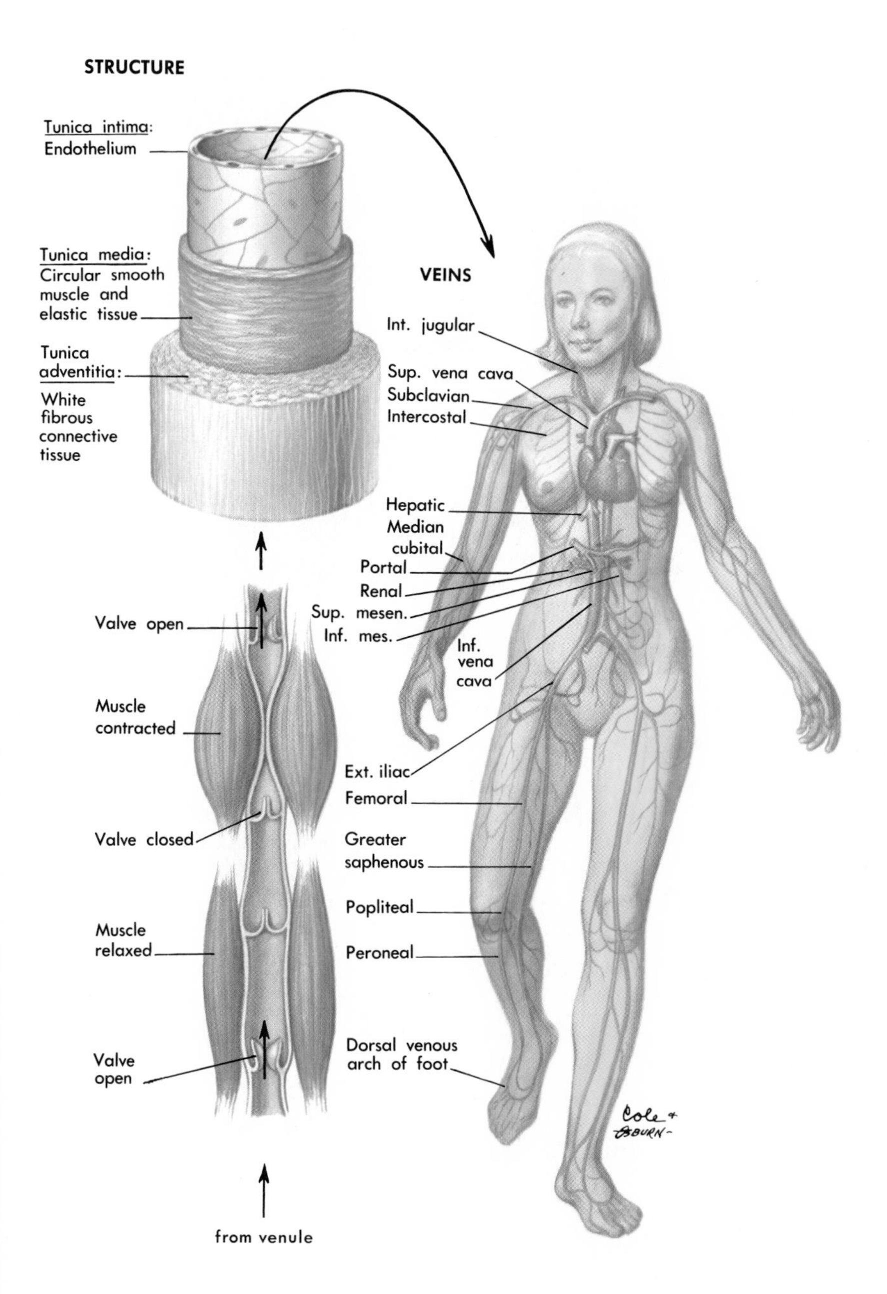

ARTERIES

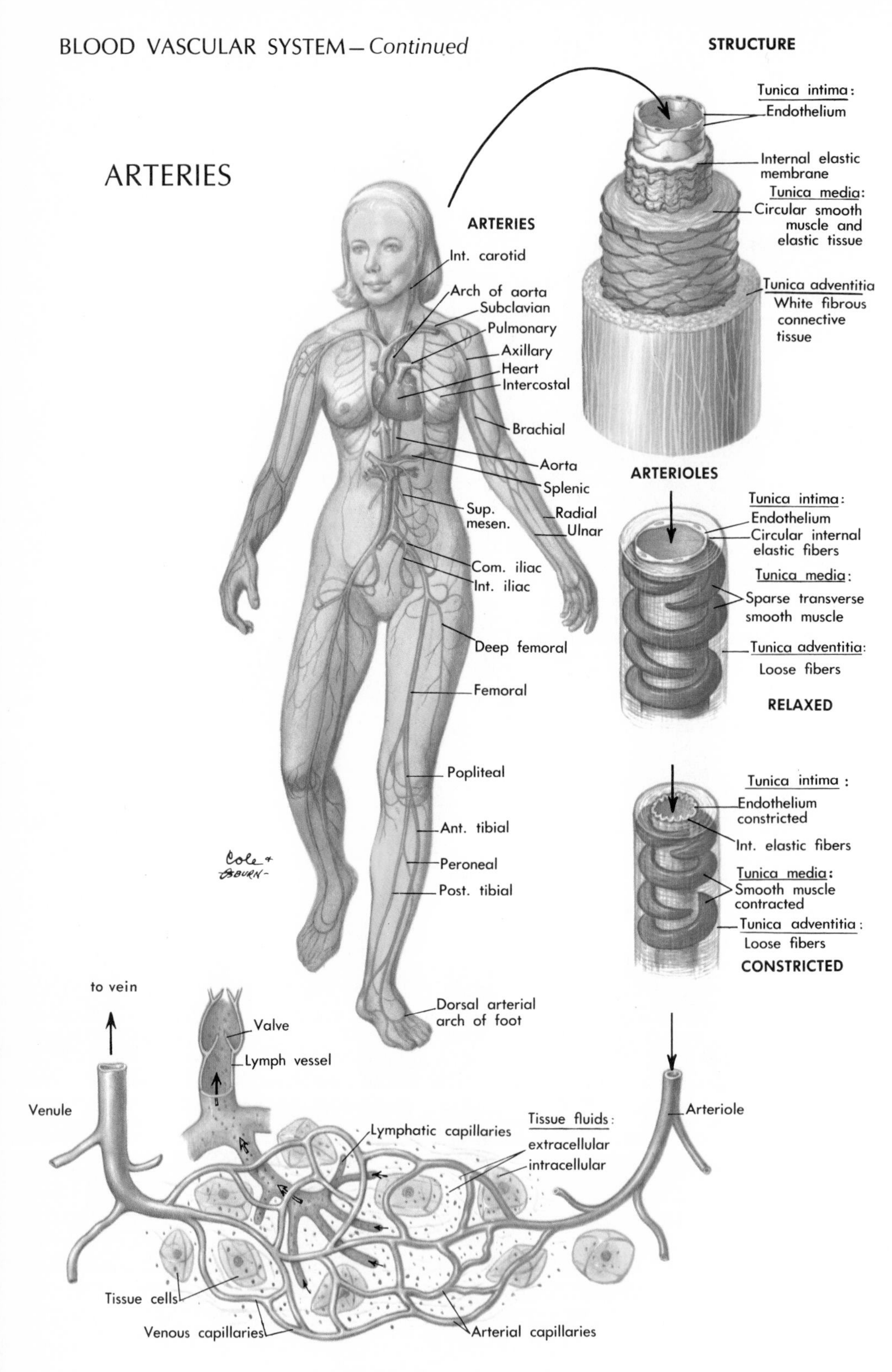

A CAPILLARY BED

DIGESTIVE SYSTEM

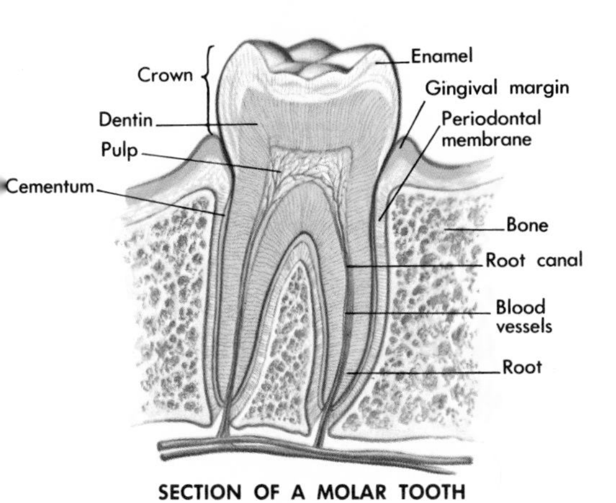

SECTION OF A MOLAR TOOTH

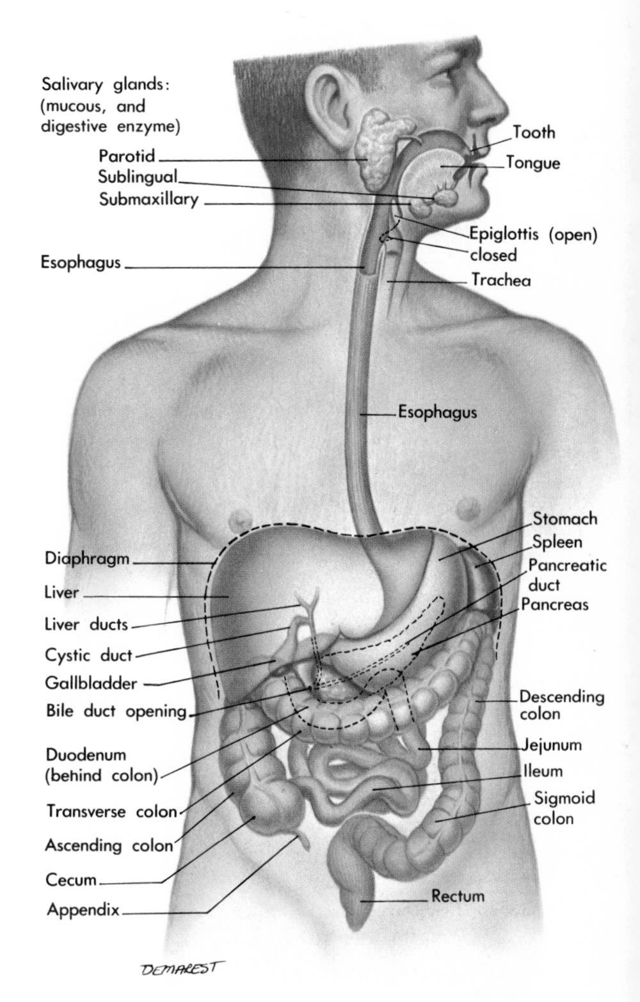

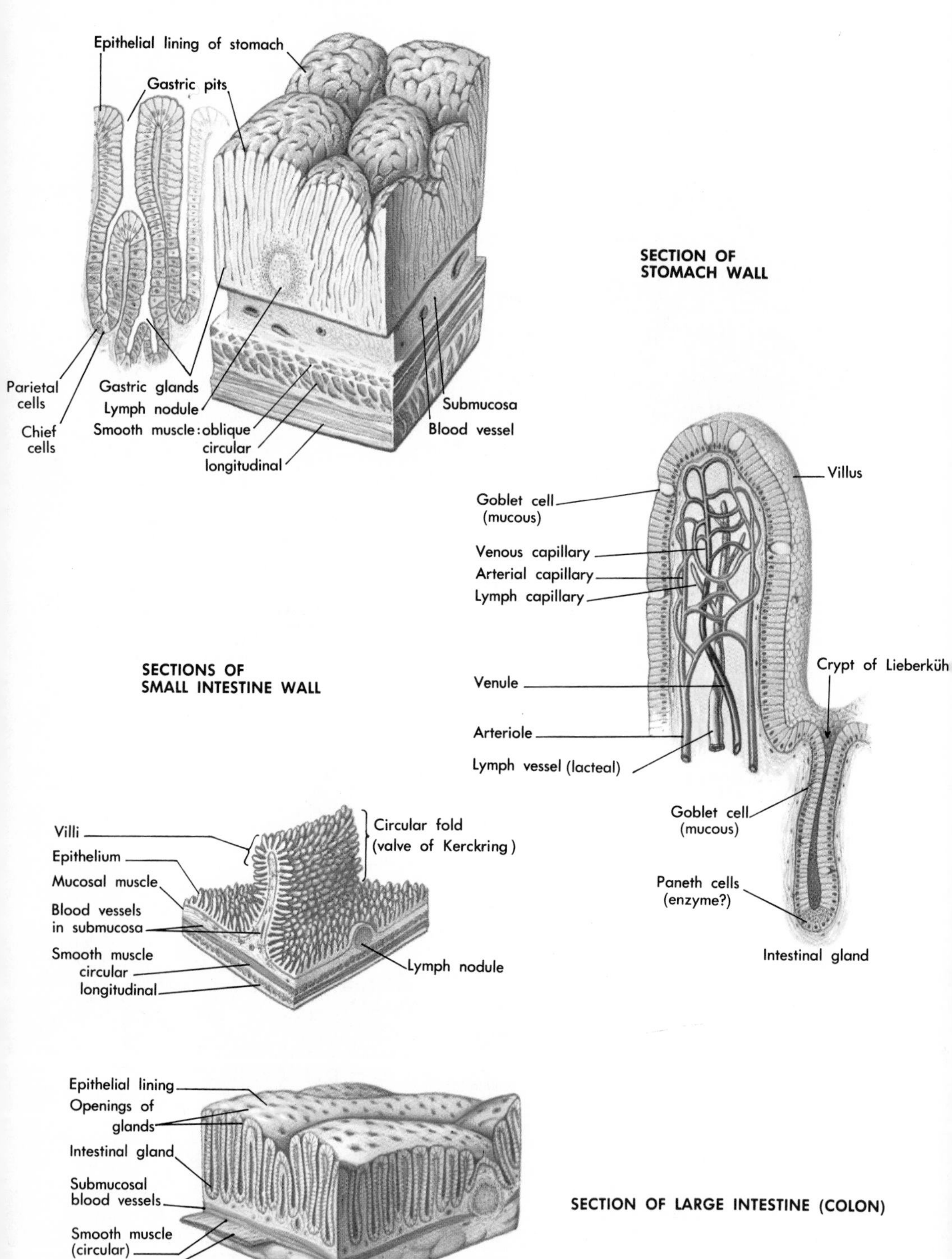
Epithelial lining of stomach
Gastric pits
Parietal cells
Chief cells
Gastric glands
Lymph nodule
Smooth muscle: oblique
circular
longitudinal
Submucosa
Blood vessel
SECTION OF STOMACH WALL
Villus
Goblet cell (mucous)
Venous capillary
Arterial capillary
Lymph capillary
Venule
Arteriole
Lymph vessel (lacteal)
Crypt of Lieberküh
Goblet cell (mucous)
Paneth cells (enzyme?)
Intestinal gland
SECTIONS OF SMALL INTESTINE WALL
Villi
Epithelium
Mucosal muscle
Blood vessels in submucosa
Smooth muscle
circular
longitudinal
Circular fold (valve of Kerckring)
Lymph nodule
Epithelial lining
Openings of glands
Intestinal gland
Submucosal blood vessels
Smooth muscle (circular)
Longitudinal muscle band
DEMAREST
SECTION OF LARGE INTESTINE (COLON)

GENITOURINARY SYSTEM

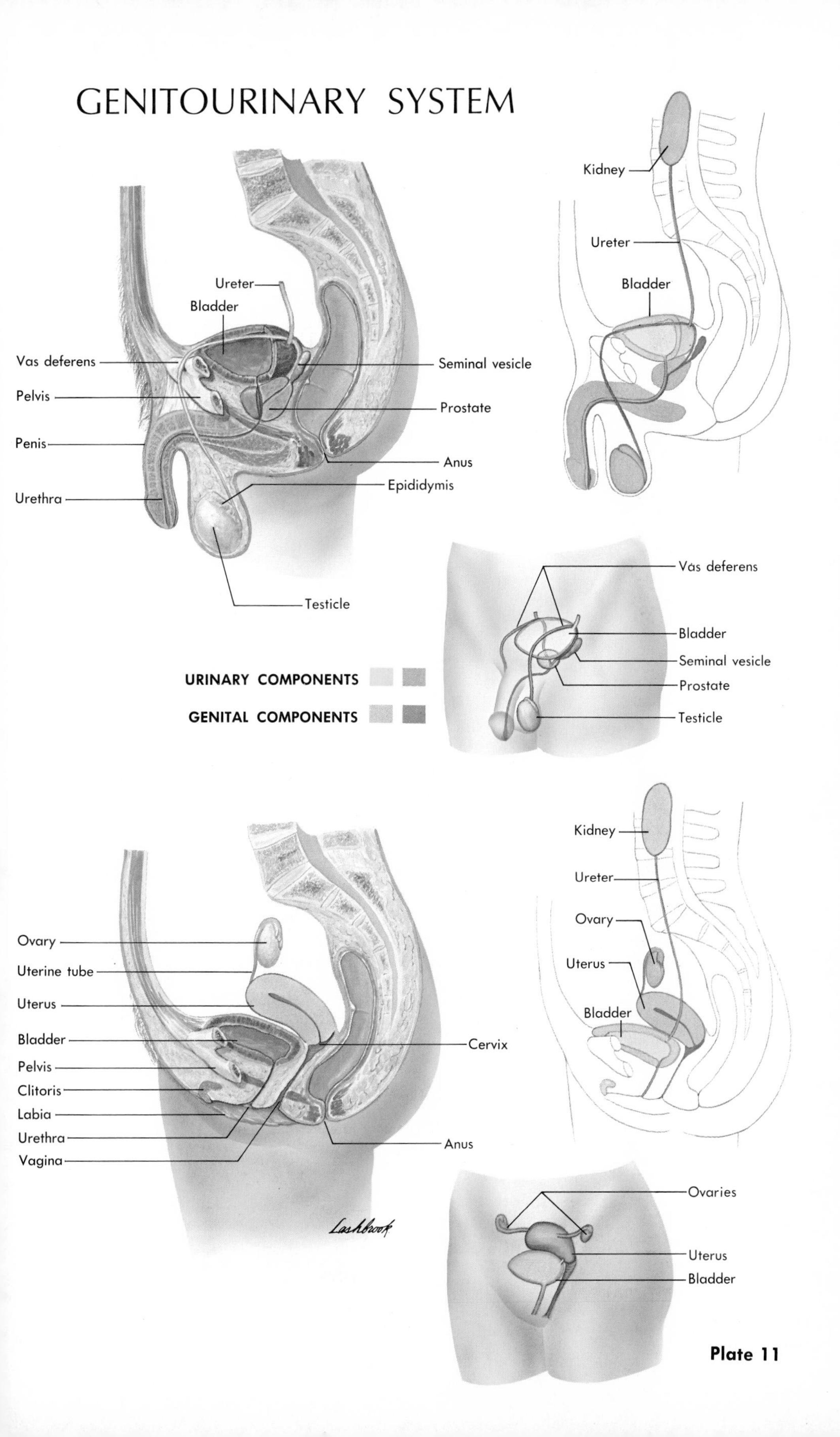

Plate 11

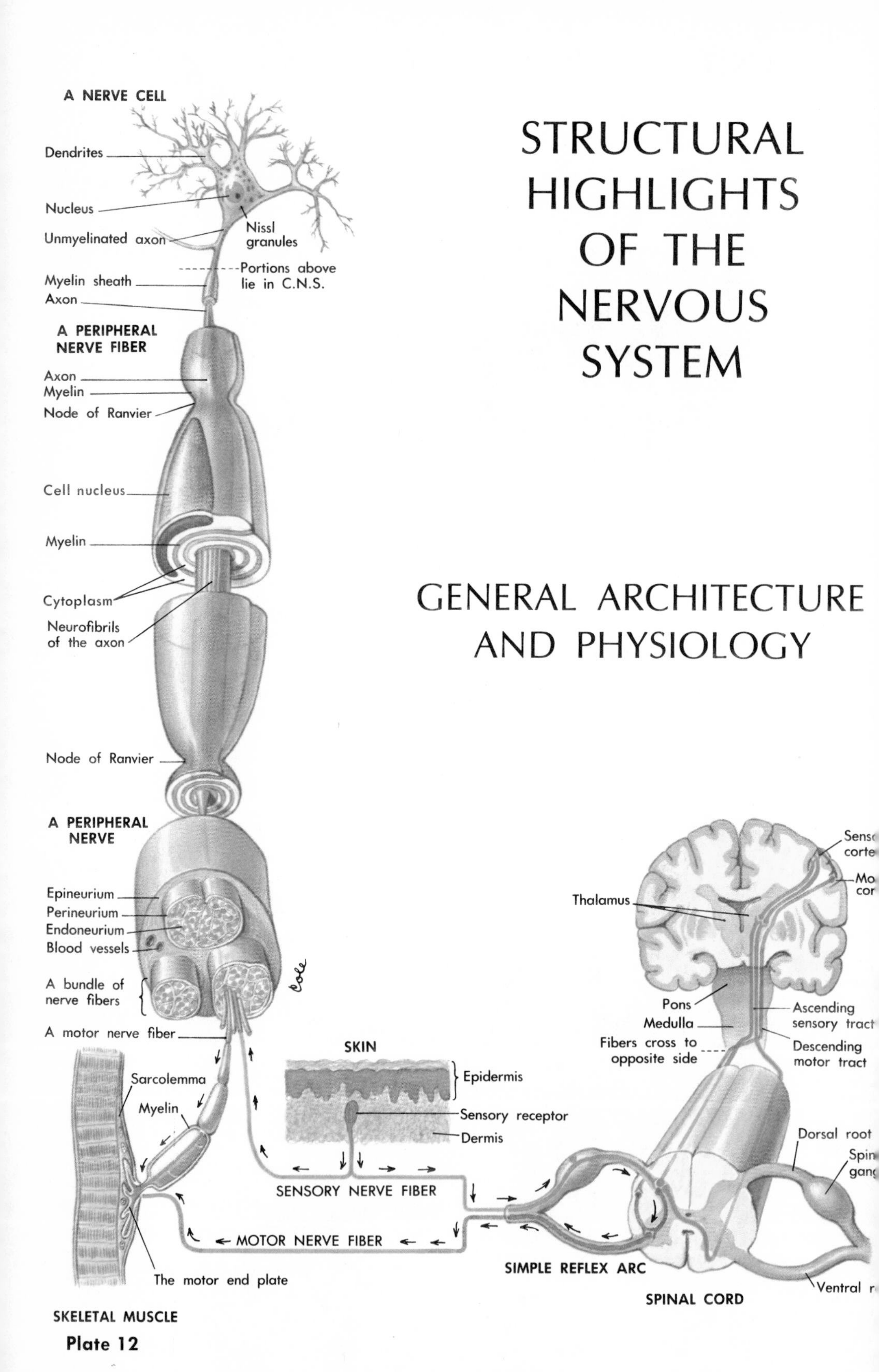
STRUCTURAL HIGHLIGHTS OF THE NERVOUS SYSTEM
GENERAL ARCHITECTURE AND PHYSIOLOGY
A NERVE CELL
Dendrites
Nucleus
Unmyelinated axon
Nissl granules
Portions above lie in C.N.S.
Myelin sheath
Axon
A PERIPHERAL NERVE FIBER
Axon
Myelin
Node of Ranvier
Cell nucleus
Myelin
Cytoplasm
Neurofibrils of the axon
Node of Ranvier
A PERIPHERAL NERVE
Epineurium
Perineurium
Endoneurium
Blood vessels
A bundle of nerve fibers
A motor nerve fiber
Cole
Sarcolemma
Myelin
The motor end plate
SKELETAL MUSCLE
SKIN
Epidermis
Sensory receptor
Dermis
SENSORY NERVE FIBER
MOTOR NERVE FIBER
Thalamus
Pons
Medulla
Fibers cross to opposite side
Ascending sensory tract
Descending motor tract
Dorsal root
SIMPLE REFLEX ARC
SPINAL CORD
Plate 12

BRAIN AND SPINAL NERVES

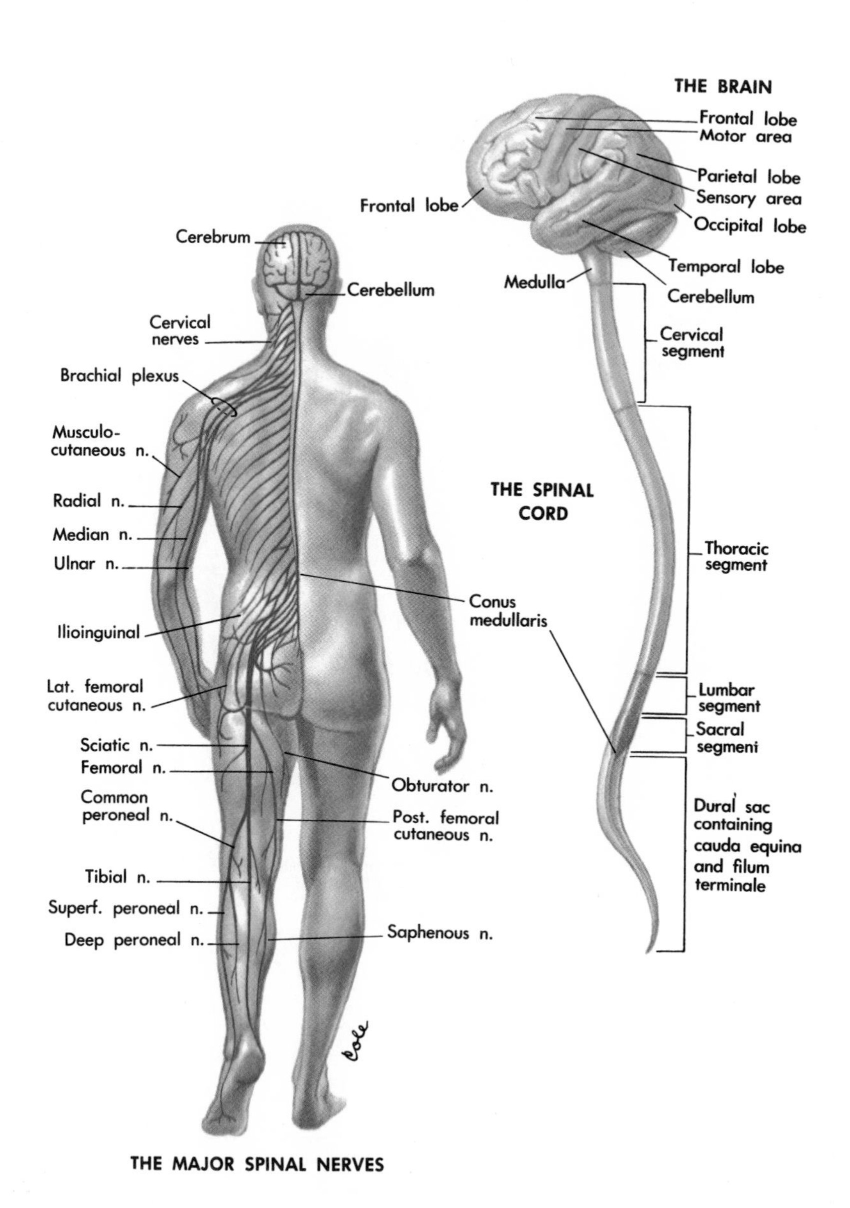

THE MAJOR SPINAL NERVES

AUTONOMIC NERVES

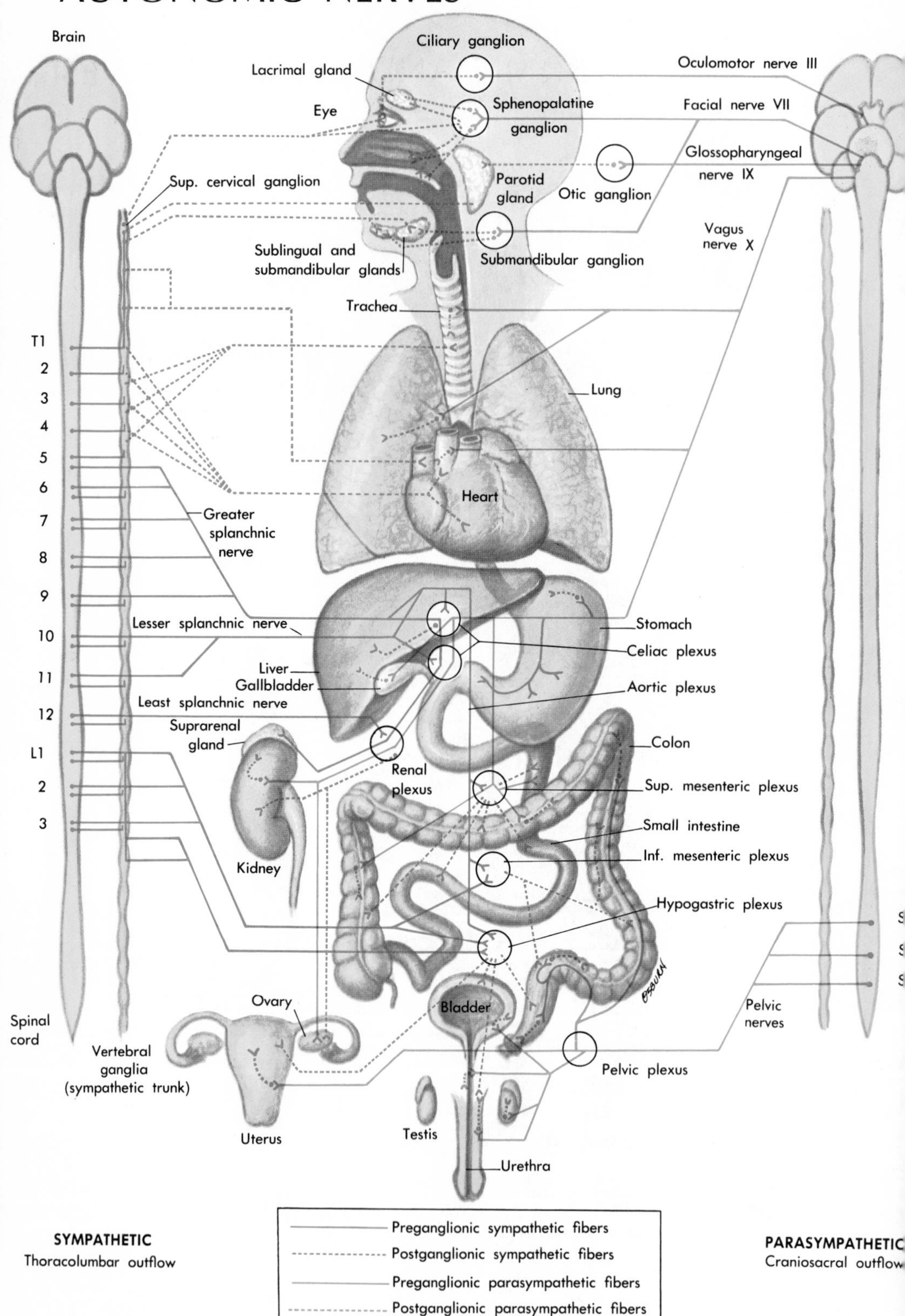

SYMPATHETIC
Thoracolumbar outflow

PARASYMPATHETIC
Craniosacral outflow

Plate 14

ORGANS OF SPECIAL SENSE

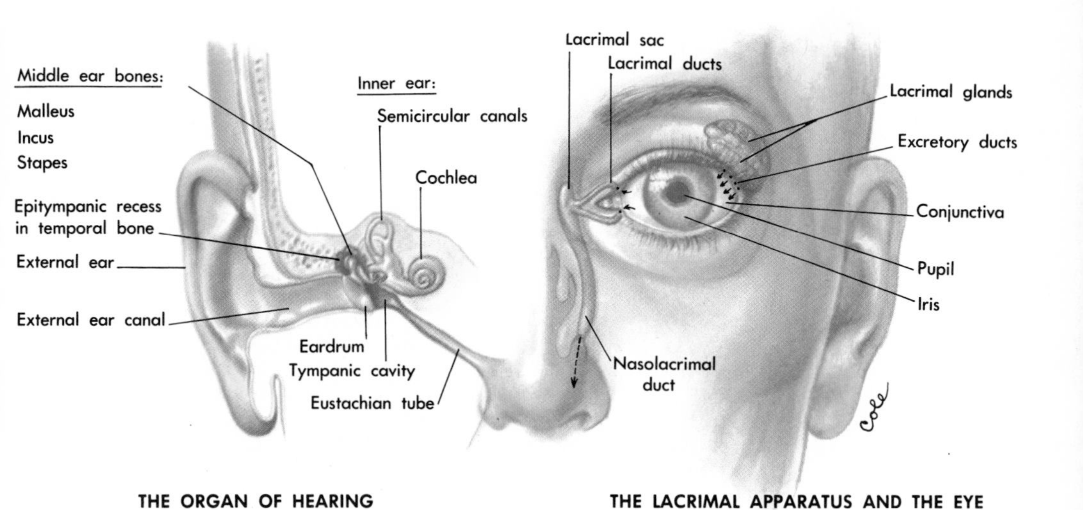

THE ORGAN OF HEARING

THE LACRIMAL APPARATUS AND THE EYE

HORIZONTAL SECTION OF THE EYE

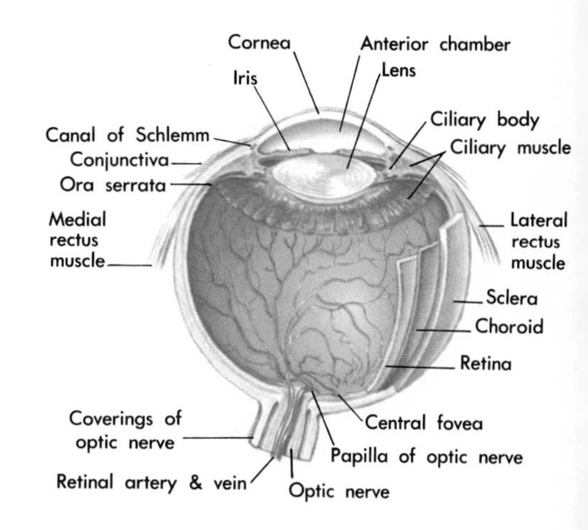

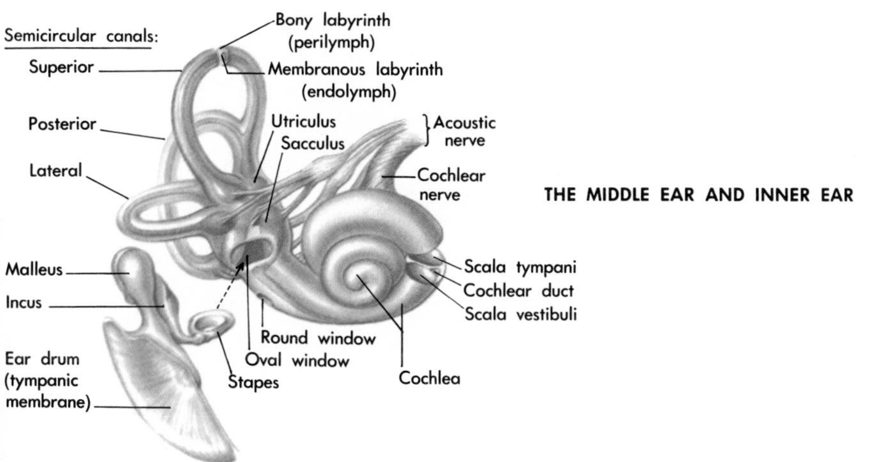

THE MIDDLE EAR AND INNER EAR

PARANASAL SINUSES

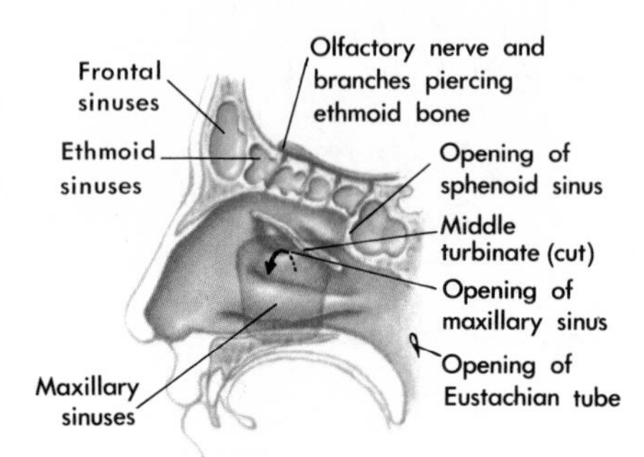

SAGITTAL SECTION OF THE NOSE

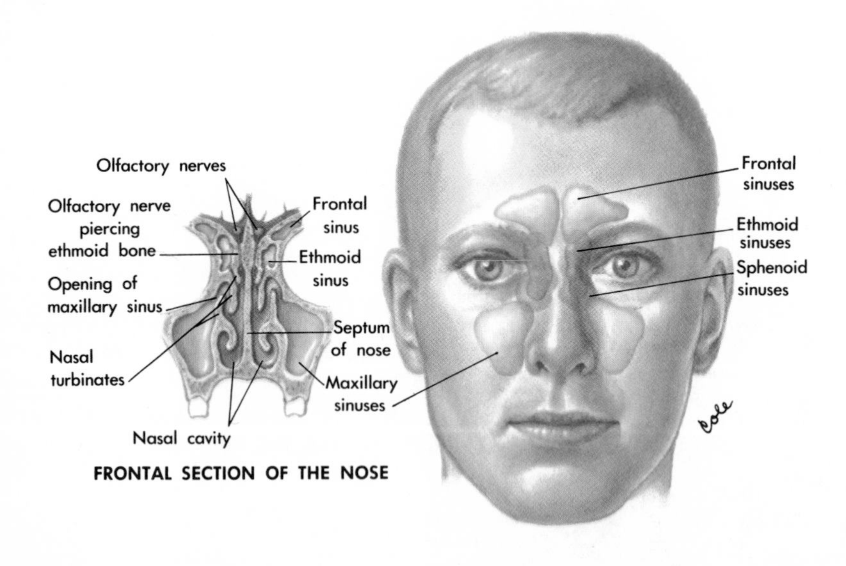

FRONTAL SECTION OF THE NOSE

Plate 16

Table 6-3. ANTIGENS AND ANTIBODIES

Blood Type	*Antigen Present in Red Cells**	*Antibodies Present in Serum*
A	A	b
B	B	a
AB	AB	neither a nor b
O	O	both a and b

RH FACTOR

Another blood antigen of clinical importance is the Rh factor. Named after the rhesus monkey, it is found in about 85 per cent of the population. Such people are said to be Rh positive, and the 15 per cent lacking the factor are called Rh negative.

The presence or absence of the Rh factor is important in the case of an Rh positive fetus with an Rh negative mother. This incompatibility between mother and fetus may lead to the destruction of the baby's red blood cells by the mother's antibodies to the factor which cross the placenta. The condition is called *erythroblastosis fetalis* or hemolytic disease of the newborn.

THE HEART

LOCATION AND STRUCTURE

The heart is the pump of the vascular system. Its job is to pump enough blood under sufficient pressure to meet the needs of the body cells and to keep the blood constantly circulating in the vessels. It is a four-chambered muscular organ, about the size of a man's fist, and it lies in the chest cavity. About two thirds of the heart lies to the left of the midline, with the pointed end, or apex, resting on the diaphragm. The apical beat can usually be felt between the fifth and sixth ribs on a line with the midpoint of the left clavicle in the reclining subject. The base, or broad end, of the heart extends just above the third rib (Fig. 6–3).

A loose-fitting sac of tough white fibrous tissue, the *pericardium*, encloses the heart. It is lined by a double layer of serous membrane, the inner layer forming the outer layer of the heart itself, the *epicardium*. Between these two layers is 30 to 50 cubic centimeters (cc.) of clear fluid, which lubricates the surfaces and prevents friction from rubbing as the heart contracts.

Under the epicardium is the main layer of the heart, the *myocardium*. It is composed of cardiac muscle tissue. (See p. 51). The lining of the heart, a thin connective tissue membrane, is called the *endocardium*.

The heart is separated into two parts by a lengthwise septum or wall of tissue. Each part is composed of an upper chamber, the atrium, and a lower chamber, the ventricle. The two atria receive blood coming into the heart. The right atrium is larger than the left, but it has thinner walls. The larger, thicker-walled ventricles pump blood out of the heart. Each has a capacity of about 85 milliliters (ml.) of blood.

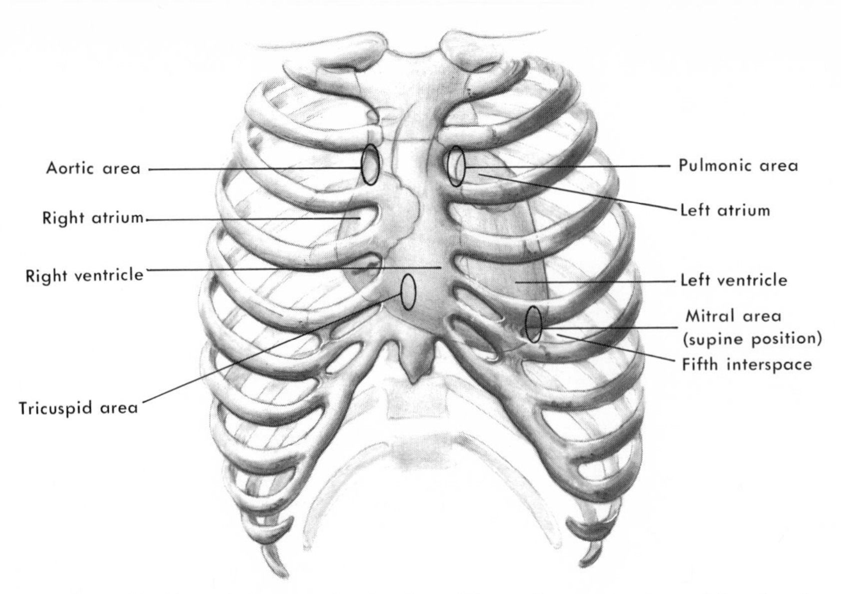

Figure 6–3. Position of the heart in the chest. The surface projections of the chambers and the auscultatory (listening) valve areas are indicated. (The anatomic valve areas are more closely grouped than the auscultatory areas.)

VALVES AND OPENINGS (Fig. 6–4)

There are three openings into the right atrium. One is the *coronary sinus*, which returns venous blood collected from the veins draining the heart muscle itself. The other two are the superior vena cava and the inferior vena cava openings. These vessels return all the venous blood from the organs of the body. There are no valves guarding these openings. Between the right atrium and the right ventricle is an opening guarded by the *tricuspid valve* (three flaps or cusps). This valve opens to allow blood to flow from atrium to ventricle, but then closes to prevent any backflow from ventricle to atrium. The underside of each flap is anchored to the walls of the ventricle by strong cords, the *chordae tendineae*.

The pulmonary artery leaves the right ventricle carrying venous blood from the heart to the lungs to be reoxygenated. Its opening is guarded by the *pulmonary semilunar* valve, which closes after blood is pumped into the artery.

Four pulmonary veins, two from each lung, bring freshly oxygenated blood into the left atrium. Like the venae cavae openings, there are no valves at the entrance of these veins. The *bicuspid* or *mitral* valve guards the opening between the left atrium and the left ventricle. It prevents leakage of blood back into the atrium. Like the tricuspid valve, its leaflets are anchored to the walls of the ventricle by the chordae tendineae, which arise from the papillary muscles.

Oxygenated blood leaves the left ventricle of the heart through the aorta. Its opening is guarded by the *aortic semilunar valve*, which closes as a result of the pressure of the blood in the aorta to prevent any backflow into the ventricle.

If any of the four valves (tricuspid, mitral, pulmonary, or aortic semilunar valves) becomes diseased, as from rheumatic fever, it may not close properly. Blood may then leak back into the wrong chamber and cause a heart murmur.

FUNCTION

The thick, muscular wall of the heart contracts and relaxes in rhythmical fashion, receiving and pumping out blood continuously. The two atria are filled with blood from their respective veins and send it through the atrioventricular openings into the ventricles. When the walls of the ventricles contract, this blood is sent out under pressure into the aorta and pulmonary artery. When the tricuspid and mitral valves close, they make the low-pitched first heart sound. The snapping shut of the two semilunar valves causes the high-pitched second heart sound. If one listens to the heart (auscultation), he will hear these two sounds which resemble the syllables "lubb-dup."

CONTROL MECHANISMS

The rhythmical nature of cardiac muscle contraction is an inherent property of this tissue—that is, the heart beat originates in and is conducted through the heart without extrinsic stimulation. This cardiac *conduction system* (Fig. 6–5) consists of specialized muscle located in certain regions of the heart. A small mass or *node* of such tissue is the sinu-atrial or *S-A node* situated in the posterior wall of the right atrium. Another mass, the atrioventricular or *A-V node* is located in the interatrial septum, close to the opening of the coronary sinus into the right atrium. From the A-V node, a bundle of fibers, the *A-V bundle* (formerly called the bundle of His), extends down into the interventricular septum where it

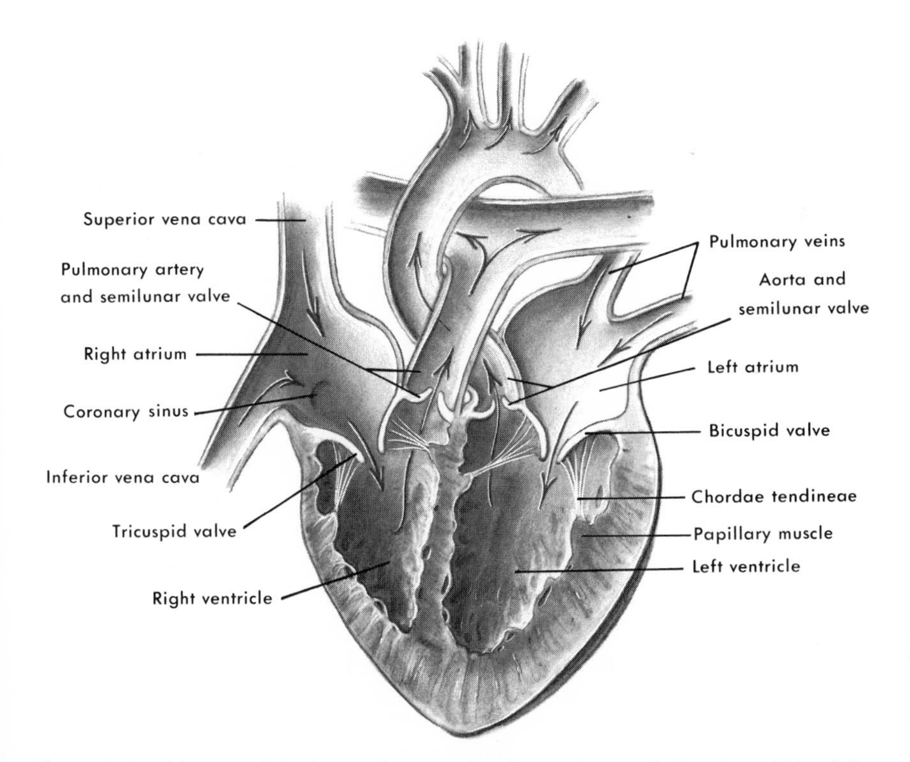

Figure 6–4. Diagram of the heart, showing chambers, valves, and direction of blood flow.

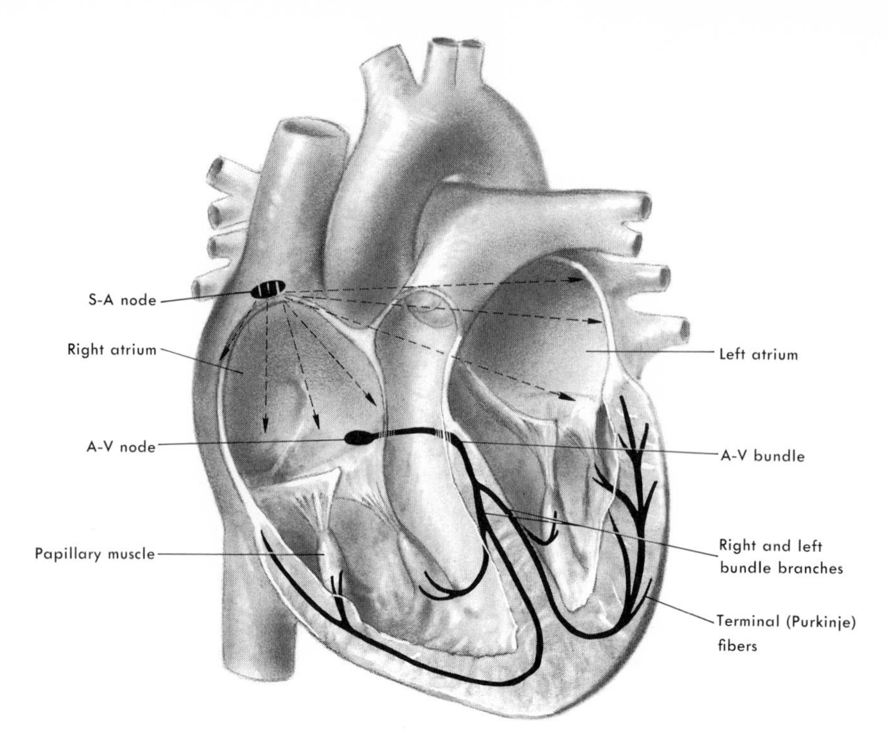

Figure 6–5. Diagram of the conduction system of the heart.

divides into right and left branches. The terminal portions of these bundle branches, the *Purkinje fibers*, spread throughout the ventricular musculature, coming into contact with the regular cardiac muscle fibers.

Evidence shows that the heart beat originates in the S-A node, and that alterations in heart rate are governed by this node. Hence it has been named the "pacemaker" of the heart. The contraction thus initiated spreads through the muscle of both atria. The wave of atrial contraction is then picked up by the A-V node and spreads to the A-V bundle. From here, via the bundle branches and Purkinje fibers, the wave of contraction spreads throughout the ventricular walls, including the papillary muscles.

The heart is supplied by the autonomic nervous system, but these nerves serve to alter the heart rate and are not responsible for the heartbeat itself. Sympathetic nerve endings supply the S-A node, the A-V node, the atria, and the ventricles. Parasympathetic fibers of the vagus nerve end near the S-A node and in the atria, but are absent from the ventricles. Stimulation of the parasympathetic fibers slows the heart rate and the strength of atrial contraction, while sympathetic stimulation causes an increase in the rate and strength of atrial and ventricular contraction. When the heart is beating at a resting rate of approximately 70 beats per minute, the elapsed time for one complete wave of contraction is 0.8 of a second. Exercise, emotions, and body temperature changes all influence the heart rate.

Because of disease, transmission of contraction through the heart may be blocked at one or more points along the conducting system. This can result in an *ectopic* heartbeat—that is, one originating somewhere other than in the S-A node. For example, if the block should occur somewhere between the S-A node and the atrial muscle, some part of the atria themselves may become the pacemaker for these chambers as well as for the ventricles. When the block occurs between atria and ventricles, the ventricular muscle beats with its own independent rhythm. If heart-block is severe and prolonged, the person may require implantation of an artificially powered pacemaker which assumes the function of maintaining normal conduction of the heartbeat.

The heartbeat is also affected by the body concentration of two chemicals, potassium and calcium. Alterations in the amounts of these ions can seriously affect the rate and strength of cardiac muscle contraction. Each chemical produces the opposite effect to that produced by the other, so it is essential that the proper ratio between the two be relatively constant in the body fluids in order for the heart to perform properly.

THE FETAL HEART

Since before birth the lungs of the fetus do not function, blood cannot be pumped to them for oxygenation. The inferior vena cava, in addition to carrying venous blood from the lower part of the body, also carries the freshly oxygenated blood from the placenta (via the umbilical vein) to the right side of the heart. This placental blood must then be sent over to the left side of the heart so it can be pumped out into the systemic circuit. To allow this transfer, an opening, the *foramen ovale*, is present between the two atria. Thus, incoming blood passes from the right to the left atrium and then down into the left ventricle.

Normally, the foramen ovale closes shortly after birth. As the lungs become inflated and breathing begins, pressures in the two atria equalize, allowing the atrial walls to join together and form an interatrial septum. The site of the foramen ovale is marked in the adult heart by an oval-shaped depression, the *fossa ovalis*. If the foramen ovale fails to close properly after birth, it will allow the mixing of blood in the heart. This failure results in what is called a *patent foramen ovale*, one of the conditions causing a so-called "blue baby."

BLOOD VESSELS

There are three kinds of blood vessels: arteries, which carry blood away from the heart; veins, which carry blood to the heart; and capillaries, which connect arteries and veins. The order of blood flow then is:

Heart----> arteries----> capillaries----> veins----> heart

ARTERIES

The arteries have thicker walls than the other vessels. They are composed of three layers of tissue, the middle layer containing smooth muscle and elastic tis-

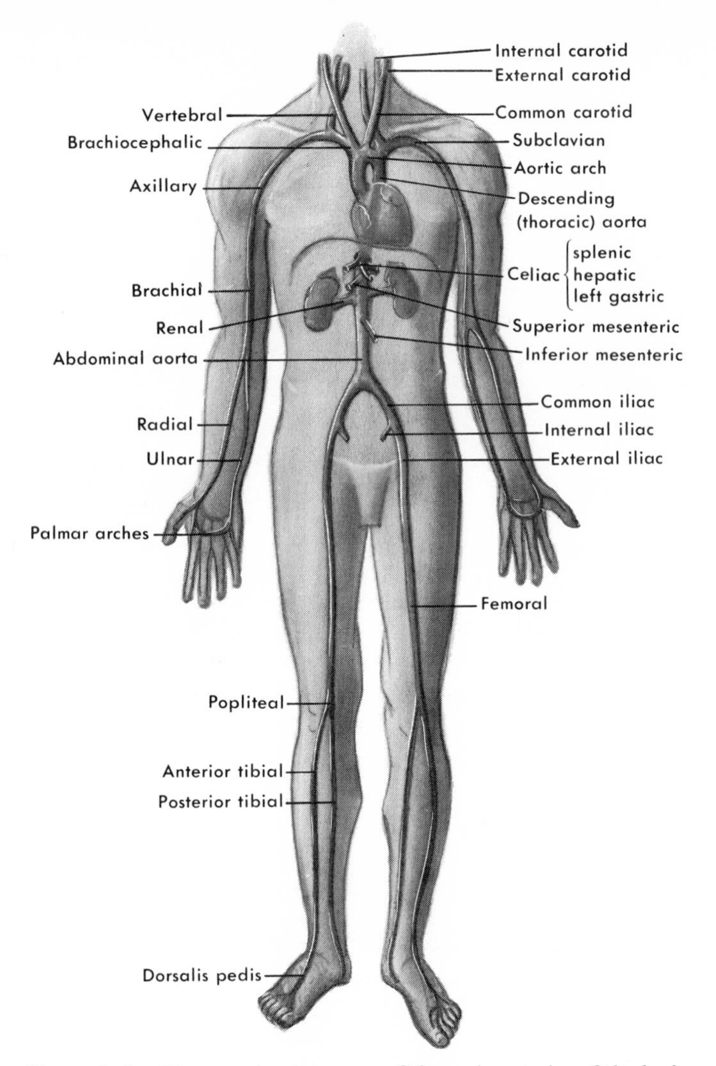

Figure 6–6. Diagram showing some of the main arteries of the body.

sue. The aorta, which leaves the left ventricle of the heart, is the largest artery in the body, being 1 to 1½ inches in diameter. Slightly smaller arteries branch off from the aorta, and these in turn give rise to still smaller branches. This branching continues until the smallest of the arteries, the arterioles, are reached. Blood flows out of them into the capillaries. Figure 6–6 shows some of the main arteries in the body.

CAPILLARIES

The wall of a capillary consists of only one layer of cells. Many capillaries are only wide enough to let red blood cells pass through in single file. There are literally miles of capillaries in the body; if all of them were laid end to end, they would form a tube 62,000 miles long. Capillaries are functionally the most important part of the circulatory system because it is through their walls that all the oxygen, nutrients, and waste products pass between the blood and the body cells. Blood flows slowly through the capillaries to allow these exchanges to take place. At the ends of the capillary beds, blood flows into the smallest of the veins, the venules.

VEINS

Veins are also composed of three layers of tissue, but unlike the arteries, their walls are thin. The venules empty blood into larger veins, and these in turn drain into still larger veins. Thus, compared to the arterial system, the venous system exhibits a kind of reverse branching. Because the pressure in the veins is low, these vessels are equipped with valves to prevent backflow of blood. If these valves break down, varicose veins result. Figure 6–7 shows some of the main veins.

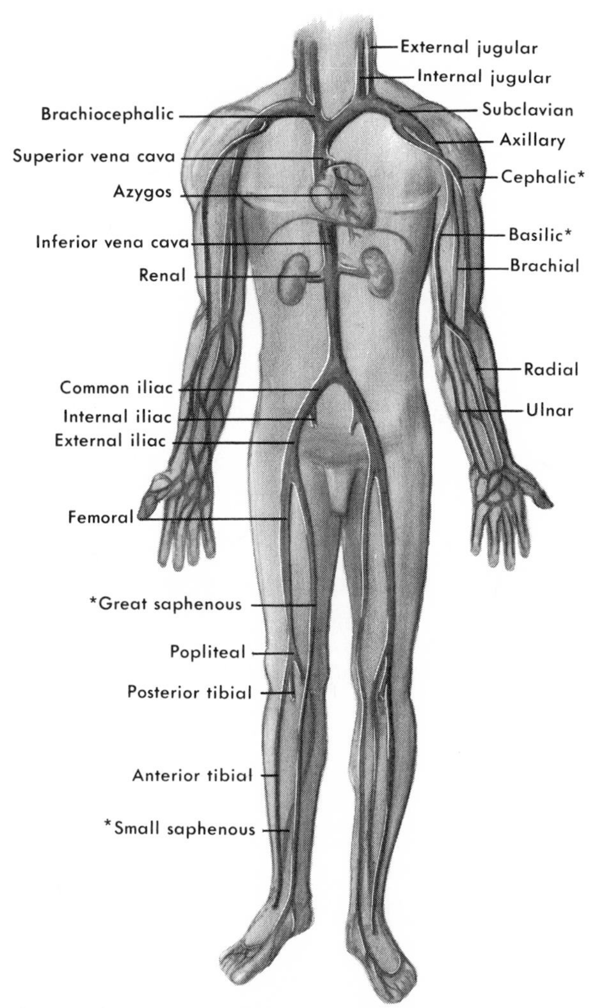

Figure 6–7. Diagram showing some of the main veins of the body. Asterisks indicate superficial veins.

VASCULAR CIRCUITS

Blood is circulated through the body in two main circuits: the pulmonary circuit and the systemic circuit. These can be represented thus:

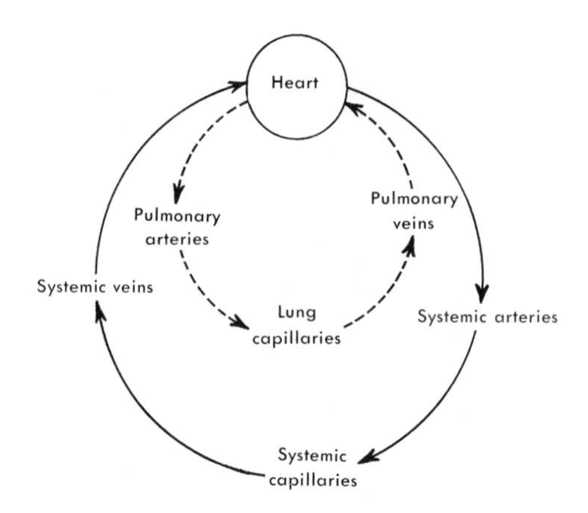

In the pulmonary circuit, blood leaves the right ventricle of the heart through the pulmonary artery. This vessel divides into right and left branches, which carry the blood to the lungs. Here the vessels progressively branch until by the time the capillaries are reached, blood is flowing close to the air spaces in the lungs. Now only two layers of cells—one the wall of the capillary and the other the wall of the air sac—separate blood from air. Oxygen and carbon dioxide are exchanged, and the blood flows on into the venules and then into larger and larger veins. Finally, it drains into the four pulmonary veins that carry it to the left atrium of the heart.

The systemic circuit is much larger. Many branches come off the aorta after it leaves the left ventricle, each bringing freshly oxygenated blood to an area of the body. For example, the first two branches are the right and left coronary arteries, which bring blood to the heart wall itself. The two carotid arteries carry blood to the head, the bronchial arteries to the lungs, the renal arteries to the kidneys, and so on. The aorta itself terminates at the level of the fourth lumbar vertebra by dividing into the two common iliac arteries. In each case, blood is circulated throughout a particular organ by progressively smaller arteries, arterioles, capillaries, and finally into the veins. Veins which drain structures above the heart empty into the superior vena cava; all others empty into the inferior vena cava. Both venae cavae open into the right atrium of the heart.

The circulation of blood in some areas of the body varies from this general pattern and so is given special names. The term *portal circulation* refers to venous blood that is collected from the digestive organs and shunted through the liver before being emptied into the inferior vena cava. This scheme allows for the deposition of the products of digestion (sugar, proteins, vitamins, etc.) in the liver cells (Fig. 6–8).

The brain receives its blood by way of four arteries, the two vertebral arteries and the two internal carotid arteries. Branches from these vessels form a circle at the base of the brain; hence it is named the *arterial circle of Willis* (Fig. 6–9).

FETAL CIRCULATION (Fig. 6–10)

Since the uninflated lungs of the fetus need comparatively little blood, most of the blood pumped from the right ventricle is shunted from the pulmonary artery into the aorta by a short connecting vessel, the *ductus arteriosus*. As in the case of the foramen ovale, this connection normally closes after birth and becomes a cord-like remnant, the *ligamentum arteriosum*. An open or patent ductus will allow mixing of unoxygenated blood in the pulmonary artery with oxygenated blood in the aorta, thus causing another type of "blue baby."

In addition to the ductus arteriosus, there is another fetal vessel called the *ductus venosus*. This vessel acts as a shunt to direct oxygenated blood from the umbilical vein into the inferior vena cava, thus by-passing the liver. After birth, this vessel closes and becomes the *ligamentum venosum* of the liver, and the umbilical vein becomes the *ligamentum teres* (round ligament) of the liver.

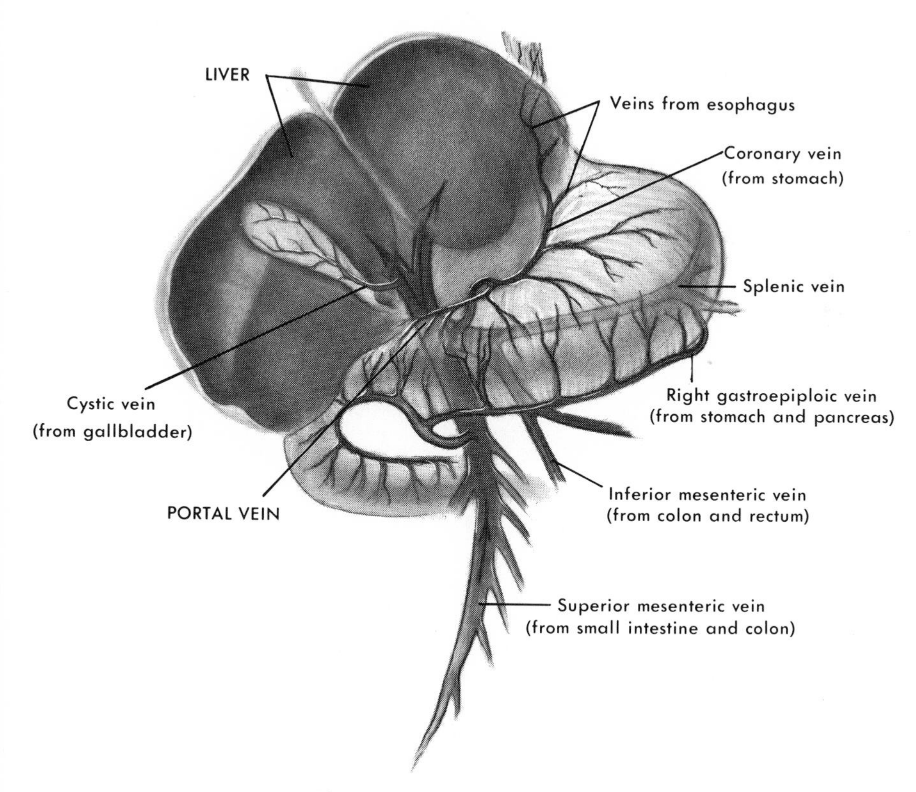

Figure 6–8. Portal circulation.

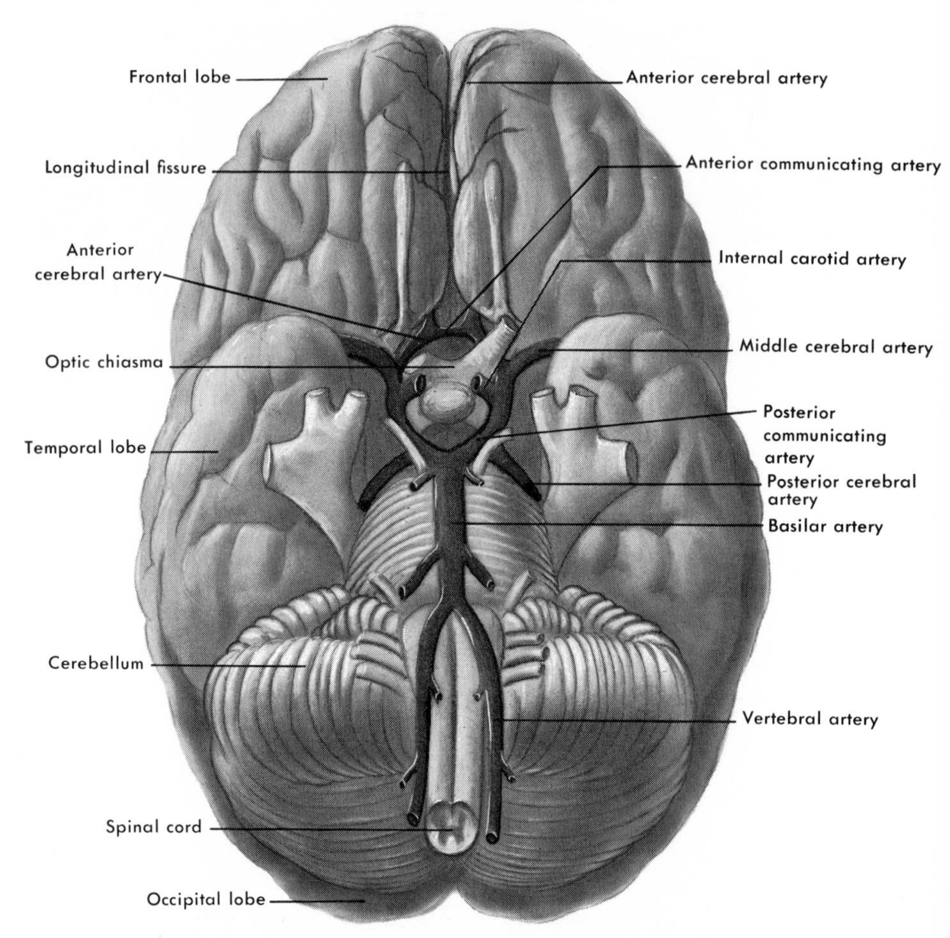

Figure 6–9. Arterial circulation at the base of the brain (circle of Willis).

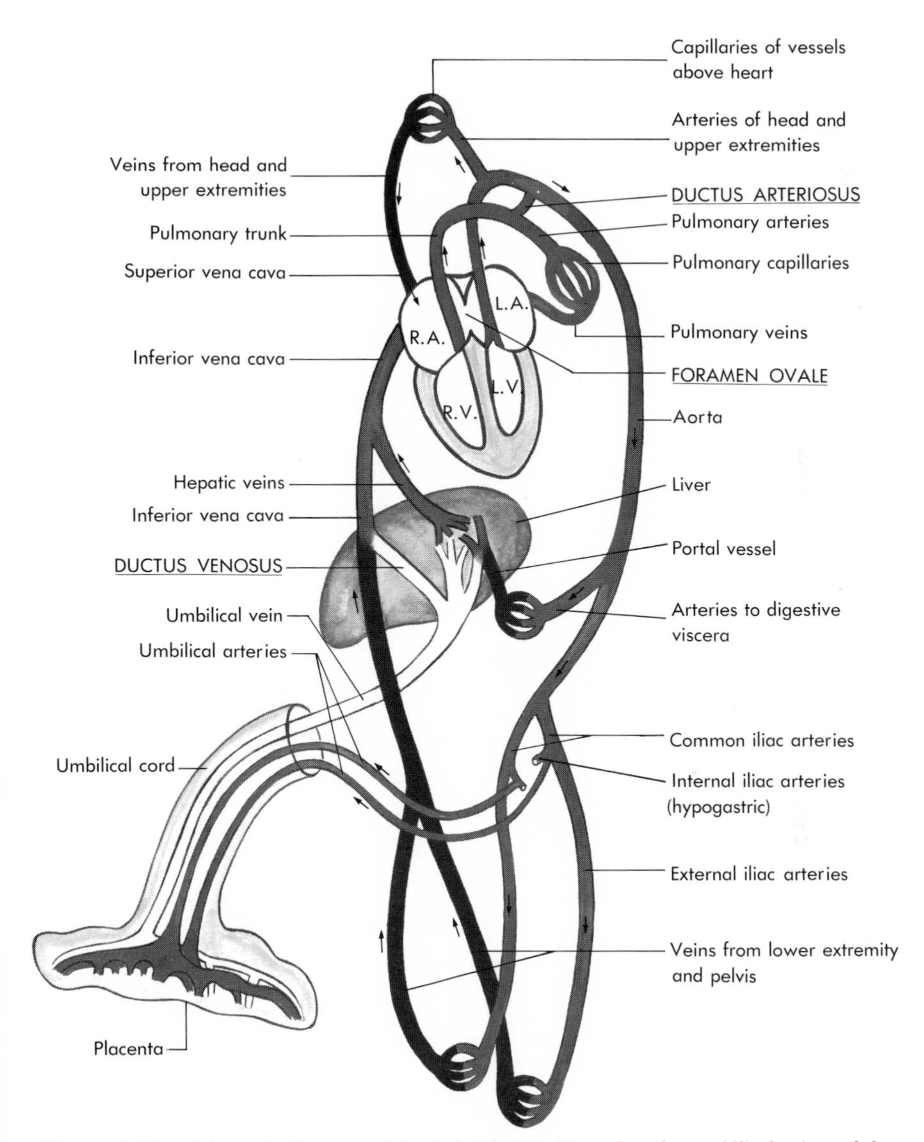

Figure 6–10. Schematic diagram of fetal circulation. Note that the umbilical vein and ductus venosus are the only vessels containing fully oxygenated blood.

PHYSIOLOGY OF CIRCULATION

THE CARDIAC CYCLE

Each complete heartbeat consists of two phases, contraction (systole) and relaxation (diastole). When the heart rate is 72 beats per min., one cardiac cycle occurs approximately every 0.8 second. During that time, the following sequence of events takes place (Fig. 6–11):

1. Ventricular systole – 0.3 second. The ventricular muscle contracts, causing the pressure of the blood in the ventricles to rise sharply – that in the left ventricle to approximately 120 millimeters of mercury and that in the right ventricle to about 26 millimeters of mercury. (The use of the term *millimeters of mercury* in measuring blood pressure will be explained on p. 131.) The A-V valves are closed before ventricular systole begins, since atrial pressure drops below that in the ventricles before the latter begin to contract. This closure is necessary to prevent any backflow of blood into the atria.

2. The semilunar valves open when the ventricular pressure becomes greater than that in the aorta and pulmonary artery. Blood is then ejected into the two arteries, most of it during the first one third of ventricular systole.

3. Ventricular diastole – 0.5 second. After ejection, the ventricular pressure drops sharply as the muscle enters its relaxation phase. As the pressure in the ventricles drops below that in the aorta and pulmonary artery, the semilunar valves snap shut, preventing backflow into the ventricles.

4. Meanwhile, the atria have filled with blood from the veins, and the pressure in these chambers begins to rise during the latter part of ventricular systole. As the ventricular pressure drops below the atrial pressure, the A-V valves open, and the ventricles fill rapidly with blood.

5. The atria then enter systole, and their contraction completes the filling of

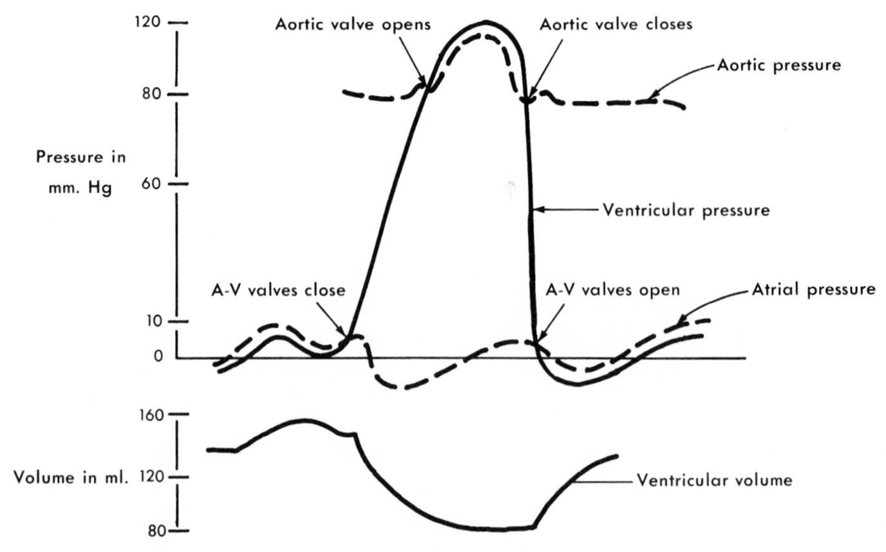

Figure 6–11. Volume-pressure relationships in the cardiac cycle.

the ventricles. Then atrial diastole begins, the A-V valves close, and another cycle begins.

There is a period of 0.4 second in the cycle during which ventricular diastole and atrial diastole overlap. This period is the so-called "rest period" during the work cycle of the heart muscle.

It should be emphasized that the ventricles are two-thirds full of blood by the time the atria enter systole. That is, rapid filling of the ventricles occurs during the first one third of ventricular diastole, with a small additional filling occurring during atrial contraction. During the period of rapid filling a third heart sound may be heard.

The duration of the cardiac cycle varies with the heart rate; as the rate increases, both systolic and diastolic phases become shorter. The amount of blood ejected from the heart during each beat is called the *stroke volume* and is usually about 70 cc., but this volume can vary under certain physiological conditions. The *cardiac output* is determined by multiplying the beats per minute by the volume of blood ejected per beat, or the heart rate times the stroke volume. Thus, with a heart rate of 70, the cardiac output is 4900 cc. Cardiac output may rise more than five times this amount in a well-trained athlete exercising heavily.

Although Figure 6–11 refers to left ventricular pressure and volume, the right heart shows the same relationships. Because the thinner-walled right ventricle sends blood through a low pressure area (the pulmonary circuit), pressures involved will be lower than for the left heart.

THE ELECTROCARDIOGRAM (Fig. 6–12)

The electrocardiogram, or ECG, is a recording of the electrical potentials generated by the heart. A small number of these electrical impulses, which are initiated in the S-A node and travel throughout the heart muscle via the conduction system (see p. 119), are conducted to the surface of the body by tissue fluids. Thus, by placing electrodes on the skin on opposite sides of the heart, these currents can be recorded. This recorded tracing is called the ECG.

A normal ECG tracing is shown in Figure 6–12. It consists of three waves, designated by the letters P, QRS, and T. (The QRS complex is actually three separate waves.) Each wave represents an event in the cardiac cycle as follows:

1. The P wave is caused by electrical potentials from the atria just prior to their contraction; this occurs during *depolarization* of the atria.
2. The QRS complex occurs just prior to ventricular contraction or depolarization of the ventricles.
3. The T wave occurs during the recovery or *repolarization* of the ventricles after contraction.

In a normal cardiac cycle, one PQRST recording will take the same amount of time that the cycle does—that is, about 0.8 second.

Abnormal cardiac rhythms, or cardiac *arrhythmias*, may be detected by the ECG. There are several types of arrhythmias. Some of them are manifested as *tachycardias* or fast heart rates, and others as *bradycardias* or slow rates. Many abnormal rhythms are caused by a partial or complete block somewhere in the conduction system of the heart. For example, a block at the S-A node (sinoatrial block) prevents impulses from entering the atrial muscle. An atrioventricular

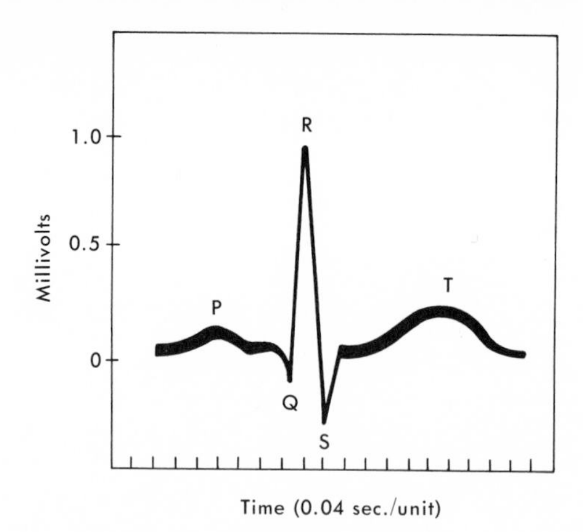

Figure 6–12. Normal electrocardiogram recording.

block prevents impulses from passing from the atria to the ventricles via the A-V bundle. The various alterations from the normal, which can be detected by the ECG, are the result of abnormal function of the heart muscle, especially that of the ventricles.

BLOOD PRESSURE

The force with which blood pushes against the blood vessel walls is called the blood pressure. It is produced by the contraction of the heart muscle. When the ventricles contract, they raise the pressure of the blood within them so that the semilunar valves are forced open and blood rushes into the aorta and pulmonary artery. The pressure of the blood entering these vessels causes their walls to stretch, and the pressure in the arterial system rises.

Pressure is highest in the aorta and progressively diminishes as the blood passes through the whole vascular system. It is lowest of all in the venae cavae (Fig. 6–13). Thus, since blood can flow only from a point of higher pressure to a point of lower pressure, this pressure *gradient* must be maintained to insure continuous circulation of blood.

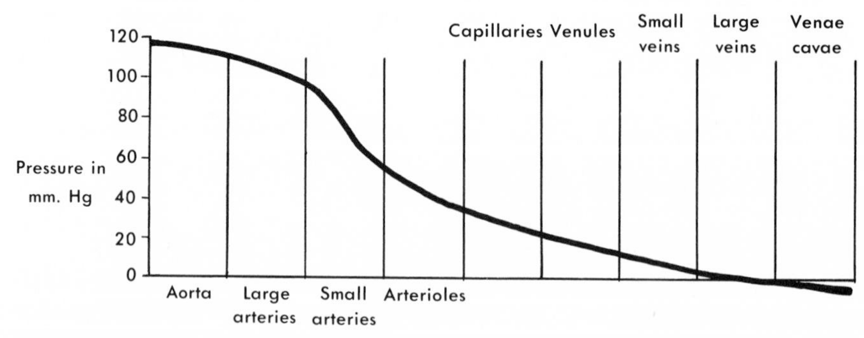

Figure 6–13. Graph showing relative pressure gradients in different parts of the circulatory system.

Although blood comes out of the heart in spurts, its flow is continuous throughout the system of vessels because the elastic walls of the vessels recoil at the end of each heart contraction and press upon the blood inside them, forcing it along. In older people the vessel walls lose some of their elasticity. The heart can compensate for this loss by raising the pressure under which it sends blood out of the ventricles.

Measuring Blood Pressure. Blood pressure is measured in terms of millimeters of mercury (mm. Hg) by means of a mercury manometer. When the pressure in a vessel is 75 mm. Hg, this means that the force exerted by the blood is enough to push a column of mercury to a height of 75 millimeters (3 inches).* The usual method of determining blood pressure is to wrap a rubber cuff around the arm (the leg also can be used) and inflate the cuff with enough air to compress the brachial artery. Then, with the stethoscope placed on the skin just above the bend in the elbow, the air pressure in the cuff is slowly released. The first sounds are heard as the brachial artery begins to open, and they are caused by the blood coming into the forearm in spurts and hitting the stationary column of blood there. The level of the mercury column on the manometer gauge at this point is recorded as the *systolic* blood pressure. This number, in mm. Hg, represents the force with which blood is pushing against the artery walls when the ventricles are contracting, that is, when they are in *systole*.

The sounds get louder as the spurts of blood get larger, becoming loudest just before that pressure at which the artery is again fully open. Because the blood is now moving freely once again, spurts of blood will no longer be hitting a stationary column and no sounds will be heard. The level of the mercury column at which the sounds become faint and die away is therefore recorded as the *diastolic* blood pressure. This number represents the force of the blood when the ventricles are relaxed, or in *diastole* (Fig. 6–14).

*If water were used instead of mercury, the column would be almost 40 inches high.

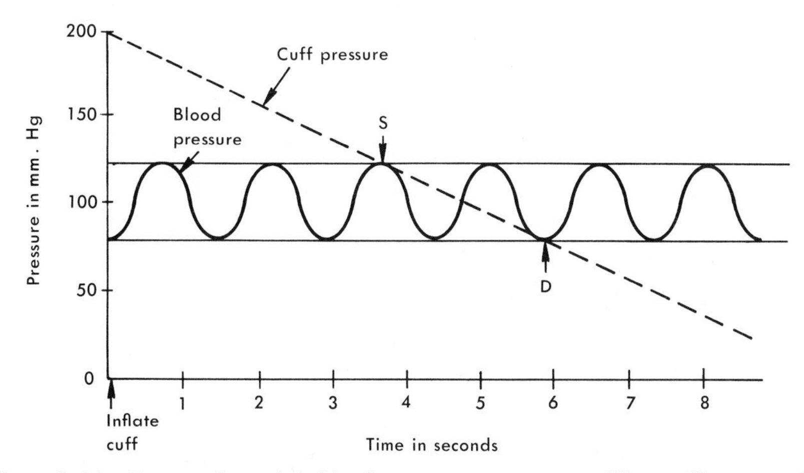

Figure 6–14. Source of sounds in blood pressure measurement. When cuff pressure drops to point S (systolic level), the first sounds are heard; at point D (diastolic level), sounds disappear as blood flow becomes continuous.

The average normal blood pressure for a young adult male is 120 mm. Hg systolic and 80 mm. Hg diastolic, usually written as 120/80. The difference between these two readings is called the *pulse pressure*. Variations from the normal pulse pressure are often a sign of diseased heart valves or hardening of the arterial walls.

BLOOD FLOW AND PERIPHERAL RESISTANCE

Blood pressure is closely related to two other factors, blood flow and peripheral resistance. Blood flow as used here refers to the volume of blood passing through the entire body every minute, or the *cardiac output*. Peripheral resistance refers to the opposition to flow offered by the blood vessel walls. The interrelationships of these three—blood pressure, blood flow, and resistance—are responsible for the maintenance of blood supply to all body tissues. Such a relationship may be expressed as

Blood Pressure (BP) = Blood Flow (BF) × Resistance (R)

This simply means that if blood flow or resistance changes, the blood pressure also changes. In other words, blood pressure is dependent upon both blood flow and resistance. The formula can be expressed in three ways to show the relationship of each factor to the other two.

Thus if (1) $BP = BF \times R$, then

(2) $BF = \frac{BP}{R}$, and

(3) $R = \frac{BP}{BF}$

This shows in (1) that, if resistance is not changed, blood pressure rises or falls with increased or decreased blood flow. Furthermore, (2) shows that, if blood flow is not changed, blood pressure will rise if resistance is increased, and it will fall if the resistance is decreased.

The second formula indicates that blood flow will rise or fall with blood pressure increase or decrease, provided the resistance is not changed. It also shows that blood flow changes in the opposite direction from alterations in resistance—that is, flow increases when resistance decreases, and flow decreases when resistance increases.

The third formula indicates that resistance is simply another way of expressing the relationship between blood pressure and blood flow. The body can bring about changes in each of these three factors by means of its reflex mechanisms, and when one of these factors is altered, it affects the other two. It has been noted (p. 129) that blood flow of cardiac output can increase to more than five times the resting amount during strenuous exercise.

Resistance to flow of blood through the vessels is caused by several factors. One such factor is the *viscosity* of the blood, or the ease with which blood flows. Water, our most abundant liquid, has been assigned a relative viscosity value of 1. Thus other fluids are given viscosity numbers relative to 1, depending upon how fast they flow when compared to water. Whole blood has a viscosity of approximately 4, which means it has a resistance to flow that is four times greater than water. The viscosity of blood is due largely to its red blood corpuscle content, although the plasma proteins are also partly responsible. Normally, the viscosity of the blood remains constant, but if the total RBC count is decreased, as in some types of anemia, then blood viscosity would also decrease.

Another factor affecting resistance to blood flow is the radius of the vessel through which it flows; the greater the radius the greater the flow. Thus the resistance to flow decreases as the vessel's radius increases. The *length* of the blood vessel also helps determine resistance to flow; the longer the vessel, the more friction between the vessel wall and the blood, and thus the greater the resistance.

About half of all resistance to flow is due to resistance by the arterioles, the constriction and dilatation of which are controlled by the sympathetic nervous system. Indeed, the main function of the arterioles (other than delivering blood to the capillaries) is to regulate peripheral resistance and thus the blood pressure. The center for this control is in the medulla of the brain (p. 86) and, upon stimulation, it will cause general vasoconstriction of the arterioles. Certain drugs, such as *epinephrine*, will produce the same effect as sympathetic vasoconstriction.

Control of Arterial Blood Pressure. The degree of exercise, changes in body posture, rapid loss of blood, and other stressful situations stimulate mechanisms to prevent major changes in blood pressure. The two main mechanisms for such minute-to-minute control are found in the nervous system and in the capillaries, while a third is handled by the kidneys.

Nervous control involves a series of reflexes in which information is transmitted to the *vasomotor center* in the brain (medulla) which, in turn, sends messages for the control of heartbeat and blood vessel constriction. This center is influenced by a number of factors, one of the most important being the level of blood pressure in the carotid artery.

In the capillary, an increase in permeability of the vessel walls leads to a shift of fluid from the body tissues into the blood vessels and vice versa. For example, with severe hemorrhage, the blood pressure tends to fall. When this occurs, fluid is reabsorbed from the interstitial spaces through the capillary walls into the vessels. Thus volume is restored and the blood pressure rises. This response, which takes several minutes, is not as rapid as that of the vasomotor center, which can respond in a matter of seconds. However, it has been shown that even in the absence of nervous system controls, the fluid shift response alone can control blood pressure level reasonably well.

The third mechanism of blood pressure control is handled by the kidneys. The nature of the mechanism itself is poorly understood; it may be that the kidney's ability to control the output of water and salt from the body is the key to this mechanism. While it is an effective control, it responds the most slowly of the three and usually takes hours to become effective.

PULSE

The pulsation felt when the fingertips are placed over an artery close to the body surface represents the alternate expansion and recoil of the elastic arterial wall. With each ventricular contraction of the heart the pressure of the suddenly ejected blood expands the walls of the aorta and is transmitted as a wave over the entire aorta and its branches. The wave lessens as it travels along the vessels. Although it cannot be felt, or palpated, in the capillaries and veins because of the low pressure in these vessels, it can be seen in them where they are close to the surface, as for example in the neck veins.

The pulse is an index of the action of the heart, the elasticity of the larger blood vessels, the viscosity of the blood, and the resistance in the arterioles and capillaries. There are several places where the arterial pulse can be taken: (1) the radial artery at the wrist, (2) the common carotid artery at the anterior edge of the sternocleidomastoid muscle, (3) the facial artery along the lower margin of the mandible, (4) the brachial artery in the arm, (5) the femoral artery in the groin, (6) the popliteal artery behind the knee, and (7) the dorsalis pedis artery on the dorsal surface of the foot.

The pulse should be described according to its *rate* (fast or slow), *size* (large or small), *type of wave* (abrupt or prolonged), and *rhythm* (regular or irregular). The average resting pulse rate in normal adults is 60 to 80 per minute and in children, 80 to 140. An increase in the pulse rate is normal during and after exercise and after eating. It is decreased during sleep. In most diseases associated with fever the pulse rate is increased, usually an average of five beats for every degree Fahrenheit. An increased pulse rate is usually present in severe anemias and becomes markedly increased after severe hemorrhage.

THE LYMPHATIC SYSTEM

In a sense, the lymphatic system is an assistant to the venous part of the vascular system. It helps return tissue fluid from the spaces between the cells to the blood from which it originated. When this tissue fluid passes from the spaces into the lymphatic vessels, it is called *lymph*.

The system is composed of a series of vessels which start as tiny, closed tubes located in the intercellular spaces throughout the body. These *lymphatic capillaries* lead into larger and larger lymphatic vessels. Finally, all the lymph empties into two main vessels, the *thoracic duct* and the *right lymphatic duct*.

The right lymphatic duct collects lymph from the upper right quarter of the body and empties it into the right subclavian vein at its junction with the right internal jugular. Lymph from the rest of the body is carried by the thoracic duct into the junction of the left subclavian vein and the left internal jugular (Fig. 6–15).

The lymphatic vessels resemble veins in structure, except that they have thinner walls and contain more valves to prevent the backflow of lymph. Lymph moves very slowly through the vessels under a relatively low pressure gradient.

Lymph nodes are located along the lymphatic vessels at intervals. A lymph node is a mass of lymphoid tissue separated into compartments by connective tissue and surrounded by a dense connective tissue capsule. Nodes vary in size from that of a pinhead to that of a lima bean. Although there are single nodes scattered throughout the body, most of them are arranged in clusters in certain areas: the floor of the mouth, the neck, the axilla, the inguinal region, the bend of the elbow, and along the main arteries (Fig. 6–16).

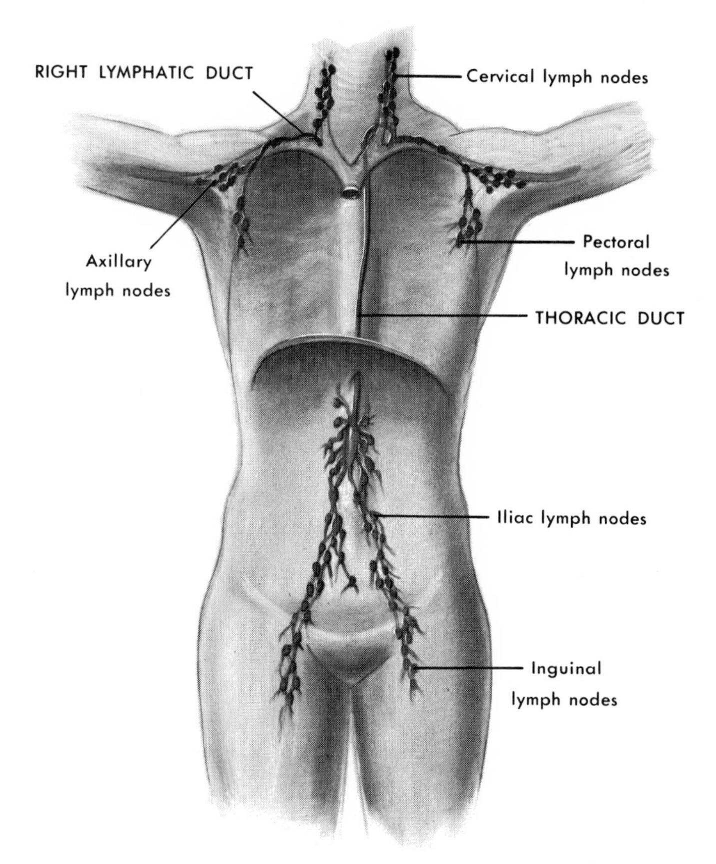

Figure 6–15. Lymph drainage of the body.

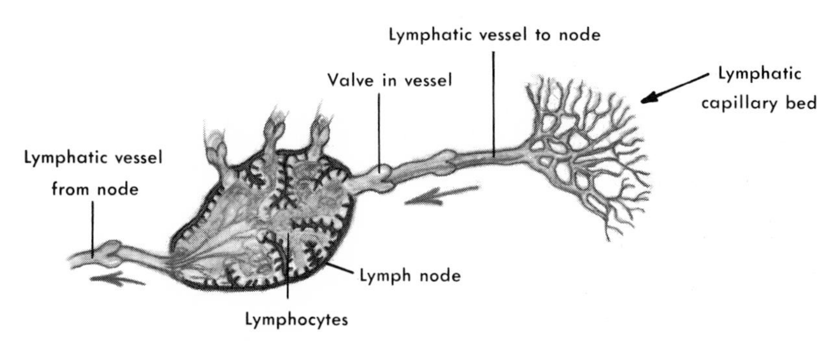

Figure 6–16. Lymphatic vessels and a lymph node.

Lymph nodes filter out bacteria and other foreign particles from the lymph as it is strained through them. Nodes also manufacture lymphocytes and perhaps also make antibodies and monocytes. Thus the lymphatic system, in addition to returning tissue fluid to the blood, is also an important part of the body defense. On the other hand, in a case of cancer or an overwhelming infection, the lymphatics may serve as a pathway for the spread of cancer cells or bacteria.

Edema, or collection of fluid in the tissues, may be caused by blockage of large lymphatic channels or by their surgical removal. An example of the latter is the swelling of the arm which sometimes occurs after removal of the breast and axillary lymph nodes in cases of breast cancer.

SPLEEN

The spleen is included here because it is made up largely of lymphoid tissue. It is located on the left side of the upper abdominal cavity, below the diaphragm and above the left kidney. The lymphoid part or white pulp of the spleen acts much like lymph nodes in filtering the blood. The white pulp also manufactures lymphocytes and monocytes. The red pulp of the spleen contains wide channels which act as a "blood bank" containing several hundred milliliters of blood. Although the evidence is not clear, the spleen may squeeze this blood out into the circulation after a severe hemorrhage has occurred.

The spleen is an important organ in the destruction of worn-out red blood corpuscles. It takes the iron out of the hemoglobin molecule so it can be reused by the bone marrow to make new RBC's. Another function assigned to the spleen is that of making antibodies. Although an important organ, the spleen is not essential to life. If it has to be surgically removed (splenectomy), its functions are taken over by other organs, such as the liver.

OUTLINE SUMMARY

Blood

1. Formed Elements
    a. Red blood cells (erythrocytes) – carry oxygen combined with hemoglobin
    b. White blood cells (leukocytes) – body defense by phagocytosis
        1) Neutrophils
        2) Eosinophils
        3) Basophils
        4) Lymphocytes
        5) Monocytes
    c. Platelets (thrombocytes) – blood clotting
    d. Normal counts (adult male)
        1) Red cells – 4½ to 5 million/cu.mm.
        2) White cells – 5000 to 10,000/cu.mm.
        3) Platelets – 250,000 to 500,000/cu.mm.

2. Plasma and Serum
   a. Plasma
      1) Liquid part of blood
      2) Composed of 90 per cent water with small amounts of dissolved protein, sugar, fat, salts, etc.
   b. Serum – plasma minus the fibrin clot
3. Blood Clotting
   a. Proceeds in three main stages (see p. 000 for sequence of events)
   b. 11 coagulation factors – deficiencies cause prolonged bleeding
4. Abnormal Blood Clots – thrombus and embolus
5. Blood Typing – important in persons needing blood transfusions
   a. Four blood types according to antigens present
      1) Type A – has A antigen and b antibodies
      2) Type B – has B antigen and a antibodies
      3) Type AB – has both A and B antigens and no antibodies
      4) Type O – has neither A nor B antigen and both a and b antibodies
   b. Rh (Rhesus factor)
      1) Present in 85 per cent of population – called Rh positive
      2) Danger to Rh negative fetus – erythroblastosis fetalis

Heart

1. Location
   a. Chest cavity; two thirds is left of midline
   b. Apex rests on diaphragm; base just under third rib
2. Structure
   a. Four chambers
      1) Two atria – receive blood into heart
      2) Two ventricles – pump blood out of heart
   b. Pericardium – outer loose sac of fibrous tissue
   c. Epicardium – outer layer of heart; inner layer of lining of pericardium
   d. Myocardium – middle layer of heart; cardiac muscle
   e. Endocardium – inner layer of heart; thin connective tissue
3. Valves
   a. Tricuspid – guards opening between right atrium and right ventricle
   b. Bicuspid (mitral) – guards opening between left atrium and left ventricle
   c. Aortic semilunar – guards entrance to aorta
   d. Pulmonary semilunar – guards opening to pulmonary artery
4. Heart Sounds – "lubb-dup" caused by valves closing
5. Control mechanisms
   a. Conduction system – nodal tissue
   b. Chemical control – calcium, potassium

Fetal Heart

1. Foramen ovale – opening between the atria; becomes fossa ovalis
2. Failure to close – mixing of blood causes "blue baby"

Blood Vessels

1. Kinds
   a. Arteries – carry blood away from heart
   b. Veins – carry blood to heart
   c. Capillaries – connect arteries and veins

2. Main Arteries and Veins of Body (see Figs. 6–5 and 6–6)
3. Two Main Circuits
    a. Pulmonary circuit—carries blood from right heart to lungs for oxygenation and back to left heart
    b. Systemic circuit—larger one; circulates blood to all body organs
4. Special Circulation
    a. Portal system—shunts blood from digestive organs through the liver
    b. Arterial circle of Willis—circle of blood vessels at base of brain
    c. Fetal circulation
        1) Ductus arteriosus—shunts blood from pulmonary artery into aorta; becomes ligamentum arteriosum
        2) Ductus venosus—shunts blood from umbilical vein into inferior vena cava; becomes ligamentum venosum
        3) Umbilical vein—brings oxygenated blood from placenta; becomes ligamentum teres

Physiology of Circulation

1. Heart Beat
    a. Starts in S-A node (pacemaker) and sweeps over heart via A-V node and A-V bundle
    b. Normal adult—72 to 80 beats/minute
2. Cardiac Cycle
    a. Two phases—systole and diastole
    b. One every 0.8 second at 72 beats/minute
    c. Sequence of events—see p. 128
    d. Stroke volume—amount of blood pumped per beat
    e. Cardiac output—multiply stroke volume by heart rate
    f. Electrocardiogram—recording of heart's electrical potentials
3. Blood Pressure
    a. Definition—force with which blood pushes against vessel walls
    b. Pressure gradient—pressure highest in arteries and lowest in veins
    c. Measurement—by mercury manometer in mm.Hg
        1) Systolic—during contraction phase (systole) of heart
        2) Diastolic—during relaxation phase (diastole) of heart
    d. Normal adult—120/80
    e. Pulse pressure—systolic minus diastolic
    f. Blood flow and peripheral resistance
    g. Control of arterial pressure
4. Pulse
    a. Definition—alternate expansion and recoil of elastic arterial wall
    b. Index of heart action, vessel elasticity, blood viscosity, and small vessel resistance
    c. Seven points in body where pulse can be taken (see p. 134)
    d. Describe according to rate, rhythm, size, and type of wave
    e. Increased in exercise and eating; decreased during sleep
    f. Normal values:
        1) Adult—60 to 80 beats/minute
        2) Child—90 to 140 beats/minute

Lymphatic System

1. Lymph
    a. Tissue fluid collected into lymphatic vessels
    b. Comes originally from blood

2. Lymphatic Vessels
   a. Return lymph to blood
   b. Thinner-walled vessels than veins with more valves
   c. Begin as blind capillaries in tissues
   d. All lymph finally collected into two main vessels:
      1) Thoracic duct
      2) Right lymphatic duct
3. Lymph Nodes
   a. Pinhead size to lima bean size
   b. Located at intervals along lymphatic vessels
   c. Clusters of nodes in certain regions—neck, axilla, groin, etc.
   d. Strain bacteria and foreign particles from lymph; make lymphocytes
   e. Route for spread of cancer cells and infections
4. Spleen
   a. Located in left side of upper abdomen
   b. White pulp—acts as filter; makes lymphocytes and monocytes
   c. Red pulp—reservoir for blood
   d. Destroys worn-out red blood corpuscles

REVIEW QUESTIONS

1. When taking the blood pressure, how can you tell when there is enough air pressure in the cuff to shut off the brachial artery?
2. If a patient's blood pressure reading is 140/95, what is his pulse pressure?
3. Why would a person with a low hemoglobin or a low RBC count complain of feeling tired all the time?
4. What is phagocytosis? Would a person with a low white cell count be likely to have more frequent and longer-lasting infections? Why?
5. What does hemopoiesis mean?
6. Trace the passage of blood through the heart, naming all the chambers, openings, and valves in their proper order.
7. Follow a drop of blood from the heart to the right hand and back to the heart, naming the main blood vessels in correct order.
8. Name the three blood vessels that join to form the portal vein.
9. Define systole and diastole.
10. Why do you think mercury instead of water is in the manometer used to take blood pressure?
11. What are polys? What is their chief function?
12. Give the three stages of blood clotting in proper order.
13. What do lymph nodes do?
14. List the functions of the spleen.
15. True or false: The pulmonary artery contains venous blood. Explain.
16. Why does a patient's blood have to be typed and crossmatched before he can be given a blood transfusion?
17. What is the difference between a thrombus and an embolus?
18. What important substance do mast cells secrete? What does this substance do?
19. What is an "ectopic" heartbeat?
20. What is the relationship of blood vessel length and radius to the rate of blood flow?
21. In the ECG, what do the letters P, QRS, and T signify?

ADDITIONAL READING

Adolph, E. F.: The heart's pacemaker. Sci. Amer. (March), 1967.
Clarke, C. A.: The prevention of "rhesus" babies. Sci. Amer. (Nov.), 1968.
Mayerson, H. S.: The lymphatic system. Sci. Amer. (June), 1963.
Wiggers, C. J.: The heart. Sci. Amer. (May), 1957.
Wiggers, C. J.: The Physiology of Shock. New York, Commonwealth Fund, 1950.
Wood, W. B., Jr.: White blood cells v. bacteria. Sci. Amer. (Feb.), 1951.

Chapter 7

THE RESPIRATORY SYSTEM

GENERAL CONSIDERATIONS

In order to carry on their necessary metabolic activities, all body cells must have a ready supply of oxygen available to them. In addition, there must be a mechanism for the removal of carbon dioxide, the main waste product resulting from chemical processes taking place in the cells. In man, these functions are carried out by the respiratory system, which consists of a series of passageways, tubes, and thin-walled membranes.

The mechanism of moving air into and out of the system is called *breathing*. The term *respiration* refers to three processes: (1) the actual exchange of gases between the air in the system and the blood in the vessels surrounding the thin membranes; (2) the exchange of these gases between the blood and the body cells; and (3) the utilization of oxygen (O_2) and the production of carbon dioxide (CO_2) by the cells. The first two processes comprise *external respiration*, whereas the third is called *internal respiration*.

ANATOMY OF THE RESPIRATORY SYSTEM (Fig. 7–1)

Structures composing the respiratory system may be divided into conducting passages and respiratory passages. The conducting passages simply provide a means by which air can pass into and out of the system. Respiratory passages are those parts directly concerned with the exchange of gases between the air and the blood.

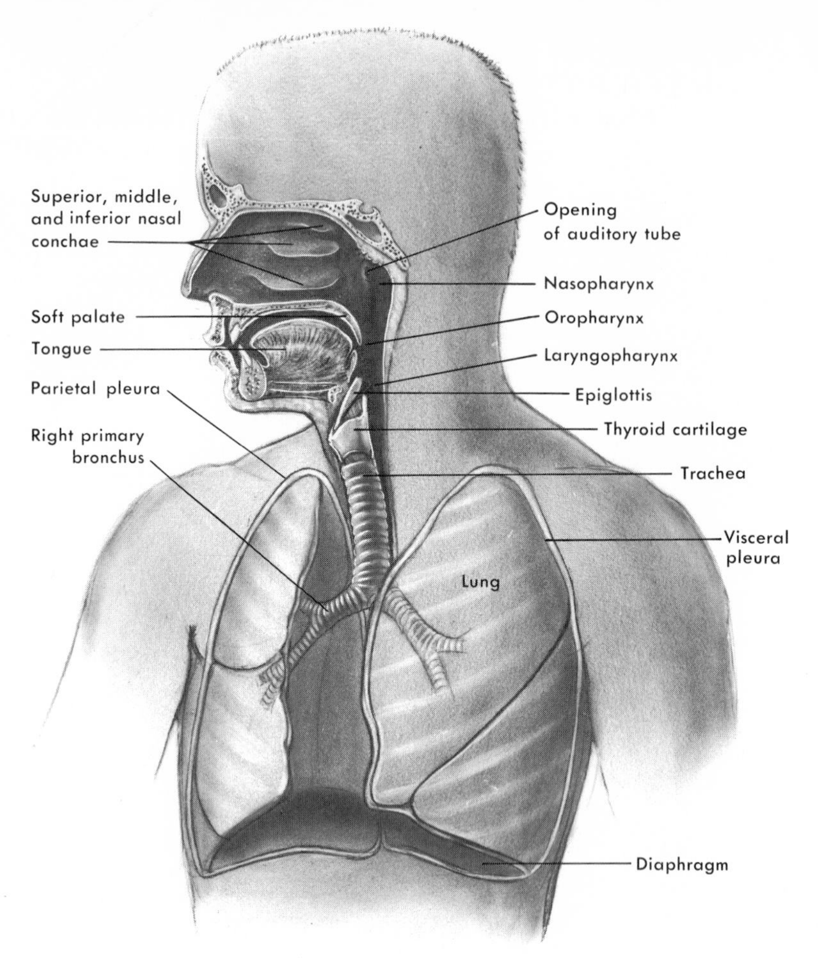

Figure 7–1. The respiratory system.

CONDUCTING PASSAGES

These passages consist of the nasal cavities, pharynx, larynx, trachea, primary bronchi, secondary bronchi, segmental bronchi, and terminal bronchioles.

Nasal Cavities. In addition to serving as a passageway for air, the nasal cavities moisten, filter, and warm air as it passes through them on its way to the lungs. The nose also contains the organ of *olfaction* for the sense of smell and participates in *phonation* (speech production).

The external portion of the nose is a bony and cartilaginous framework covered with skin. It is separated into two cavities by the nasal septum, which is formed by cartilage, and the vomer and ethmoid bones. The roof is formed by the nasal bones and parts of the ethmoid, frontal, sphenoid, vomer, and palatine bones. The maxilla and palatine bones form the floor of the cavity. Each lateral wall has three bony projections, the nasal conchae, which subdivide each cavity into groove-like passageways called *meatuses* (Fig. 7–2). The openings for drainage of the paranasal sinuses and the lacrimal duct are located in these grooves.

The nasal cavity is lined with pseudostratified, ciliated columnar epithelium with goblet cells. The nostrils, or openings to the outside of the body, are called the *anterior nares*. Posteriorly, the *posterior nares* open directly into the pharynx, the next portion of the system.

Pharynx. The pharynx is a funnel-shaped passageway extending from the base of the skull to a point opposite the lowest cartilage of the larynx (Fig. 7–3). For purposes of description, it can be divided into three regions according to location: nasopharynx, oropharynx, and laryngopharynx. The pharynx is a common passageway to both the respiratory and digestive systems.

The *nasopharynx* is the upper portion lying directly posterior to the nasal cavities and above the soft palate. Its posterior wall abuts against the base of the occipital bone and the first two cervical vertebrae. There is an opening in each lateral wall for the *auditory tubes*, which connect the nasopharynx with the middle ear cavities. Two masses of lymphoid tissue lie at the upper end of the posterior wall. These masses are the *pharyngeal tonsils*, commonly referred to as the adenoids.

The *oropharynx* extends from the soft palate down to the level of the hyoid bone in the neck. Thus, it lies posterior to the oral cavity and the tongue. The *tonsillar pillars* are two folds of tissue on the lateral walls. The spaces between these folds contain the *palatine tonsils*, usually called simply the tonsils.

The *laryngopharynx* is the lowermost portion of the pharynx. It continues from the level of the hyoid bone and ends at the lower border of the cricoid cartilage. At this point the pharynx becomes continuous with the esophagus. Thus the laryngopharynx lies posterior to the larynx, which is the continuation of the respiratory tract. This point marks the place, then, where the food pass-

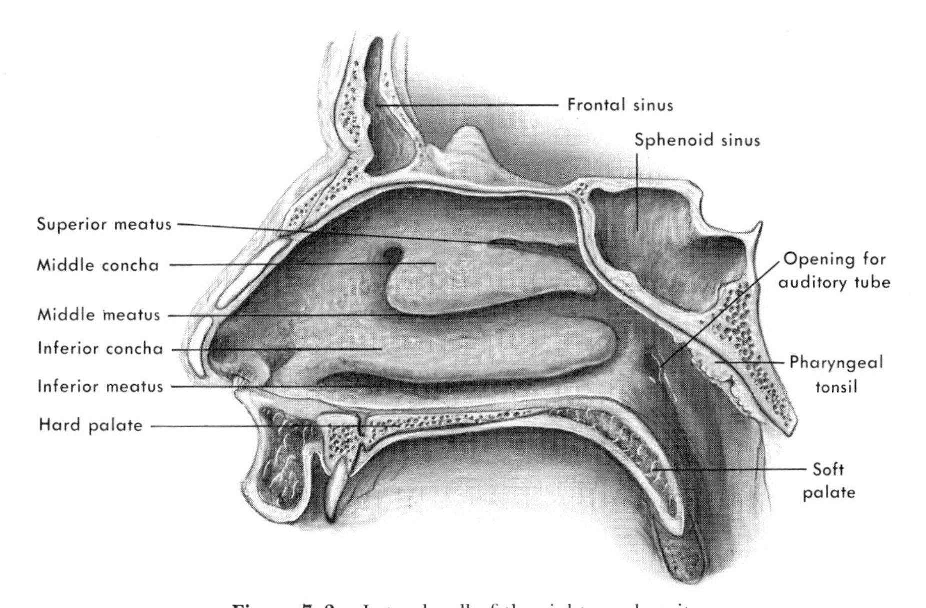

Figure 7–2. Lateral wall of the right nasal cavity.

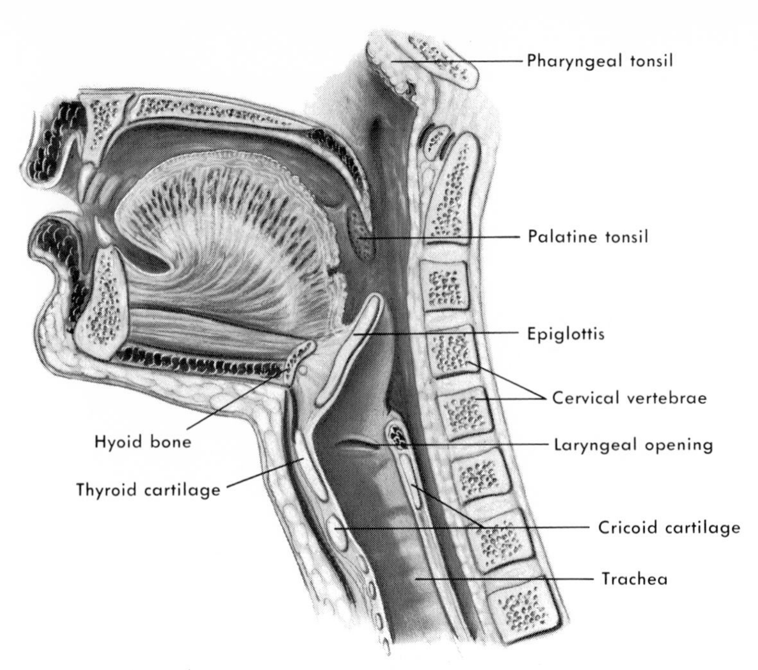

Figure 7–3. The pharynx and larynx – median section.

ageway and the airway cross one another. Food is pushed backward into the esophagus, whereas air moves anteriorly to enter the larynx.

The wall of the pharynx has a thick muscular coat. It is lined by mucous membrane with an underlying layer of strong fibrous tissue. There are three pairs of external circular muscles called the *constrictors*, which overlap and blend with each other. The muscle fibers of the inferior constrictors become continuous with the musculature of the esophagus. In addition, three pairs of longitudinal muscles run from the styloid process, the soft palate, and the auditory tube to insert into the pharyngeal wall.

In the act of *swallowing*, the tongue pushes a ball of food backward into the pharynx. Constriction of the longitudinal pharyngeal muscles then drives it downward as far as the upper limit of the larynx. From here, contraction of the pharyngeal constrictors forces it backward and down into the esophagus. In this way, the opening into the larynx is closed off so that swallowed material is pushed forcibly down into the esophagus instead of passing into the airway.

Larynx. In addition to passing air from the pharynx above into the trachea below, the larynx controls the expulsion of air from the lungs in the production of sound. Thus, it is sometimes referred to as the voice box. Situated between the root of the tongue and the trachea, it lies in the anterior portion of the neck in front of cervical vertebrae 4, 5, and 6.

The larynx is composed of nine cartilages; three are single and six are paired. The largest of these is the *thyroid* cartilage, which is composed of two lateral plates fused together in the anterior midline. This line of fusion forms the *laryngeal prominence*, commonly referred to as the Adam's apple. The spoon-shaped *epiglottic* cartilage, lying behind the root of the tongue, has a free, rounded upper end. In swallowing, the epiglottis furnishes additional protection against the entrance of food or liquid into the airway. The inferior boundary of the larynx is formed by the ring-shaped *cricoid* cartilage (Fig. 7–4).

The larynx is the tone-producing organ, furnishing sound which is then modified into the human voice by chambers above and below the larynx. *Phonation*, or speech, results from action on this sound by the pharynx, tongue, lips, and palate. Thus the larynx is not the organ of speech, but the organ of sound production. It is possible for one whose larynx has been removed to speak because he can produce sound by swallowing air and expelling it through the esophagus. Speech produced in this manner, however, lacks pitch and volume control.

Certain of the laryngeal muscles are important in sound production. The *vocal folds*, known collectively as the *glottis*, stretch across the laryngeal opening from anterior to posterior. The free edges of the folds contain the *vocal ligaments*, and the more lateral portions are formed by the *vocalis* muscle. The space between the folds, the *rima glottidis*, can be varied in size by action of the laryngeal muscles. The rima widens during breathing and narrows during the production of sound (Fig. 7–5).

Trachea. The trachea, or windpipe, is a rigid tube about $4\frac{1}{2}$ inches (12 cm.) long that conducts air from the larynx down into the bronchi. It lies in the midline, anterior to the esophagus and opposite the sixth cervical to fourth thoracic vertebrae. The upper half of it lies in the neck and the lower half lies in the chest.

The walls of the trachea are formed by 16 to 20 **C**-shaped cartilage rings, the posterior ends of which are joined by fibrous tissue and smooth muscle. These cartilages prevent the tracheal walls from collapsing, thus insuring that the airway is open at all times. The mucous membrane lining the trachea has cilia which sweep dust and foreign particles up into the pharynx so they can be

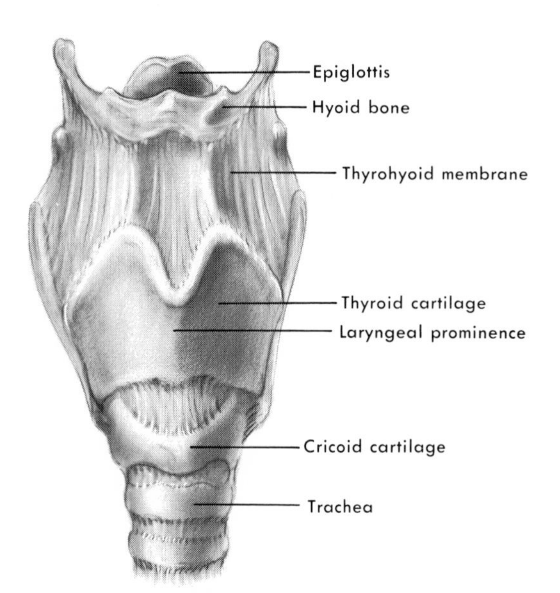

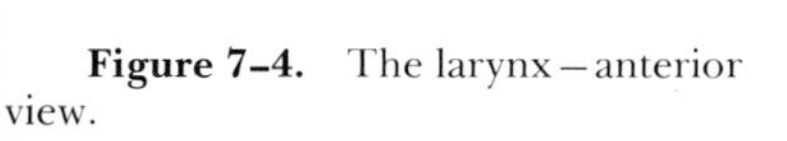

Figure 7–4. The larynx—anterior view.

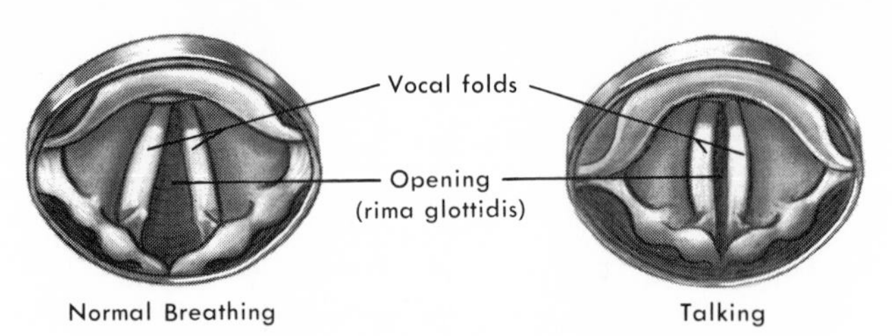

Figure 7–5. The voice box as viewed from above.

removed by action of the cough reflex. The presence of many mucous glands serves to moisten air as it passes through.

Bronchial Tree (Fig. 7–6). The bronchial tree constitutes a branching set of tubes with progressively smaller diameters. These tubes deliver air from the trachea to all parts of the lung tissue. The trachea ends at the level of the sternal angle (upper border of the fifth thoracic vertebra) behind the manubrium of the sternum, by dividing into right and left *primary bronchi.* Like the trachea, both bronchi have cartilage rings in their walls to keep them open. The right bronchus, being the more direct continuation of the trachea, is straighter, shorter, and larger than the left.

At the level of the fifth thoracic vertebra, the right bronchus enters the right lung. It then divides into three *secondary bronchi,* supplying the superior, middle, and inferior lobes. Each secondary bronchus then further divides into two to four smaller *segmental bronchi.*

The left bronchus divides into two secondary bronchi, one to each lobe of the left lung. These also further divide into smaller segmental bronchi. Within the lobes of both lungs, the area supplied by one segmental bronchus is called a *bronchopulmonary segment.*. There are 18 such segments composing the bronchial tree.

The bronchial tree continues to branch, dividing into more and smaller tubes called *bronchioles.* As these are reached, the amount of cartilage and the number of mucous glands in the walls gradually decrease, but the amount of

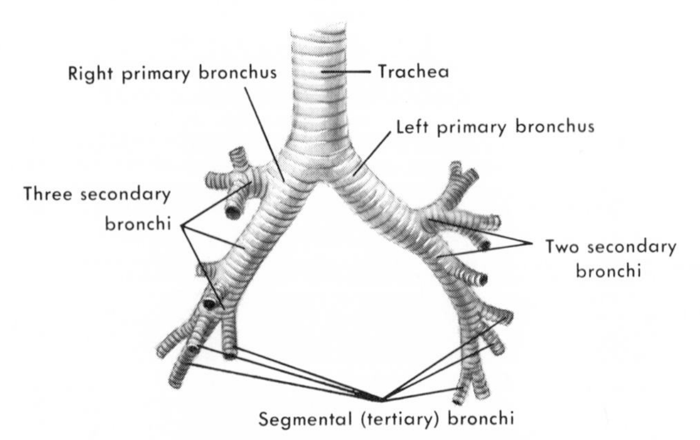

Figure 7–6. Branches of the bronchial tree.

smooth muscle increases. When a diameter of about 1 mm. (about 1/8 inch) is reached, the cartilage disappears.

Each lung lobe is subdivided into *lobules* of various sizes and shapes. A bronchiole enters each lobule and gives off many *terminal bronchioles*. These number from 50 to 80 in each lobe and mark the functional end of the air-conducting passageways of the respiratory system.

RESPIRATORY PASSAGES

A functional lung lobule (Fig. 7–7) consists of a *respiratory bronchiole*, its branching *alveolar ducts*, which in turn lead into *alveolar sacs*, and the thin-walled, cup-shaped *alveoli*, which form the walls of the sacs. No respiratory function takes place until this point in the system is reached.

The final branches of the bronchioles within a lobule are the respiratory bronchioles. Each has a diameter of about 0.5 mm. and opens into several alveolar ducts. These in turn open into spaces called alveolar sacs. The alveoli of each sac are composed of a thin layer of epithelium. Alveoli also appear as outpouchings from the sides of the respiratory bronchioles and alveolar ducts.

The walls of the alveolar sacs are surrounded by pulmonary capillaries, which are supported by delicate fibrous and elastic tissue. Thus, air in the alveoli and blood in the capillaries are separated by only a very thin partition, which is readily crossed by diffusing gases.

It has been estimated that there are approximately 14 million alveolar ducts and 700 million alveoli contained in the tissue of the lungs. Thus, the total respiratory surface area is 50 to 100 square meters, or 25 to 50 times the surface area of the entire body.

THE LUNGS (Fig. 7–8)

The two cone-shaped lungs are the organs of respiration. They normally lie free within the pleural cavities of the thorax except where they are attached at

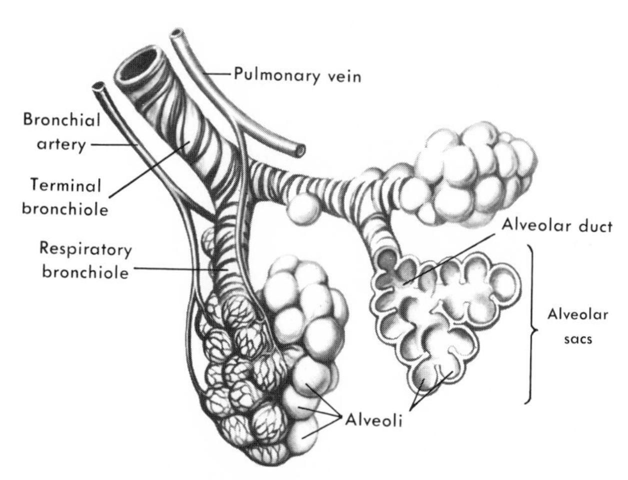

Figure 7–7. Functional lung lobule.

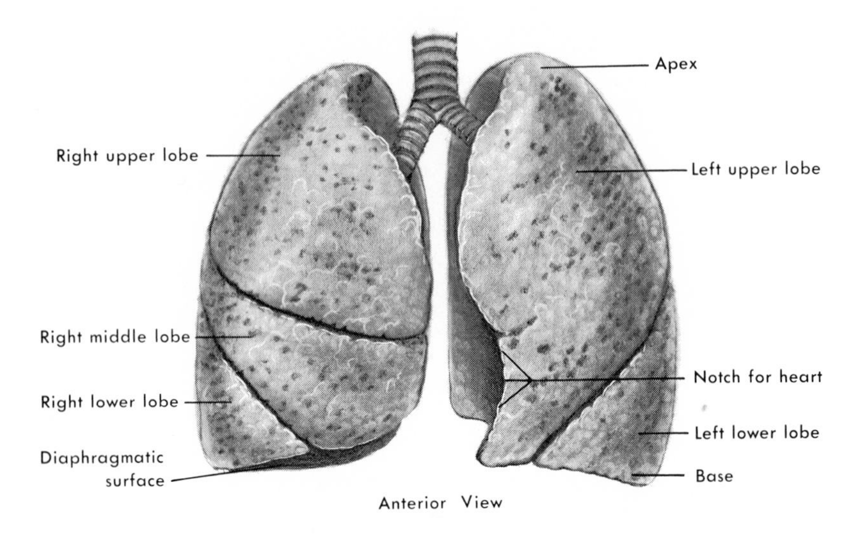

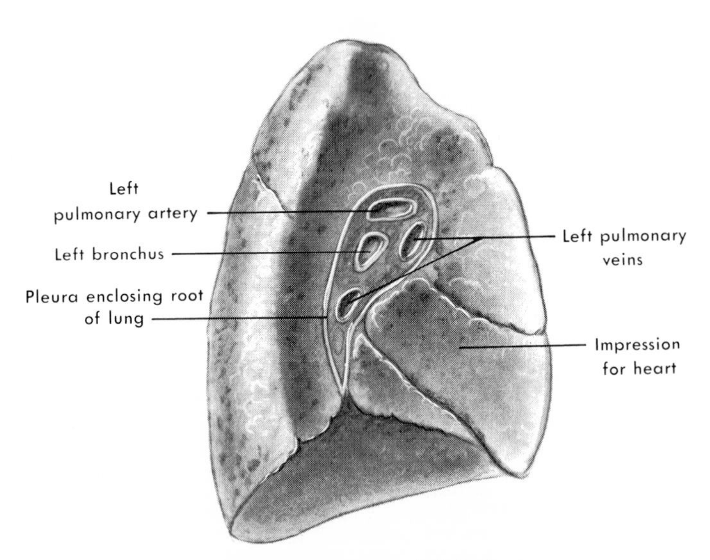

Left Lung, Medial Surface

Figure 7–8. The lungs.

their roots. The right lung has three lobes and the left has two. Because the liver forces the diaphragm up, the right lung is shorter than the left. However, because the heart occupies more of the left side of the thorax, the volume of the left lung is smaller.

The lungs are light, spongy, elastic organs. The superior end or *apex* of each rises above the sternal end of the clavicle, and the shape of the broad, concave *base* is molded by the diaphragm. A slit on the medial surface of each lung is the *hilum*. At this point all structures—arteries, veins, nerves, and bronchi—enter or leave the lung substance, forming its *root*.

PLEURA (Fig. 7-9)

The pleurae are paired, closed, double-layered serous sacs which enclose the lungs. The inner *visceral layer* adheres closely to the lung substance, following all the indentations of the lobes. It is separated from the outer *parietal layer*, which lines the thoracic walls. A small amount of serous fluid between the two prevents them from rubbing together as the lungs move during breathing.

To visualize the relationship of the pleurae to the lungs, imagine a closed fist plunged into an inflated balloon. The fist, representing the lung, forms a two-layered cavity in the balloon, which represents the pleura. The reflection of the balloon around the wrist represents the circular reflection of the pleural layers around the roots of the lungs; it is called the *pulmonary ligament*.

BLOOD SUPPLY

Venous blood is conducted from the right ventricle of the heart to the lungs by the pulmonary artery. Its branches deliver the blood to the capillary beds sur-

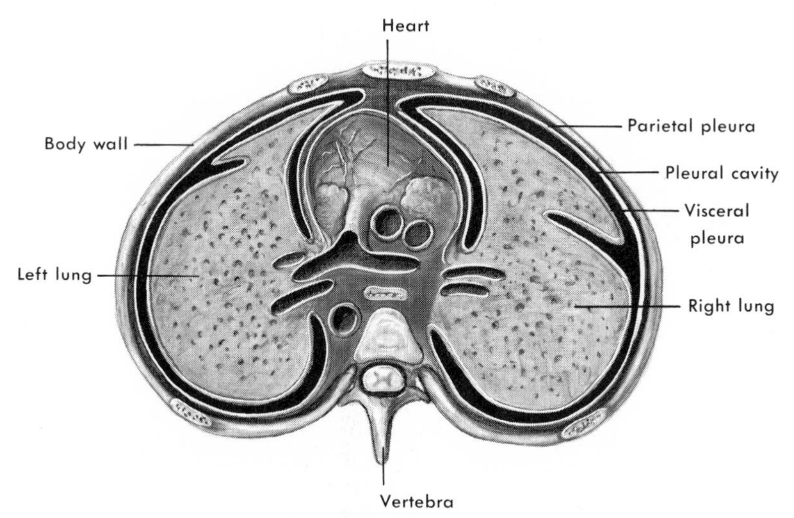

Figure 7-9. Cross section of the thorax as viewed from above, showing the relationship of the lungs and pleurae.

rounding the alveolar sacs. Each segmental bronchus is accompanied by a branch of the pulmonary artery. After the blood is oxygenated, it is collected by venules and then by branches of the pulmonary veins, which deliver it to the left atrium of the heart.

Oxygenated blood is supplied to the lung tissue itself by three bronchial arteries. Two of these come directly off the thoracic aorta, and the third is usually a branch of the first intercostal artery. These three arteries supply the lower trachea and bronchi as far as the respiratory bronchioles. Venous blood is returned by bronchial veins, which drain into larger veins in the thorax. However, because there are connections between the bronchial capillary network and the pulmonary capillary network, some of the blood carried to the lung by the bronchial arteries is returned by way of the pulmonary veins.

PHYSIOLOGY OF THE RESPIRATORY SYSTEM

MECHANICS OF BREATHING

Some terms used in the discussion of breathing can be defined as follows: *Eupnea* is normal, resting breathing. *Dyspnea* is labored or difficult breathing. *Tachypnea* or polypnea is breathing at an increased rate. *Hyperpnea* is breathing with an increased depth. *Apnea* is the cessation of breathing. The space inside the lungs is referred to as the *intrapulmonic* space, and the space inside the thorax but outside the lungs is called the *intrathoracic* space.

The volume of the closed thorax is changed by the muscles of inspiration and expiration. In normal inspiration, the muscles of importance are the *diaphragm* and the *intercostal* muscles (muscles between the ribs). During deep or forced inspiration, additional muscles are used. Quiet expiration is accomplished simply by the relaxation of the inspiratory muscles. In forced expiration, contraction of the abdominal muscles occurs to assist in expelling more air from the lungs.

When the diaphragm contracts in inspiration, it moves downward, enlarging the thorax in its vertical diameter. Contraction of the intercostal muscles produces an increase in the lateral diameter of the chest. The anteroposterior diameter is increased during forced breathing when additional muscles are used. Therefore, all diameters of the thorax can be increased, resulting in an enlargement of the thoracic cavity.

As the thorax becomes larger, the pressure of air inside the lungs (intrapulmonic pressure) decreases and falls below that of the air outside the body (atmospheric pressure). This pressure difference, or gradient, causes air to move in through the open airways and into the lungs until the two pressures are equal again. This equalization occurs at the end of inspiration when the inspiratory muscles have reached their maximum point of contraction.

Expiration occurs when the inspiratory muscles relax. The diaphragm moves upward, and the ribs return to their normal position. The elastic lungs are then able to recoil, increasing the pressure on the air inside them. When the intrapulmonic pressure rises above the atmospheric pressure, air is expelled from the lungs.

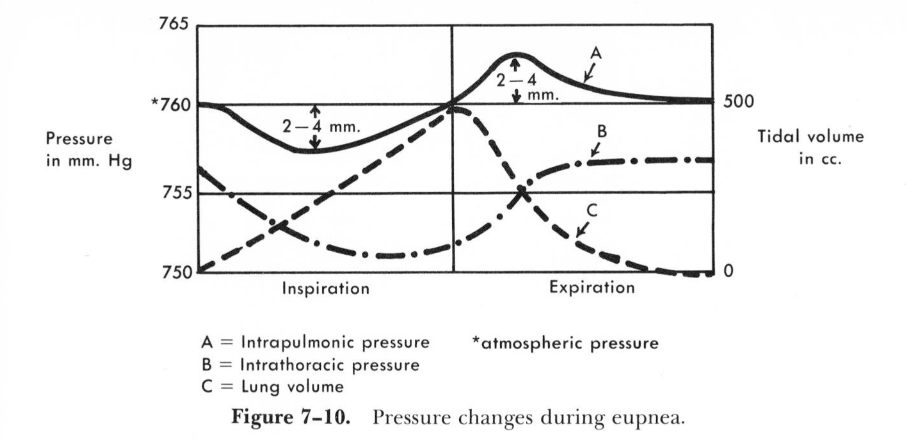

Figure 7–10. Pressure changes during eupnea.

In eupnea the magnitude of the pressure gradients during inspiration and xpiration vary from 2 to 4 mm. Hg. During forced breathing these changes are nuch larger (Fig. 7–10).

Note that intrathoracic pressure is negative during the entire breathing ycle; that is, it is always below atmospheric pressure. (There are some excepions to this which will be discussed later.) However, it is more negative during nspiration when the lungs are stretched to a greater extent than it is during exiration when they recoil. If enough air is deliberately or accidentally introduced nto the thoracic cavity, the lungs will collapse from the increased pressure. This ondition is called *pneumothorax*, and before the use of surgery and drugs, it was nduced to collapse the diseased lung of tuberculosis patients to allow it to rest. Because the air so introduced is gradually absorbed, the procedure had to be epeated at intervals in order to keep the lung in a collapsed state.

Intrathoracic pressure becomes positive when the opening to the larynx is losed and a forced expiratory effort is made. Called the Valsalva maneuver, it is ommonly executed when an individual strains in performing functions such as rination, defecation, and during childbirth. If one continues this maneuver for oo long a time, dizziness and even fainting may result. Fainting occurs because vhen the positive intrathoracic pressure is prolonged, there is a decrease in the mount of blood entering the right atrium of the heart. The result is a drop in he amount of blood the heart pumps out with each beat, which in turn leads to a all in arterial blood pressure. Coughing and sneezing, which are protective eflexes of the respiratory system, produce positive intrathoracic pressure in nuch the same way as the Valsalva maneuver.

UNG VOLUMES (Fig. 7–11)

The volume of air moved into the lungs in a single breath is called the *tidal olume*. In the normal young adult male this amounts to about 500 cc. (All volmes given here are for the average young adult male.) At the end of an average esting expiration, about 2.5 liters of air remains in the lungs. This volume is eferred to as the *functional residual capacity*.

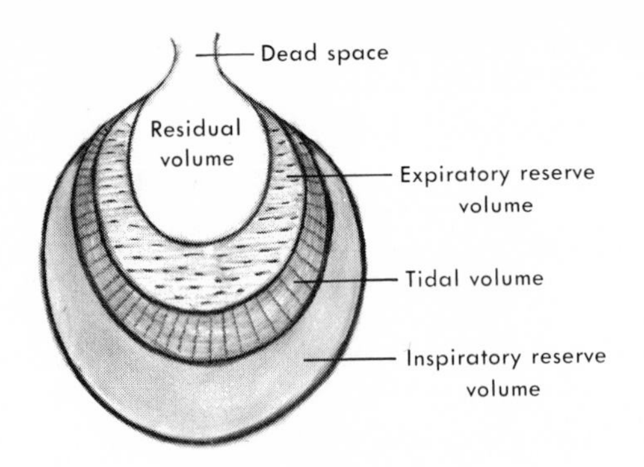

Figure 7–11. Relationships of lung volumes. (Residual volume is the volume of air remaining in the lungs at the end of a maximal expiration.)

Inspiratory capacity is the volume of air that can be forcefully inhaled after a normal expiration. The sum of this capacity plus the functional residual capacity comprises the *total lung capacity*. The volume of air that can be inhaled in addition to the total volume is the *inspiratory reserve volume*. Similarly, the *expiratory reserve volume* is that volume which can be exhaled after the exhalation of the tidal volume.

One of the clinical tests of lung function is the measurement of the *vital capacity*. It is calculated as follows:

Tidal volume + inspiratory reserve volume +
expiratory reserve volume = vital capacity

The average value for vital capacity is about 4500 cc.

Even when the lungs are collapsed, a small volume of air remains trapped in the air sacs. This air is called *minimal air*, and it accounts for the fact that any lungs that have ever contained any air will float in water. Thus, a stillborn baby's lungs, which have never been inflated, will not float in water.

Not all of the tidal volume air reaches the areas of the lung where oxygen and carbon dioxide are exchanged with blood. The conducting passages—nose, pharynx, larynx, etc.—are the parts of the system containing such air. This air is known as *dead space volume* and amounts to about 150 cc.

It is important to remember that during a normal breathing cycle there is not a complete exchange of gas between the lungs and the outside air. Instead, the large volume of gas in the lungs (functional residual capacity) is diluted with a small volume of atmospheric air (tidal volume).

In order to determine the effectiveness of respiratory ventilation, the *effective minute volume* is calculated. This volume represents the total amount of new air entering the alveoli each minute and is figured as follows:

Effective minute volume = (tidal volume − dead space) × rate

If the tidal volume is 500 cc., the dead space 150 cc., and the breathing rate is 14 times per minute, the effective minute volume is 4900 cc. Consequently, during each minute, 4900 cc. of air is actually participating in gas exchange with the blood. It is obvious that anything which increases dead space volume or decreases tidal volume will decrease effective minute volume. Such a decrease will occur in any condition which results in nonfunctioning lung tissue—surgical removal, atelectasis, emphysema, etc.

GAS EXCHANGE AND TRANSPORT

The composition of outside air, given in terms of dry gas, is remarkably constant. Air in the alveoli is always saturated with water vapor. If one breathes on a cold window pane, the vapor condenses into droplets of moisture. Thus the body loses water and heat by the expiration of moist, warm air. Because dogs have very few sweat glands, they pant heavily when an increase in heat loss is required. In panting, the large volume of air moved over the respiratory passages is largely dead space air.

Panting differs from *hyperventilation*, which is an increase in the quantity of air breathed (minute volume) as a result of an increase in the rate and/or depth of respiration. Hyperventilation can cause dizziness and fainting. Carbon dioxide is excreted at a faster than normal rate, and thus its concentration in the blood is decreased. The alkalosis caused by the loss of carbon dioxide from the arterial blood either interferes with cerebral metabolism directly or produces its effect by slowing the cerebral blood flow.

Expired air is a mixture of inspired air and alveolar air. Approximately four fifths of the air we breathe is nitrogen, which is an inert gas and at sea level ordinarily plays no role in gas exchange. As inspired air is passed through the lungs, it loses about 5 volumes per cent oxygen and gains almost the same amount of carbon dioxide. Because a gradient for oxygen exists between the outside air and the body tissues, oxygen diffuses from the alveolar air to the blood and from the blood to the tissues. There is a reverse gradient for carbon dioxide, and thus it diffuses from the tissues to the blood and from the blood to the alveolar air.

Transport of Oxygen. Oxygen is present in the blood in two forms: part is physically dissolved in the plasma, and the remainder is chemically combined with hemoglobin. The concentration of hemoglobin is usually 14 to 15 grams per 100 cc. of blood. One gram of hemoglobin combines with 1.34 cc. of oxygen, and when the blood is saturated, almost 20 volumes per cent of oxygen is present in this form. About 60 to 70 times as much oxygen in arterial blood is combined with hemoglobin as is physically dissolved in the plasma.

The equation for the combination of oxygen with hemoglobin can be expressed as follows:

$$\text{Hb (hemoglobin)} + O_2 \text{ (oxygen)} \rightleftharpoons HbO_2 \text{ (oxyhemoglobin)}$$

This reaction is readily reversible, as indicated by the arrows in both directions. Therefore, hemoglobin easily gives up its oxygen when the blood reaches the tis-

Table 7-1. APPROXIMATE COMPOSITION OF RESPIRATORY GASES AS VOLUMES PER CENT

	Nitrogen	*Oxygen*	*Carbon Dioxide*
Inspired air	79	21	0.04
Expired air	79	16	4.40
Alveolar air	80	14	5.60

sues. Carbon monoxide is a deadly poison because when it is present in the inspired air, even in small amounts, it combines with hemoglobin in place of oxygen, thus depriving the tissues of an adequate oxygen supply. Furthermore, the resulting compound, *carboxyhemoglobin*, is very stable and so renders hemoglobin inactive for oxygen transport.

Carbon Dioxide Transport. Carbon dioxide diffuses into the blood of the tissue capillaries. Some of it remains physically dissolved in the plasma. Some combines with water to form a weak acid (carbonic acid), which in turn reacts with blood salts to form *bicarbonate*. Still another portion combines with hemoglobin to form *carbaminohemoglobin*.

Bicarbonate is the principal form in which carbon dioxide is transported by the blood and amounts to about 89 per cent of the total present. When venous blood reaches the lungs, about 65 per cent of this amount is given up to the alveolar air.

CONTROL OF BREATHING

Nervous Control (Fig. 7–12). Two groups of nerve cell bodies in the medulla of the brain compose the *inspiratory center* and the *expiratory center*. Nerve impulses from the inspiratory center pass down the phrenic and intercostal nerves to the diaphragm and intercostal muscles. Impulses from the expiratory center discharge through nerves to the muscles of expiration. These two centers act reciprocally; that is, when one is stimulated and discharging, the other is inhibited. Although impulses from higher brain centers can influence the respiratory center, as in the voluntary control of breathing, the centers can function in carrying on rhythmical breathing even after the medulla has been separated from the rest of the brain.

When the inspiratory center discharges, the diaphragm and intercostal muscles contract and inspiration occurs. Inflation of the lungs causes stimulation of *stretch receptors*, located mostly in the visceral pleurae. These nerve endings

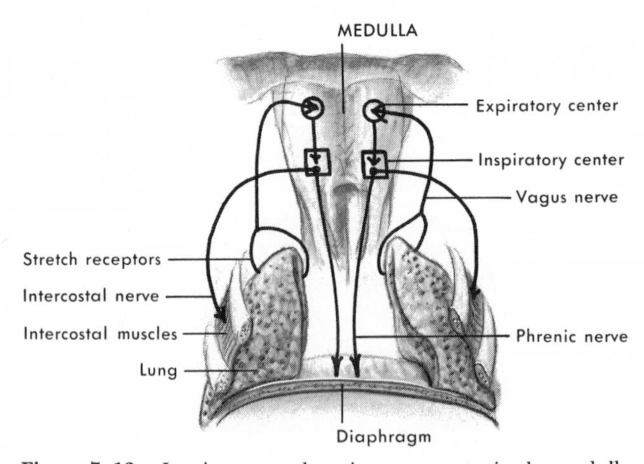

Figure 7–12. Inspiratory and expiratory centers in the medulla.

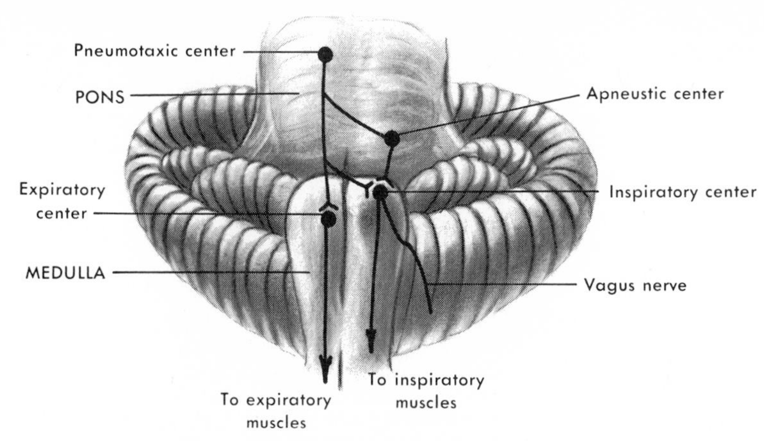

Figure 7–13. Location and relationships of respiratory centers.

elay impulses through fibers of the vagus nerve to the medulla, which stimulates he expiratory center. Stimulation of the expiratory center causes inhibition of he inspiratory center and passive expiration occurs. This reflex inhibition of inspiration is called the *Hering-Breuer reflex.*

Two other respiratory centers are contained in the pons (Fig. 7–13). These are the *apneustic center* and the *pneumotaxic center,* both of which modify and conrol the medullary centers' activities. When both vagus nerves carrying impulses o the expiratory center are cut, the Hering-Breuer reflex is abolished. This loss ordinarily would result in *apneustic* (gasping) breathing. However, under these conditions, the pneumotaxic center takes over and inhibits apneustic breathing by stimulating the expiratory center, thus promoting periods of normal expiraion. Stimulation of the apneustic center results in forceful inspirations and weak expirations, thus enhancing signals from the inspiratory center to the inspiraory muscles.

If the impulses from these two centers to the inspiratory and expiratory enters are blocked, respirations become shallow and irregular in rhythm. Basic espiratory rhythm, however, persists, showing that the medulla is the major oördinating area for respiration.

Chemical Control. Carbon dioxide is the major chemical factor regulating alveolar ventilation. It acts directly on the respiratory center in the medulla to precisely control involuntary breathing. Very small increases in the carbon dioxde content of the blood produce marked increases in respiratory rate. Above the evel of 10 per cent, however, carbon dioxide becomes a depressant, and concenrations of 30 per cent or higher are anesthetic. When carbon dioxide in the blood stimulates the medullary center, alveolar ventilation increases, which in urn decreases alveolar carbon dioxide. This results in a decrease of tissue carbon dioxide to normal levels. In this way, the respiratory center maintains a relaively constant level of carbon dioxide in the body.

Although of lesser importance, changes in oxygen level also have some inluence on respiration. A decrease in oxygen tension produces an increase in respiratory ventilation. This increase is a result of an indirect effect on the respiratory center. Nerve endings are located in the walls of the common carotid arteries and the aortic arch. If the oxygen level of the blood drops low enough (hemoglobin saturation below 93 per cent), these afferent nerve endings are stimulated to send impulses to the respiratory center with a resulting increase in ventilation rate. This reflex is weak, however, and it probably plays very little part in the usual regulation of respiration.

ANESTHETICS

One of the most prevalent causes of respiratory depression results from the use of general anesthetics. These, in contrast to local anesthetics, affect the central nervous system, causing loss of sensation and consciousness. The deeper the anesthesia, the more depressed the central nervous system becomes. Fortunately, in most cases the respiratory centers in the medulla are spared until the deepest anesthetic plane is reached.

Ether, one of the oldest general anesthetics in terms of use (it was introduced in 1540), remains one of the safest agents in regard to sparing the respiratory centers. It is such an irritating substance that it cannot be "pushed" on the patient too fast. Furthermore, it reflexly stimulates respiration until the deepest anesthetic plane is reached. Thus, ether depresses the respiratory centers the *least*, but it depresses the cerebral cortex the *most*—an important feature of a good general anesthetic.

HYPOXIA

Hypoxia is any condition in which there is a deficiency in delivery or utilization of oxygen by the body tissues. It is obvious that this condition can occur at any point in the respiratory system from the air that is inspired to the tissue cells themselves. A classification of the causes of hypoxia can be based, then, on the point of breakdown in the delivery of oxygen to the tissues. There are six such categories:

1. Atmospheric deficiency. This condition can result from either (1) a decrease in total barometric pressure, as in breathing air at high altitudes, or (2) a decrease in the amount of oxygen present in the air breathed, as in inhaling natural gas, which contains no oxygen. In either case, increasing the percentage of oxygen in the inhaled air alleviates the hypoxia.

2. Ventilatory deficiency. In the normal lung this deficiency can result from an obstruction in the airway or a failure of the respiratory muscles. In the abnormal lung, an insufficient amount of functioning lung tissue could produce this condition. Examples are surgical removal, pneumothorax, and atelectasis. (Atelectasis is a collapsed or airless state of all or part of a lung from one of many causes.)

3. Diffusion deficiency. Poor diffusion of gases across capillary membranes in the lungs commonly results from pulmonary fibrosis. The alveolar walls are thickened by fibrous tissue and by pulmonary edema, in which accumulation of fluid into the air sacs obstructs the passage of oxygen.

4. Hemoglobin deficiency. This deficiency can result from either a reduction in total hemoglobin, as in anemia, or the formation of inactive hemoglobin, as in the inhalation of carbon monoxide.

5. Circulatory deficiency. A circulatory deficiency results from the mixture of arterial and venous blood, as in the case of a defect in the wall separating the heart ventricles. The result is that the arterial blood reaching the tissues has reduced oxygen content. Another cause is reduced blood flow, either to all organs, as in cardiac failure, or to specific organs because of local obstruction of the blood supply by a clot, severe vasoconstriction, etc.

6. Inadequate tissue oxygenation. Even with an adequate amount of oxygen being delivered to the body tissues, it is still possible for them to suffer hypoxia if there is interference with oxygen diffusion from blood to tissue cells. Such interference can result from interstitial edema, the collection of excess fluids between the cells, which increases the distance that oxygen must diffuse. Cyanide, a deadly poison, produces tissue hypoxia by interfering with the utilization of oxygen by the cells.

OUTLINE SUMMARY

General Considerations
1. Respiratory system supplies cells with oxygen and removes carbon dioxide produced by them.
2. Breathing moves air into and out of the system.
3. External respiration involves gas exchange between air in the system and blood and between blood and the tissue cells.
4. Internal respiration is the utilization of oxygen and the production of carbon dioxide by the cells.

Anatomy of the Respiratory System
1. Conducting Passages
   a. Two nasal cavities – warm, filter, and moisten air; sense of smell; phonation
   b. Pharynx – funnel-shaped passageway connecting nasal cavities with larynx
      1) Nasopharynx – opening for auditory tube; pharyngeal tonsils
      2) Oropharynx – palatine tonsils
      3) Laryngopharynx – continuous with esophagus; airway and food passage cross here
   c. Larynx – organ of sound production; nine cartilages, including thyroid, epiglottis, and cricoid
   d. Trachea – rigid windpipe formed by **C**-shaped cartilage rings
   e. Bronchial tree
      1) Two primary bronchi – right and left
      2) Five secondary bronchi – three right and two left
      3) Segmental bronchi – supply a total of 18 bronchopulmonary segments
      4) Bronchioles – 50 to 80 terminal bronchioles in each lung lobe
2. Respiratory Passages
   a. Functional lung lobule – respiratory bronchiole, alveolar ducts, alveolar sacs, alveoli
   b. Two lungs – cone-shaped organs of respiration
      1) Upper apex, lower base; hilum on medial surface
      2) Root – all structures entering or leaving lung
   c. Pleurae – paired, closed, double-layered serous sacs enclosing lungs and lining thoracic cavity

3. Blood Supply
   a. Pulmonary artery—blood to capillaries surrounding alveolar sacs; exchange oxygen and carbon dioxide
   b. Bronchial arteries—supply walls of bronchial tree with oxygenated blood
   c. Pulmonary and bronchial veins—return blood to heart

Physiology of the Respiratory System

1. Mechanics of Breathing
   a. Inspiration
      1) Diaphragm and intercostal muscles enlarge thoracic cavity
      2) Intrapulmonic pressure decreases below atmospheric pressure and air moves into lungs
   b. Expiration
      1) Inspiratory muscles relax; thorax decreases in size; elastic lungs recoil
      2) Intrapulmonic pressure rises above atmospheric levels; air expelled from system
   c. Intrathoracic pressure—always negative except during Valsalva maneuver, coughing, and sneezing
2. Lung Volumes
   a. Tidal volume—amount of air moved into lungs in one breath; averages 500 cc.
   b. Functional residual capacity—about 2.5 liters of air remaining in lungs after resting expiration
   c. Inspiratory capacity—air forcefully inhaled after normal expiration
   d. Total lung capacity—equals the sum of the functional residual capacity and the inspiratory capacity
   e. Inspiratory reserve volume—air inhaled in addition to tidal volume
   f. Expiratory reserve volume—air expired after exhalation of tidal volume
   g. Vital capacity—equals tidal volume + inspiratory reserve volume + expiratory reserve volume (averages 4500 cc.)
   h. Minimal air—that remaining in lung even after it is collapsed
   i. Dead space volume—air occupying conducting passages; average 150 cc.
   j. Effective minute volume—equals (tidal volume − dead space) × rate
3. Gas Exchange and Transport
   a. Composition of outside air (see Table 7-1, p. 153)
   b. Oxygen transport
      1) Some dissolved in blood plasma
      2) Most combined with hemoglobin as unstable compound, oxyhemoglobin
   c. Carbon dioxide transport
      1) Some physically dissolved in plasma
      2) Some combined with hemoglobin to form carbaminohemoglobin
      3) Most combines with water to form carbonic acid and then bicarbonate
4. Control of Breathing
   a. Nervous control
      1) In medulla—inspiratory center and expiratory center
         a) These act reciprocally and are the major coördinating areas
         b) Hering-Breuer reflex inhibits inspiration
      2) In pons—apneustic center and pneumotaxic center

a) apneustic – reinforces signals from inspiratory center
b) pneumotaxic – stimulates expiratory center when Hering-Breuer reflex is abolished

b. Chemical control
1) Carbon dioxide – major chemical factor regulating alveolar ventilation; acts directly on medullary centers
2) Oxygen – less important chemical factor; indirect effect on medullary centers

5. Anesthetics
a. Danger – general anesthetics depress respiratory centers
b. Ether – one of safest agents because it spares respiratory center

6. Hypoxia – state of oxygen deficiency in body
a. Atmospheric deficiency – low barometric pressure; inhalation of foreign gases
b. Ventilatory deficiency – obstructed airway; insufficient active lung tissue
c. Diffusion deficiency – pulmonary thickening; fluid in lungs
d. Hemoglobin deficiency – amount of hemoglobin reduced below normal level; formation of inactive hemoglobin
e. Circulatory deficiency – mixing arterial and venous blood; clots; vasoconstriction
f. Inadequate tissue oxygenation – tissue edema; cell poisons

REVIEW QUESTIONS

1. Explain the difference between breathing and respiration.
2. What part of the respiratory system is common to the digestive system?
3. What structures are surgically removed in a "T and A" (tonsillectomy and adenoidectomy)?
4. Describe what happens in swallowing.
5. Why is it not really accurate to call the larynx the voice box? Can a person whose larynx has been removed still speak? Explain.
6. Describe the components of a functional lung lobule.
7. Describe the relations of the pleurae to the lungs.
8. How does expiration occur?
9. What is the Valsalva maneuver?
10. What happens in pneumothorax?
11. How is the vital capacity calculated?
12. What is dead space volume?
13. If a person breathes 12 times a minute and has a dead space volume of 200 cc. and a tidal volume of 550 cc., what is his effective minute volume?
14. Why, under normal conditions, does oxygen move from the alveolar air into the blood and from blood into the tissue cells?
15. Write the reversible reaction for the combination of oxygen and hemoglobin.
16. Describe the respiratory centers in the medulla and pons.
17. How does carbon dioxide regulate alveolar ventilation?
18. How does a decrease in oxygen content of the blood affect respiratory ventilation?
19. Explain why breathing carbon monoxide in sufficient concentration causes death.

ADDITIONAL READING

Clements, J. A.: Surface tension in the lungs. Sci. Amer. (Dec.), 1962.
Comroe, J. H., Jr.: Physiology of Respiration: An Introductory Text. Chicago, Year Book Medical Publishers, Inc., 1965.
Comroe, J. H., Jr.: The lung. Sci. Amer. (Feb.), 1966.
Winter, P. M., and Lowenstein, E.: Acute respiratory failure. Sci. Amer. (Nov.), 1969.

Chapter 8

THE DIGESTIVE SYSTEM

INTRODUCTION

Just as all living cells need oxygen to survive, so too they need materials for growth, energy, and repair. These nutritive substances must be taken into the body and changed into forms which can be utilized by the cells. The digestive system is composed of organs which fulfill these needs by properly preparing such essential materials, called *food*, through a series of processes called *digestion*. *Absorption* refers to the passage of these prepared materials from the digestive tract to the blood. The utilization of the absorbed material by the body cells is called *metabolism*.

GENERAL PLAN (Figs. 8–1, 8–2)

The digestive system consists essentially of a tube extending approximately 27 feet from the lips to the anus. In addition, there are organs lying outside the tube but connected with it which are called accessory organs. These organs are the liver, the gallbladder, and the pancreas.

The wall of the tube is composed of four main layers of tissue. The *mucosa* (mucous membrane), or lining layer, contains a thin basal layer of smooth muscle which allows for local movements and folds. It also has a connective tissue layer which contains small blood vessels, lymphatic tissue, and in some areas secretory glands. The innermost layer of the mucosa is formed by epithelial tissue, the type varying in different portions of the tube according to function.

The *submucosa* lies underneath the mucosa and is a loose connective tissue layer. It contains networks of larger blood vessels and autonomic nerve fibers.

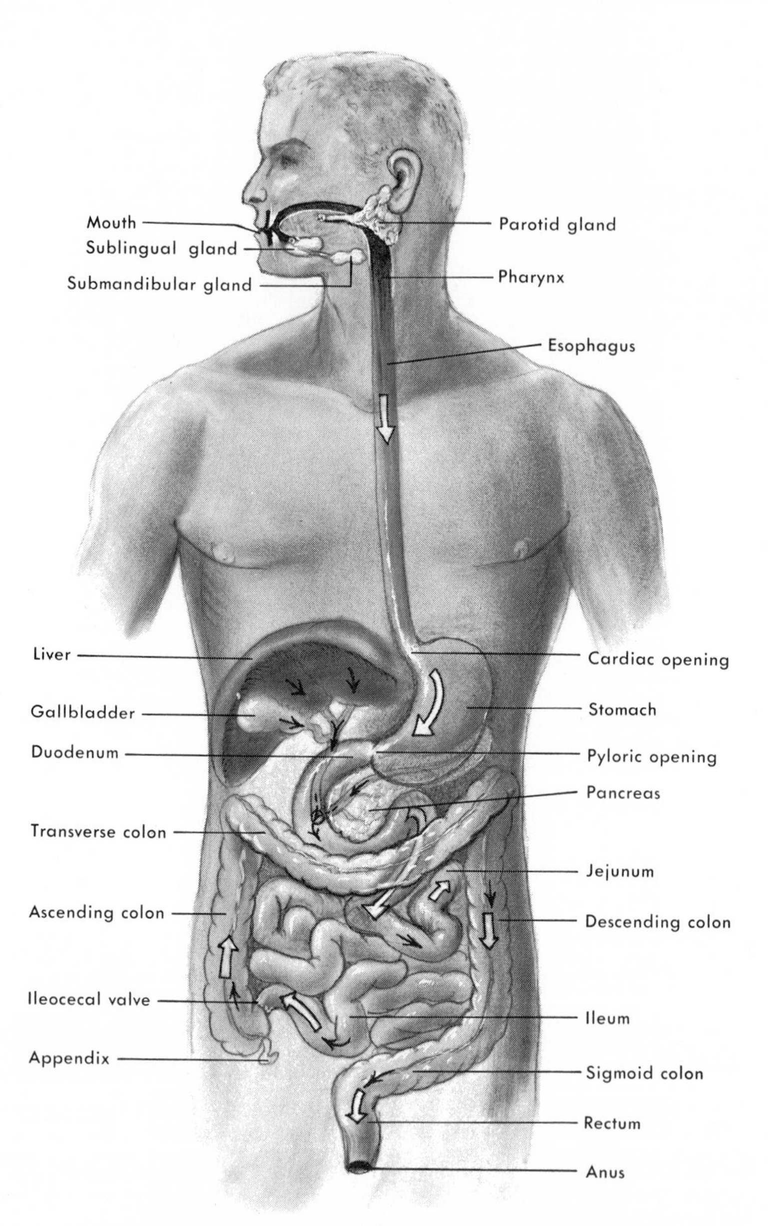

Figure 8-1. General plan of the digestive system.

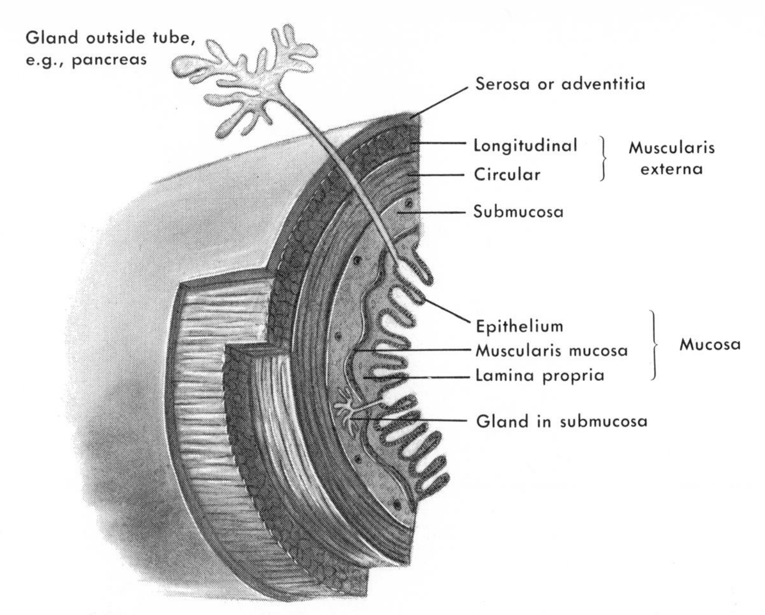

Figure 8–2. Layers of the digestive tube (as seen in cross section).

Submucosal glands are present in only two parts of the tube: in the duodenum of the small intestine and in the esophagus.

Lying under the submucosa is the main muscle layer of the tube. Called the *muscularis externa*, it usually consists of two layers of smooth muscle. The fibers of the inner layer run circularly around the tube, whereas those of the outer layer run longitudinally or lengthwise. This arrangement varies somewhat from one portion of the tube to another.

The outermost layer of the digestive tube is a thin covering of either fibrous or serous membrane. That portion of the tube lying above the diaphragm is covered with fibrous membrane called the *adventitia*. Below the diaphragm the covering layer is serous membrane and is called the *serosa*.

The *peritoneum* (Fig. 8–3) is a double-layered serous membrane which lines the abdominal cavity and is reflected over the viscera within it. The lining layer is called parietal peritoneum, and that covering the organs is called the visceral peritoneum. The latter contains varying amounts of fat, depending upon the nutritional state of the individual. Certain double folds of the peritoneum, called *ligaments* (not to be confused with ligaments which connect bones), pass between two organs or between organs and the abdominal wall. These ligaments help to hold the viscera in place and transmit nerves and blood vessels to and from them.

The *lesser omentum* is the ligament which passes from the lesser curvature of the stomach and first part of the duodenum to the inferior surface of the liver. It carries the hepatic artery, portal vein, and main bile duct in its free margin. The *greater omentum* passes between the greater curvature of the stomach and the transverse colon, hanging down like an apron over the small intestine. A fan-shaped fold of peritoneum, the *mesentery*, enfolds most of the small intestine and anchors it to the posterior abdominal wall.

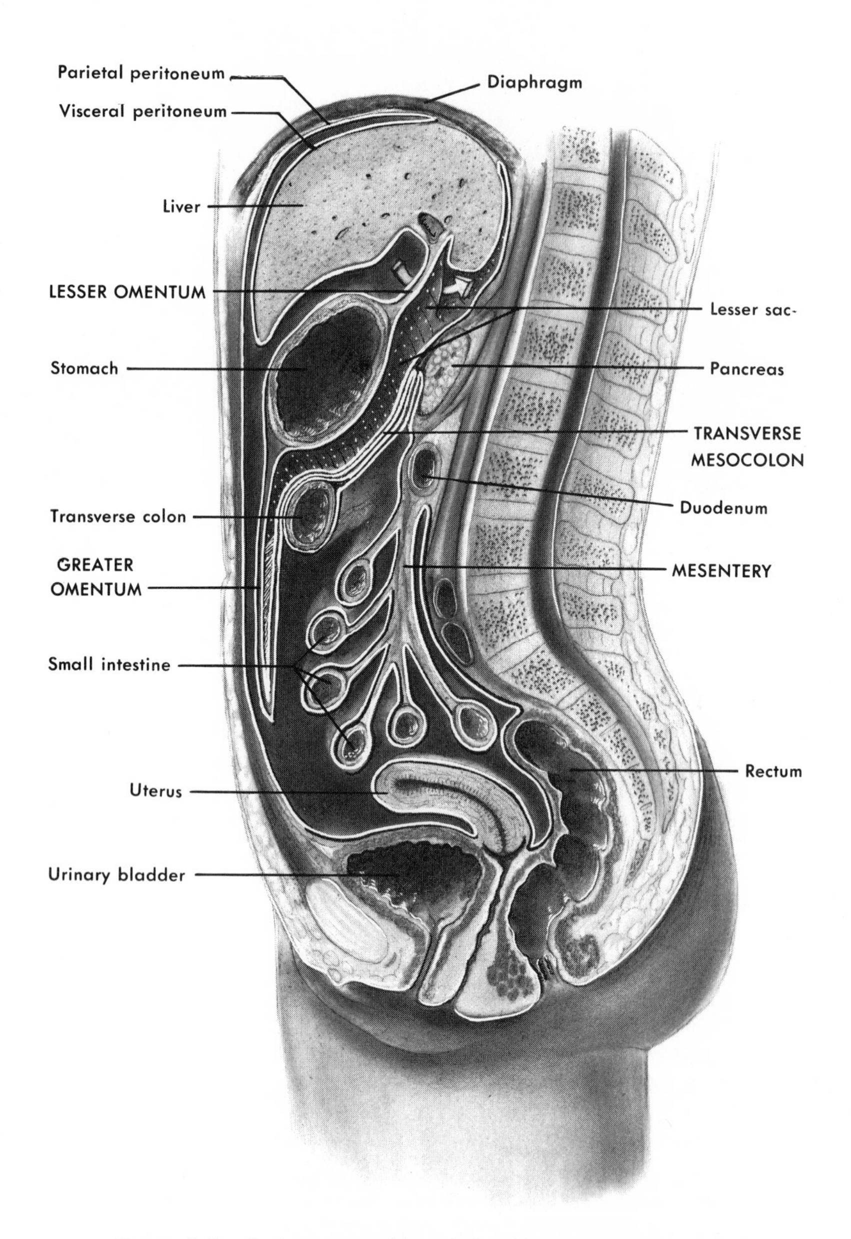

Figure 8-3. Peritoneum and its relationships – midsagittal section.

Some abdominal organs are not covered by peritoneum. The abdominal aorta, pancreas, most of the duodenum, and kidneys lie posterior to the peritoneal cavity instead of actually inside it. Therefore, these structures are said to be *retroperitoneal.*

ANATOMY OF THE DIGESTIVE SYSTEM

Mouth. The parts of the mouth to be described are the vestibule, cheeks, tongue, teeth, roof, and salivary glands.

VESTIBULE. The vestibule is that portion of the mouth between the lips and cheeks and the teeth. Skeletal muscle and fibroelastic connective tissue form the lips. Their red, free margins are covered with modified skin. Many blood capillaries close to the surface of the skin give the lips their red color. Since the skin of the lips is not as heavily keratinized as other areas of the body, it must be kept moist to prevent drying and cracking.

CHEEKS. The cheeks are lined with mucous membrane. Their main bulk is formed by the buccinator muscles, which help keep food between the teeth while chewing.

TONGUE (FIG. 8–4). The tongue is composed of interlacing bundles of skeletal muscle fibers with fibroelastic tissue between them. Other skeletal muscles pass from the hyoid bone and the styloid processes and insert into the tongue. Thus, in addition to changing shape, the whole tongue can move in a variety of directions. The fold of mucous membrane on the ventral or undersurface of the tongue is the *frenulum.* The dorsal or upper surface is covered with thick mucous membrane, which is studded with small elevations called *papillae.* The taste buds are located on these papillae. Masses of lymphoid tissue, called the *lingual tonsils,* are located at the root of the tongue. The tongue is concerned with taste, chewing, swallowing, and speech.

TEETH. Two sets of teeth are usually present in the jaws at birth. The

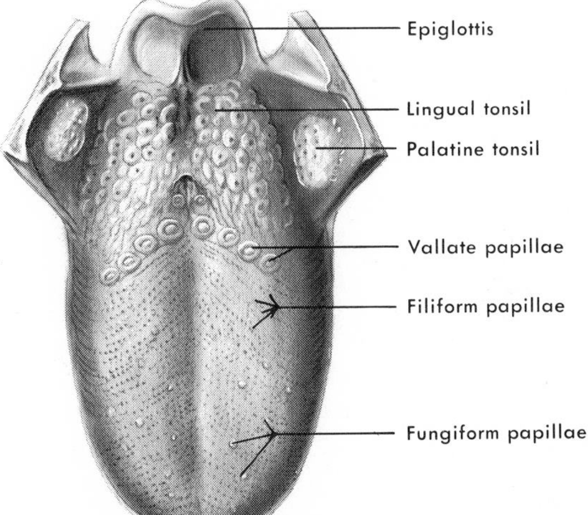

Figure 8–4. The tongue – dorsal surface.

first set to erupt are the *deciduous* or baby teeth, numbering ten in each jaw. They usually begin to appear about the age of six months. The 32 *permanent* teeth (16 in each jaw) erupt between the ages of six and 18 years. In some cases, however, the last four permanent teeth, or wisdom teeth, may fail to erupt at all (Fig. 8–5).

The *crown* is that part of the tooth that extends above the gum. It is covered with *enamel*, a very hard calcified material derived from epithelium. The rest of the tooth is embedded in a socket in the alveolar process of the mandible or maxilla and is called the *root*. The root is covered with *cementum* and is separated from the bone by *periodontal membrane*, which holds the tooth firmly in place. The *neck* of the tooth is the junction between the crown and the root. Calcified connective tissue called *dentin* forms the bulk of the tooth. In its center is the *pulp cavity*, which is expanded in the upper portion of the tooth and narrowed as it extends into the root. This space contains nerve fibers and blood vessels. (Fig. 8–6)

ROOF OR PALATE. The bony hard palate and the anterior portion of the soft palate form the roof of the mouth. The palatine processes of the maxillae and the horizontal plates of the palatine bones are covered with thick mucous membrane, part of which is actually the periosteum of the underlying bones. Interlacing skeletal muscle forming the soft palate allows it to move so that in swallowing it can close off the nasopharynx and prevent food from passing into the nose. The anterior part of the soft palate is attached to the posterior margin of the hard palate. The free posterior margin of the soft palate forms an arch from one side to the other, and this arch marks the posterior limit of the mouth. A cone-shaped extension, the *uvula*, projects downward from the center of the arch. The passage from the mouth to the pharynx is called the *fauces*.

SALIVARY GLANDS. The mucous membrane of the mouth contains several small glands which produce a secretion called *saliva*. In addition, there are three pairs of larger salivary glands which are not located in the mouth cavity but discharge saliva into it. Saliva is essentially fluid, but it usually contains some white blood cells and other cell fragments. It varies in consistency from thin and watery to thick and sticky, depending upon the type of food in the mouth.

The two *parotid* salivary glands lie below and in front of the ears, between the mastoid process of the temporal bone and the ramus of the mandible on each side of the face. Each parotid duct runs parallel to the zygomatic arch, pierces the buccinator muscle of the cheek, and opens into the vestibule of the mouth opposite the second upper molar tooth.

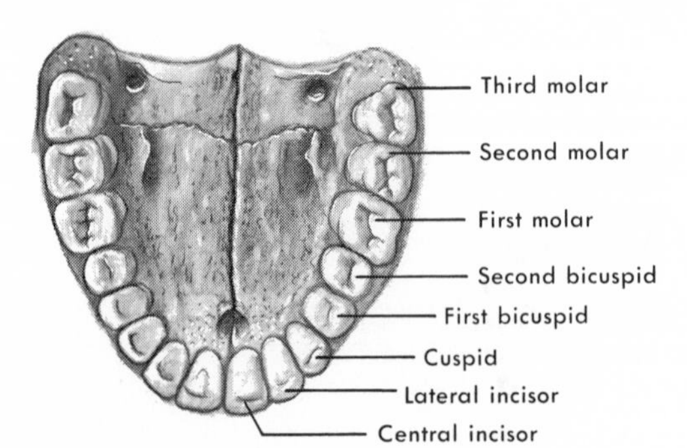

Figure 8–5. Permanent teeth – upper jaw. The third molars are also known as wisdom teeth.

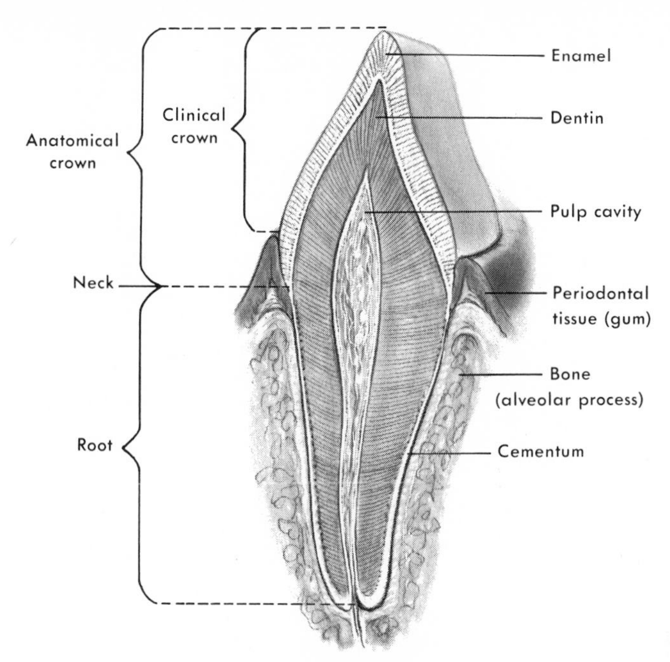

Figure 8–6. Diagram of a typical tooth.

The *submandibular* glands are in contact with the inner surface of the mandible, lying below and in front of the mandibular angle. Their ducts open into the floor of the mouth on either side of the frenulum.

The *sublingual* glands lie beneath the mucous membrane in the floor of the mouth near the midline. Their secretion empties through several ducts which either open directly into the mouth or join with the submandibular ducts.

Pharynx. This passageway is common to both respiratory and digestive systems. It was described in Chapter 7.

Esophagus. Lying posterior to the trachea, the *esophagus* is a collapsible tube about 10 inches long. Continuing from the laryngopharynx, it passes through the mediastinum, pierces the diaphragm, and opens into the stomach. There are mucous glands in the submucosal layer and cardiac glands in the connective tissue layer of the mucosa.

In the wall of the upper third of the esophagus the external muscle is skeletal. This muscle is under the control of the autonomic nervous system, which is an exception to the general rule that skeletal muscle is voluntary. In the lower third of the tube the muscle is smooth, but in the middle third the fibers are a mixture of skeletal and smooth. Most of the esophagus lies above the diaphragm, and this portion is covered with adventitia.

Stomach. The stomach (Fig. 8–7) lies in the upper part of the abdominal cavity, just below the diaphragm. The opening between the esophagus and the stomach is the *cardiac orifice*. In lower animals this opening is guarded by a ring of circular muscle called the cardiac sphincter. Although such a muscle ring is not present in the human body, the fibers of the muscularis externa in this area are

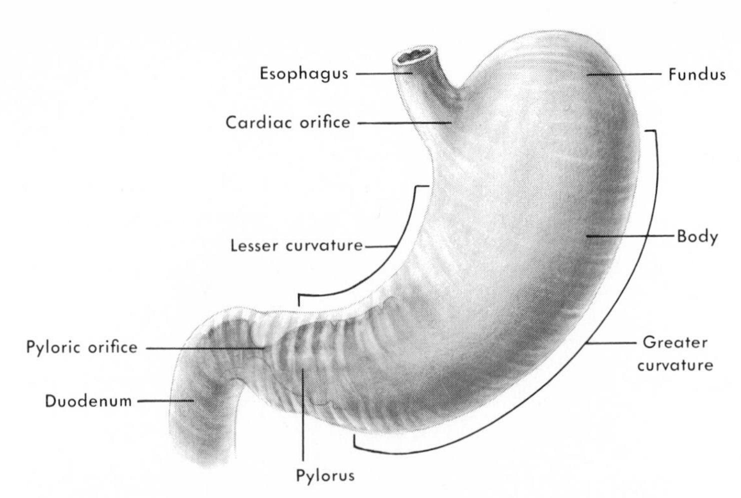

Figure 8–7. The stomach – anterior view.

in a state of contraction. Their contraction helps to prevent a reverse flow of stomach contents into the esophagus.

The *fundus* is that portion of the stomach lying above a line drawn through the cardiac orifice. Two thirds of the remainder of the organ is the *body*, and the most distal portion is the *pylorus*. The *pyloric orifice* opens into the small intestine. It is guarded by the *pyloric sphincter*, a fold of the middle layer of the muscularis externa.

The stomach presents two surfaces, anterior and posterior. The medial border is the *lesser curvature*, and the lateral border is the *greater curvature*. When the stomach is empty, the mucous membrane is thrown into longitudinal folds called *rugae*. The rugae disappear when the stomach is full.

The digestive juices of the stomach are secreted by the *gastric glands*, located in the fundus and body. Three layers of smooth muscle – inner oblique, middle circular, and outer longitudinal – make up the muscularis externa. The serosa covering the stomach is visceral peritoneum.

Small Intestine. The small intestine is about 21 feet long and 1 inch in diameter. It extends from the pyloric orifice to the ileocecal orifice, which is the entrance to the large intestine. The *ileocecal valve* guarding this orifice acts to prevent the flow of fecal material from the large intestine backward into the small intestine. The three portions of the small intestine are the duodenum, the jejunum, and the ileum.

The first 12 inches of the tube is the *duodenum* (Fig. 8–8). It forms a **C**-shaped curve around the head of the pancreas. The bile and pancreatic ducts open into the duodenum on an elevation called the *duodenal papilla*. This area is about 3 inches below the pylorus of the stomach. These two ducts usually open side by side but separately, sharing a thin, common wall between them. Most of the duodenum is retroperitoneal.

The *jejunum* forms the next two fifths (about 8 feet) of the small intestine. The *ileum* forms the distal portion of the tube and is about 12 feet long. Both parts are suspended from the posterior abdominal wall by the mesentery. There

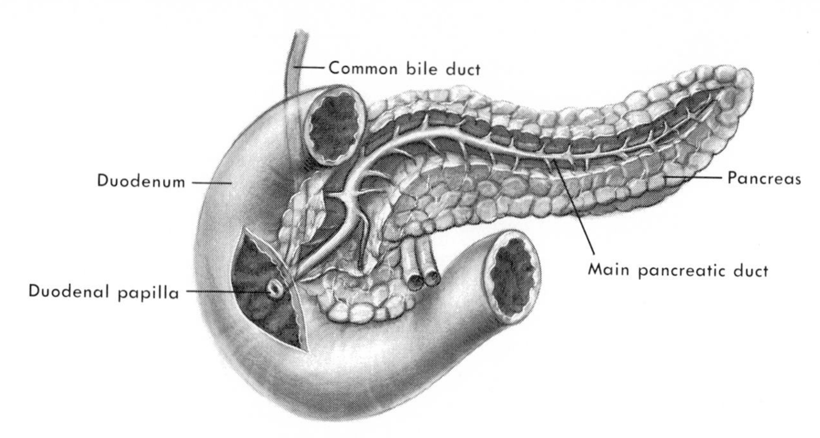

Figure 8–8. The duodenum and pancreas.

is no clear dividing line between the two portions, and they differ from each other as a result of gradual changes. The jejunum has a slightly larger diameter, thicker walls, and a deeper color than the ileum. Furthermore, the amount of fat contained in the mesentery increases from above downward.

The inner surface of the small intestine is thrown into *circular folds* (plicae circularis). These folds are formed by the mucosa and submucosa, and unlike the rugae of the stomach they are permanent foldings. Small, finger-like projections of the mucosa, called *villi*, are located on and between the folds. Each villus contains a blood capillary and a lymphatic capillary. The circular folds begin in the duodenum at a level below the duodenal papilla. They are very numerous in the jejunum and then become reduced in size and number in the ileum. The villi are present in large numbers in the duodenum but gradually decrease in number as the small intestine progresses (Fig. 8–9).

There is an abundance of aggregated lymphatic tissue present in patches along the border of the ileum, which is opposite the mesenteric attachment.

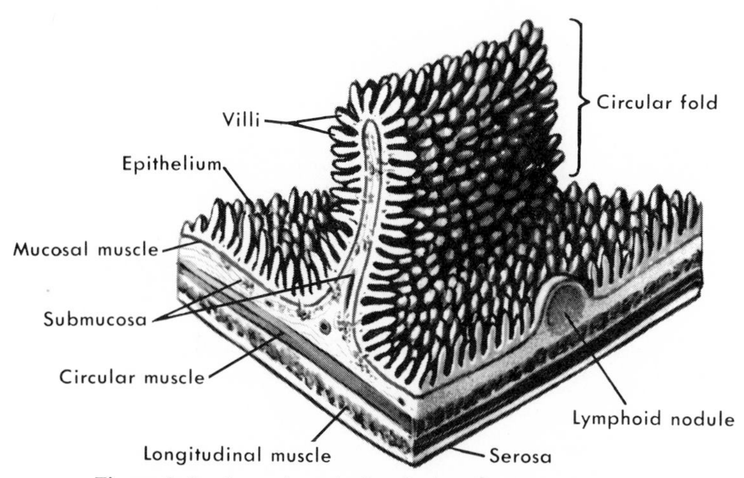

Figure 8–9. Inner intestinal wall, showing villi and plicae.

Mucosal *intestinal glands* (crypts of Lieberkühn), dipping down almost to the muscle layer of the mucosa, are present throughout the length of the small intestine. *Submucosal glands* (Brunner's glands) are found only in the duodenum.

The circular and longitudinal layers of the muscularis externa become thinner as the distal end of the tube is reached. Autonomic nerve fibers and parasympathetic ganglia lie between the two muscle layers. Visceral peritoneum forms the serosa of the small intestine.

Large Intestine. The large intestine continues from the small intestine at the ileocecal orifice and continues to the anus. It is about 4½ to 5 feet long and 2½ inches in diameter. In addition to its larger diameter, it has three main gross differences from the small intestine. First, the longitudinal muscle layer is incomplete, forming three separate bands equidistant from each other along the whole length of the tube. These bands are called the *teniae coli*. Second, since the teniae are shorter than the large intestine itself, the wall of the tube bulges between them to give a "puckered" effect. These bulges are called *haustra*. Finally, the visceral peritoneum or serosa forms small, fat-filled pouches where it attaches to the teniae. These are the *epiploic appendages* (Fig. 8–10).

The parts of the large intestine are the cecum, the colon, the rectum, and the anal canal. The *cecum* is a blind pouch about 2 to 3 inches long which lies below the junction of the ileum and colon. It opens above into the ascending colon, and its closed end is directed downward. The end of the ileum is folded into the cecum, forming the ileocecal valve.

The originally tapered tip of the cecum develops into a narrow tube, the *appendix*, which arises about 1 inch below the ileocecal valve. Although its length varies, it is usually about 3 inches long. Attached only by its proximal end to the cecum, it is completely covered by peritoneum and contains an abundance of lymphatic tissue.

The *ascending colon* extends upward from the cecum along the right posterior abdominal wall to the undersurface of the liver. Here it turns toward the midline to form the *right colic* or *hepatic flexure*. The largest part of the large intestine is the *transverse colon*. It crosses the upper abdominal cavity from right to left and then turns downward, forming the *left colic* or *splenic flexure*. It is held in position by peritoneal ligaments which attach it to the posterior abdominal wall and the greater curvature of the stomach. Another ligament holds it in contact with the spleen. Most of the small intestine lies below and behind the transverse colon.

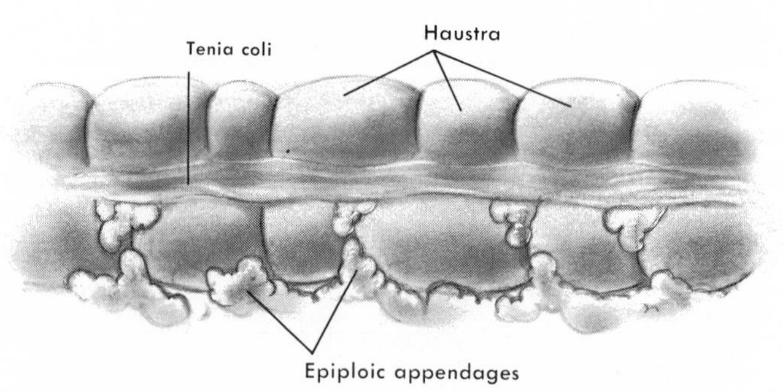

Figure 8–10. Section of the large intestine.

The *descending colon* extends from the left colic flexure down along the left side of the abdomen to the brim of the pelvis. As it passes anterior to the ileum, it turns medially and inferiorly to end in the sigmoid colon.

As its name implies, the *sigmoid colon* is **S**-shaped. It continues from the descending colon along the left pelvic wall and then transversely across the pelvis to the right. Finally, it passes back and down to continue as the rectum.

The *rectum* extends from the sigmoid colon to the anus and is approximately 5 to 7 inches long. The rectum proper goes as far as the muscles forming the floor of the pelvis, opposite the tip of the coccyx. From here it continues as the *anal canal,* which pierces the pelvic floor muscles. The canal then passes down between the anal muscles, which compress it to a slit.

The terminal opening of the digestive tube is the *anus.* A ring of involuntary muscle, the internal anal sphincter, encircles the anal canal just above the point where the mucosa of the tube changes to skin. The voluntary external anal sphincter overlaps the internal sphincter but extends down to end in the skin around the anus.

Although there are no villi in the large intestine, its walls contain an abundance of intestinal mucous glands. Solitary lymph nodules are scattered throughout the connective tissue layer of the mucosa. A dense network of veins is present in the mucosa and submucosa of the rectum. The condition in which some of these veins become large and tortuous is called *hemorrhoids.* The ascending and descending parts of the colon lie behind the peritoneum, and the rectum and anal canal lie below it.

Accessory Organs. The liver, gallbladder, and pancreas are not actually part of the digestive tube, but they are closely related to it in their functions.

LIVER (FIG. 8–11). The liver is the largest organ in the body, weighing about three pounds in the adult. Located in the upper right quadrant of the abdominal cavity, its superior surface lies under the dome of the diaphragm. Its inferior surface is molded over the stomach, the duodenum, the right colic flexure, and the right kidney. The gallbladder lies in an impression on this visceral surface. Although visceral peritoneum is reflected over most of the liver, its posterior surface is bare.

Four lobes of the liver may be distinguished externally. Anteriorly, the *right lobe* is separated from the *left lobe* by a two-layered peritoneal fold, called the falciform ligament. On the undersurface, the *quadrate lobe* lies between the gallbladder fossa and the groove for the ligamentum teres. The small *caudate lobe* lies between the depressions for the inferior vena cava and ligamentum venosum. (These last two ligaments are the remains of blood vessels present in the circulatory system of the fetus. See Chapter 6.)

The *porta* or "door" of the liver is the region where the hepatic artery and portal vein enter the organ and where the bile ducts leave it. The *common bile duct* is formed by the joining of the *hepatic duct,* which carries bile from the liver, and the *cystic duct,* which carries bile from the gallbladder. The common duct then carries the bile into the duodenum through an opening on the duodenal papilla. The *hepatic artery* furnishes arterial blood for the nourishment of the liver cells. The *portal vein* carries blood containing the products of digestion from the intestinal tract into the liver. Three *hepatic veins* carry venous blood from the liver and empty into the inferior vena cava.

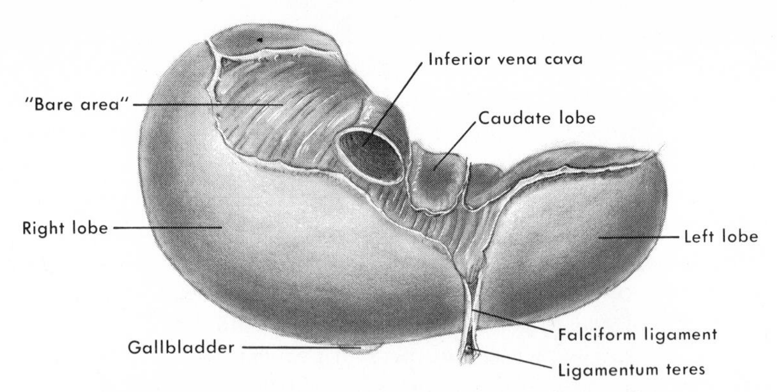

Figure 8–11. The liver—diaphragmatic (superior) surface.

Internally, the liver *lobules* are the functional units of liver substance. In the human liver these are not distinctly separated from each other by connective tissue boundaries. However, each lobule has certain anatomic features by which it can be recognized.

A liver lobule (Fig. 8–12) is a five- or six-sided area of liver substance with a *central vein* running lengthwise through its center. Liver cells, arranged in layers or plates, radiate from the central vein out to the periphery of the area. Five or six *portal areas*, each containing a branch of the portal vein, an hepatic artery, and a bile duct, are located around the periphery of the lobule. Between the branching rows of liver cells there are small communicating channels or *sinusoids*, which contain blood from the portal vein and hepatic artery. Blood flows through the sinusoids from the periphery of the lobule toward its center (cen-

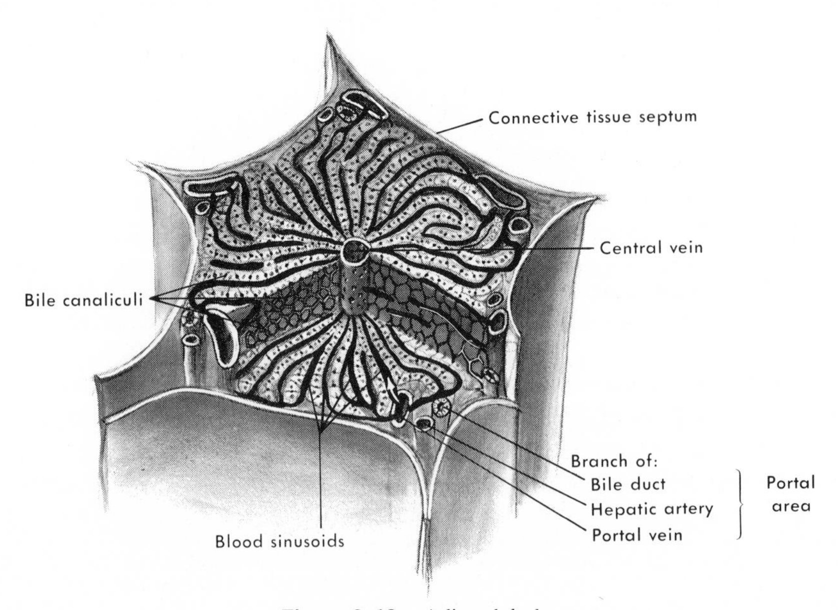

Figure 8–12. A liver lobule.

tripetal flow) and empties into the central vein. *Bile* is secreted by the liver cells into tiny canals or *canaliculi*. These run from the center of the lobule toward the periphery (centrifugal flow) where the bile empties into the bile duct located in each portal area. Phagocytic cells, called *histiocytes* (Kupffer cells), are attached to the walls of the sinusoids.

Importance of the Liver. Because the liver is so vital to health and to life itself, some of its main functions will be considered here.

One of the regulatory functions of the liver is that of controlling the blood sugar level. Sugar is the body's main energy source, and its level in the blood must be maintained within a relatively narrow range. Since the amount of sugar eaten varies with almost every meal, the liver must be able to both absorb excess sugar from the blood and liberate it to the blood. The hormone *insulin* is also needed for the proper handling of sugar in the body; it will be discussed in Chapter 10.

The liver also stores and secretes other essential nutrients. It receives most of these materials from the intestinal tract, which has prepared them for absorption into the blood. The chemical steps taking place in the body which enable cells to utilize essential materials are called *intermediary metabolism*, and the liver plays the dominant role in these processes. It is also responsible for *detoxifying*, or rendering harmless, many substances which could be harmful if allowed to accumulate in the body. Rapid synthesis of plasma proteins is an important liver function in restoring hydrostatic pressure of the blood following hemorrhage.

The liver plays a role in blood clotting (p. 115) since it produces several clotting factors, such as prothrombin and fibrinogen. Its bile salts are necessary for the absorption of vitamin K from the gastrointestinal tract, which in turn is needed for the production of prothrombin. Thus persons with liver disease may have serious bleeding problems.

Another important liver function is that of producing *bile*. A brownish yellow fluid, it is secreted continuously by the liver in amounts averaging about 600 cc. per day. Bile contains the bile salts, which are very important in the digestion of fat.

GALLBLADDER (FIG. 8–13). Located on the undersurface of the liver, the gallbladder is a sac-like organ about 4 inches long. Its closed end is bulb-like and is called the fundus. The main portion is the body, and the tapered open end is called the neck. When empty, the mucous membrane is thrown into numerous folds, which disappear as the organ becomes distended with bile.

Smooth muscle in the wall of the gallbladder resembles that of the intestine. Glands are present only in the mucosa of the neck region. The narrow *cystic duct*, which leads out of the organ, is a side branch of the hepatic duct, which comes from the liver. At the entrance to the cystic duct folds of mucous membrane are arranged spirally to form the *spiral valve*. The cystic and hepatic ducts join to form the *common bile duct*, which conveys bile into the duodenum. A weak sphincter muscle guards the duodenal opening. However, proximal to the common duct's point of entry into the duodenal wall, its muscle layer forms the *sphincter choledochus*. This sphincter is strong enough to stop the passage of bile into the duodenum.

The gallbladder stores and concentrates bile. Bile leaving the gallbladder is six to ten times as concentrated as that which comes to it from the liver. Concen-

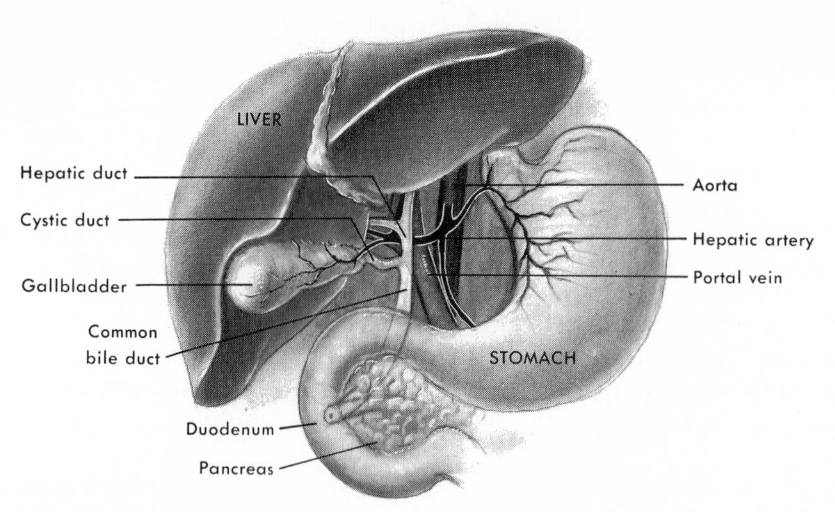

Figure 8–13. Biliary apparatus.

tration is accomplished mainly by the absorption of water from the bile into the mucosa of the gallbladder. The organ's storage capacity is 35 to 50 cc.

PANCREAS. The pancreas is a slender organ, about 6 to 9 inches long, lying in the abdomen behind and under the stomach. Its head lies in the curve formed by the duodenum. Its body extends horizontally toward the spleen, and the blunted tail often makes contact with the spleen. Because it has such a thin connective tissue capsule, the lobulated structure of the pancreas can be seen with the naked eye. The organ has been described as having the appearance of cottage cheese.

The digestive juices secreted by the pancreatic acinar (gland) cells are collected into small ducts—all of which drain into the *main pancreatic duct* (of Wirsung). The main duct begins in the tail of the pancreas and runs lengthwise through it, with the smaller ducts joining it as it passes toward the head. As it leaves the pancreas, the main duct joins with the common bile duct to pass through the wall of the duodenum to open on the duodenal papilla.

The pancreas also produces two additional substances, *insulin* and *glucagon*. These are *hormones*, which are secretions sent directly into the blood of the organs producing them rather than collected into a duct system. Although most of the pancreas consists of the acinar cells, which produce the digestive enzymes, insulin and glucagon are produced by small masses of cells called *islets*, which are scattered throughout the gland. These hormones will be discussed in Chapter 10.

PASSAGE OF FOOD THROUGH THE DIGESTIVE TRACT

Types of Movement. After food has been chewed (mastication) and swallowed (deglutition), it must of course be moved through the gastrointestinal tract for further preparation and final disposition. The two main types of gastrointestinal movements are mixing movements and propulsive movements.

PERISTALSIS. The basic propulsive movement of the digestive tract is *peristalsis*. It can be likened to squeezing toothpaste up through a tube with the fingers. A ring of contraction occurs in the circular layer of smooth muscle in the wall of the digestive tube. This ring then moves forward, pushing ahead of it any material present in the tube. All tubular structures in the body having smooth muscle will show peristalsis when they are stimulated. Thus, this movement occurs in such structures as the bile ducts, other glandular ducts, and the ureter.

The presence of material in the tube, causing it to distend or bulge, is the usual stimulus for peristaltic movement at that point. However, strong peristaltic waves are dependent on an intact nerve supply to the smooth muscle. Peristalsis in the gastrointestinal tube occurs in both directions from the point of stimulus, but it usually dies out in the direction toward the mouth. The reason why it persists in the direction toward the anus has never been satisfactorily explained (Fig. 8–14).

SEGMENTAL MOVEMENTS. In addition to peristaltic waves, rhythmical segmental contractions occur in the circular smooth muscle of the small intestine. These are small, ring-like "standing wave" contractions, which do not progress up or down the tube as the peristaltic waves do. They are most frequent in the duodenum. They aid digestion and absorption by thoroughly mixing the contents with digestive juices (Fig. 8–15).

Passage Through the Pharynx and Esophagus. Peristalsis takes place first in the pharynx. Then reflex elevation of the tongue, soft palate, and larynx closes off the mouth, nose, and trachea. Peristalsis aided by gravity moves food down the esophagus to the stomach in about five seconds. As the peristaltic wave approaches the lower end of the esophagus, the cardiac constrictor opens and the stomach is relaxed to receive the swallowed mass.

Passage Through the Stomach. The irregular but vigorous hunger contractions present when the stomach is empty disappear when a meal is eaten. Peristaltic waves then occur, which move the food down toward the pylorus of

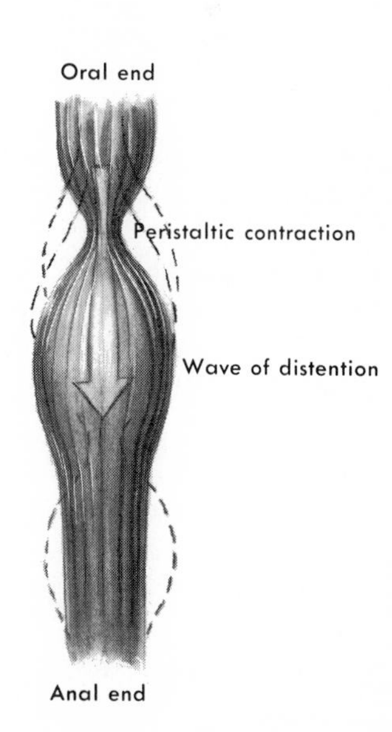

Figure 8–14. Peristaltic movement in the digestive tube.

Figure 8–15. Intestinal segmental movements.

the stomach. Digestive juices secreted by the gastric glands in the stomach wall come into contact with the food lying against the inner surface of the organ. Weak mixing waves tend to move the outer layer of food and gastric secretions toward the distal end of the stomach. These mixing movements increase in intensity as the contents approach this area.

When the stomach is empty, liquids will pass through it in a few minutes. When solid food is eaten, it usually reaches the first part of the duodenum in 20 to 30 minutes. Depending on the type of food, it takes two to four hours for the stomach to empty completely.

Passage Through the Small Intestine. Vigorous peristaltic waves, caused by contraction of the circular smooth muscle, occur in the small intestine. In addition, there are fewer vigorous contractions of the longitudinal muscle layer. These waves do not sweep over the whole intestine (as they do in the stomach) but progressively involve about 5 to 15 cm. of the intestinal wall. Rarely, as in the taking of a cathartic, a peristaltic *rush* may occur in which a wave appears at the pylorus and passes over the whole small intestine in about one minute. This rush, however, is not strong enough to move all the contents with it. As noted previously, segmental contractions serve to mix the contents and the digestive secretions in the small intestine.

The presence of a meal in the stomach causes peristalsis in the ileum whether or not food is present there. This reaction is called the gastroileal reflex. After a meal is eaten, the first portion of food appears in the terminal part of the ileum in about four hours. At this time the ileocecal valve opens and remains open as more food moves down the tract. The peristaltic rate in the ileum increases until it pushes the contents through the valve and into the colon. Complete transit time through the small intestine is from three to six hours. Thus, the time elapsed from ingestion to entrance into the colon is about six to nine hours.

Passage Through the Large Intestine. Rhythmic segmental contractions occur only in the ascending colon portion of the large intestine. Their functions are to further mix the contents and to permit absorption of water from them. Peristaltic rushes pass over the rest of the large intestine from the transverse colon to the rectum. These rushes occur only three or four times a day. The first one appears immediately after a meal is taken (gastrocolic reflex) and thus moves the contents from the previous meal into the rectum.

It takes approximately ten hours for material to move from the cecum to the distal end of the colon. The overall transit time from ingestion to the rectum is 15 to 24 hours.

Defecation. *Defecation* is the elimination of *feces* from the rectum. It is a function chiefly of the smooth muscle of the rectal wall and internal anal sphincter and of the voluntary muscle of the external sphincter. A reflex is set up by the presence of material in the rectum as a result of peristaltic rushes through the colon. Involuntary nerve centers in the spinal cord inhibit the internal anal sphincter, thus causing it to relax. This reflex can be voluntarily inhibited in the adult by action of the cerebral cortex; thus it is a learned response. On the other hand, appropriate signals can aid defecation by contracting abdominal and pelvic muscles, a process referred to as straining. This voluntary muscle contraction can cause a tenfold increase in pressure within the rectum.

DIGESTION

Digestion can be defined as the chemical breakdown of ingested food into particles (molecules) small enough to be absorbed into the body. In general, starches and sugars must be broken down to simple compounds called *monosaccharides*; proteins to their simplest elements, *amino acids*; and fats to *fatty acids* and *glycerol*. These chemical reactions go rapidly because of the action of chemicals called *enzymes*, which are secreted by the digestive glands.

An amylase is a digestive enzyme which attacks starches. Lipases break down fats, and proteinases and proteases attack the long chemical chains which make up proteins.

Action in the Mouth. Because food remains in the mouth for only a short time, very little if any digestion takes place here. Primarily, the food is broken down mechanically by mastication between the teeth. In chewing, saliva becomes mixed with the food, and this substance has important functions.

Saliva has a serous or watery component which contains an alkaline enzyme called *ptyalin* or *salivary amylase*. This enzyme may begin the breakdown of cooked starch present in the mouth if the food is well mixed with the saliva. Ptyalin's action continues after food reaches the stomach, but it soon stops when the material becomes thoroughly mixed with the acid gastric juice.

The mucus component of saliva is a sticky substance which lubricates the food, facilitating swallowing. Saliva also contains a solvent which liberates materials in the food to enhance its *taste*. About 1 to 1.5 liters of saliva are secreted per day.

The release of saliva from the glands into the mouth is entirely under nervous control. This control involves both a conditioned reflex, initiated by the sight and smell of familiar foods, and an unconditioned reflex, caused by the presence of food in the mouth. The degree of salivation can be affected by the emotions; for example, fear can cause the mouth to become very dry.

The consistency of food in the mouth determines whether the glands secrete a copious, watery saliva for dry material or a scanty, sticky saliva for moist

Table 8-1. DIGESTION IN THE STOMACH

Substance in Gastric Juice	*Substance Acted Upon*	*Products Formed*
Hydrochloric acid	Proteins	Softened protein
Gastric lipase	Emulsified fats	Fatty acids, glycerol
Pepsin	Softened protein	Proteoses, peptones

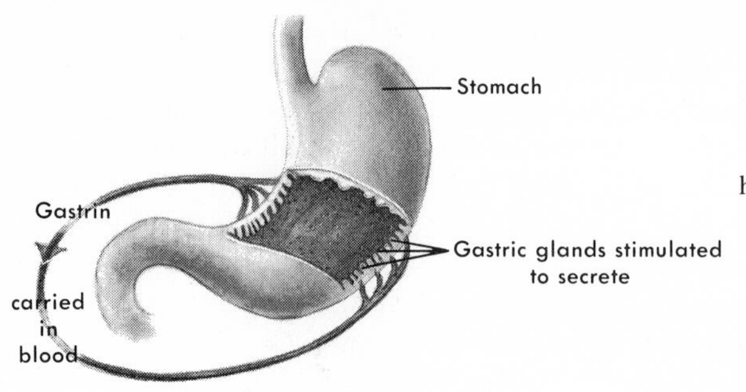

Figure 8–16. Action of the hormone gastrin.

food. The former is the result of parasympathetic stimulation, whereas the latter is a function of the sympathetic division.

Action in the Stomach. Gastric secretions contain mucus, hydrochloric acid, and digestive enzymes. Mucous cells are located in the cardiac and pyloric regions of the stomach, whereas the enzyme-producing glands are in the fundus and body. About 2 liters of gastric juice are secreted each day.

Mucus apparently protects the gastric mucosa from undergoing digestion by the acid gastric juice. It may also contain the so-called "intrinsic factor," the lack of which gives rise to a disease called pernicious anemia. Certain foods, especially meats, cause the formation of a hormone called *gastrin* in the gastric mucosa. Gastrin is absorbed into the blood and carried back to the gastric glands, stimulating them to secrete (Fig. 8–16).

HYDROCHLORIC ACID (HCl). This strong acid is responsible for the extreme acidity of pure gastric juice. It is secreted by the parietal cells in the stomach mucosa. Hydrochloric acid acts on proteins in the stomach by softening

Table 8-2. DIGESTION IN THE SMALL INTESTINE

Substance Secreted	*Substance Acted Upon*	*Products Formed*
From pancreas:		
Trypsin	Chymotrypsinogen, proteoses, peptones	Chymotrypsin, peptides, amino acids
Chymotrypsin	Proteoses, peptones	Peptides, amino acids
Carboxypeptidase	Peptides	Amino acids
Pancreatic amylase	Starches	Maltose
Pancreatic lipase	Fats	Fatty acids, glycerol
From liver:		
Bile (bile salts)	Fats	Emulsified fats
From small intestine:		
Enterokinase	Trypsinogen	Trypsin
Peptidases	Peptides	Amino acids
Maltase	Maltose	Glucose
Lactase	Lactose	Glucose and galactose
Sucrase	Sucrose	Glucose and fructose
Intestinal lipase	Fats	Fatty acids, glycerol
Intestinal amylase	Complex sugars	Double sugars

them, thus facilitating their digestion by enzymes. Because of its acidity, hydrochloric acid has a strong antiseptic action which helps prevent infection from entering the body via the gastrointestinal route.

PEPSIN. Pepsin is a powerful proteinase secreted by the chief cells. It actually is present in these cells in the inactive form, *pepsinogen,* which is converted to pepsin when it is secreted into the acid gastric juice. Pepsin, in acid solution, attacks nearly all types of protein present in food and breaks them down into simpler compounds called *proteoses* and *peptones.*

GASTRIC LIPASE. This enzyme is of little importance because normal gastric juice is too acid for its activity. It is most effective in breaking down emulsified fats such as those in milk, cream, and egg yolk. Unemulsified fats are not broken down by this relatively weak enzyme.

Some books list rennin as a gastric enzyme, but this is not correct. Rennin is found in the stomach of a calf, and its action is to clot milk. In the human stomach, however, whatever clotting occurs is probably performed by pepsin.

The partially digested food, sometimes referred to as chyme, is now ready to leave the stomach. Proteins have been partially broken down and emulsified fats have been acted upon. Starches are relatively unchanged except for the slight action of salivary amylase.

Action in the Small Intestine. The reaction of the digestive juices in the small intestine is alkaline. Chyme entering the duodenum is acid in reaction because it has been mixed with the acid gastric juice. This acidity stimulates the release of a hormone called *secretin* from the duodenal mucosa. Secretin is absorbed into the blood and carried by it to the pancreas. The presence of protein material in chyme stimulates the release of secretin as well as the release of another hormone called *pancreozymin.* These two hormones (1) provide the proper alkaline reaction in the duodenum, (2) protect the duodenal mucosa by neutralizing the hydrochloric acid present in the chyme, and (3) cause the pancreas to secrete its digestive juices.

If the meal has a high fat content, the duodenal mucosa is also stimulated to produce a hormone called *enterogastrone,* which inhibits the further production of gastric juice. Digestion in the small intestine is carried out by pancreatic juice, bile, and intestinal juice.

PANCREATIC JUICE. About 1 to 1.5 liters of pancreatic juice per day is secreted by the pancreatic cells and sent into the duodenum through the pancreatic duct. Its alkalinity is a result of its content of sodium bicarbonate (baking soda).

Pancreatic juice contains several powerful enzymes. *Trypsinogen* and *chymotrypsinogen* are two inactive substances which are changed into two active proteases. *Trypsin* is formed from trypsinogen by another enzyme, *enterokinase,* which is secreted by the intestinal mucosa. Trypsin, in turn, changes chymotrypsinogen into *chymotrypsin.* These two proteases attack the proteoses and peptones, breaking them down into *peptide chains,* which are simpler protein compounds, and *amino acids,* which are the basic building blocks of protein. Amino acids are also split off the peptide chains by another enzyme called *carboxypeptidase* (formerly called erepsin).

Pancreatic amylase (amylopsin) is the most powerful amylase in the digestive tract. Its action is similar to that of salivary amylase in that it converts the complex starches to simpler compounds. Pancreatic amylase breaks starches down to the double sugar (disaccharide) *maltose*, or malt sugar.

One fat-digesting enzyme, *pancreatic lipase* (steapsin), is present in pancreatic juice. By itself it is relatively weak, but its action is greatly increased by the presence of other substances in the intestine, especially the bile salts. The lipase breaks fats down to fatty acids and glycerol.

BILE. Bile is secreted more or less continually by the liver. It passes into the hepatic duct, through the cystic duct into the gallbladder, and down the common bile duct into the duodenum. The stimulus for bile secretion is the presence of bile salts in the duodenum. These are absorbed into the portal blood and carried back to the liver.

Before a meal, the sphincter choledochus in the common bile duct remains closed. When food, especially fat, reaches the duodenal mucosa, a hormone called *cholecystokinin* is released. The hormone is absorbed into the blood and travels to the gallbladder, causing it to contract. This contraction raises the pressure of the bile in the common duct, which causes the sphincter to relax, and bile is discharged into the duodenum.

As secreted by the liver, bile has a golden or brownish yellow color. It has a bitter taste and is slimy in consistency. The main constituents of bile are water, bile pigments, bile salts, and cholesterol.

Bile pigments give bile its color, and they come from the hemoglobin portion of worn-out red blood cells. Most of the pigments are formed elsewhere in the body and then sent to the liver for excretion. One breakdown product of bile pigments gives feces its color. Jaundice is the yellowish color given to the skin when bile accumulates in the blood circulation. It is often a sign of liver or biliary tract disease.

Bile salts are considered the most useful constituents of bile. They are not excreted from the body as are the bile pigments. After secretion into the intestinal tract, they are absorbed almost completely (90 per cent) into the portal blood and carried back to the liver for resecretion. Bile salts have several important functions:

1. They transform pancreatic lipase from a weak to a powerful digestive enzyme.
2. They aid in the absorption of fat-soluble vitamins, which is especially important in the case of vitamin K, the antihemorrhagic vitamin.
3. They stimulate the liver to secrete bile.
4. They help emulsify fats in the duodenum, thus allowing more surface area for the action of pancreatic lipase.
5. They combine with fatty acids, making them water-soluble and thus more easily absorbed.
6. They keep cholesterol in solution.

Cholesterol in the bile comes from the liver, from the decomposition of red blood cells, and from the diet. Not very soluble, cholesterol may precipitate out of solution and form gallstones. The actual mechanism of gallstone formation is not very clear, even though their composition is known.

Summary of Bile Functions. Bile (1) helps neutralize the acid chyme coming from the stomach, thus aiding the action of the alkaline pancreatic and intestinal enzymes; (2) promotes the absorption of fat and fat-soluble vitamins; (3) stimulates the liver to secrete more bile; (4) aids in fat digestion; and (5) promotes the excretion of certain drugs, some metals, the bile pigments, etc.

INTESTINAL JUICE PROPER—SUCCUS ENTERICUS. In addition to the enzymes sent into the small intestine from the pancreas, the glandular cells at the base of the crypts of Lieberkühn produce several digestive enzymes. The most important of these will be discussed.

Succus entericus secretion is stimulated primarily by the presence of chyme in the intestine. Thus, the greater the amount of chyme, the greater the secretion. Of less significance is the possibility that a hormone, *enterocrinin*, which stimulates the mucosal glands, is secreted by the intestinal mucosa.

Protein enzymes called *peptidases* complete the final stage of protein breakdown to amino acids. Three sugar-digesting enzymes called *disaccharidases*—maltase, sucrase, and lactase—split the double sugars maltose, sucrose, and lactose into simple sugars or *monosaccharides.* The intestinal cells also produce *enterokinase,* the enzyme which changes trypsinogen into trypsin. A relatively weak intestinal lipase is secreted. It converts fat to fatty acids and glycerol. Finally, a small amount of *intestinal amylase* is produced for splitting sugars to the disaccharide stage.

The submucosal glands (Brunner's) in the duodenum elaborate large amounts of mucus. This mucus protects the duodenal mucosa from digestion by the acid gastric juice. Stimulation of the sympathetic nervous system strongly inhibits these glands and may be a factor in the production of duodenal ulcers. Additional mucus is secreted by cells located in the mucosa of the whole small intestine.

Thus, digestion is completed in the small intestine, with its enzymes finishing the process of breaking down the three main food components—starches, fats, and proteins—into their simplest components. These materials are now ready to be absorbed (Table 8-3).

Table 8-3. SUMMARY OF HORMONES PRODUCED BY THE DIGESTIVE TRACT

Hormone	*Source*	*Action*
Gastrin	Gastric mucosa	Stimulates secretion of gastric juice
Enterogastrone	Duodenal mucosa	Inhibits gastric secretion
Secretin	Duodenal mucosa	Stimulates secretion of alkaline pancreatic fluid
Pancreozymin	Duodenal mucosa	Stimulates secretion of pancreatic enzymes
Cholecystokinin	Duodenal mucosa	Stimulates release of bile from gallbladder
Enterocrinin (?)	Small intestine mucosa	Possible stimulation of intestinal glands

Action in the Large Intestine. Digestion as such does not occur in the large intestine, since the process is essentially completed in the small intestine. The cells in the crypts of Lieberkühn do not secrete any digestive enzymes. They do, however, produce large quantities of alkaline mucus, which serves to lubricate the contents and to neutralize the acids formed by intestinal bacteria.

The processes that proceed in the large intestine are due mostly to the millions of bacteria which live there. Living and dead bacteria, in fact, make up one fourth to one half the dry weight of feces. Other components of feces are undigested food residues, unabsorbed secretions (bile, pancreatic juice, etc.), and cell debris. Water is absorbed from this material, and the characteristic consistency of feces results. Other substances taken into the body in varying amounts are excreted by way of the large intestine. Some of these are iron, calcium, and phosphate.

About 1 liter of gas is present in the gastrointestinal tract of an adult. Its composition varies with the diet; it is predominantly hydrogen on a high milk diet, whereas on a mixed or meat diet it will be largely nitrogen. Intestinal gas may cause distress because of distension.

Putrefaction. Proteins in the form of undigested food residue, unabsorbed amino acids, cell debris, and dead bacteria are decomposed in the large intestine by the action of live bacteria. This process involves several chemical steps which collectively are called *putrefaction*. Two substances resulting from this decomposition (indole and skatole) give feces its characteristic odor.

Auto-intoxication. An overabundance of some of the products of intestinal putrefaction may result in diarrhea. On the other hand, the question has been raised as to whether some of these breakdown products, which in themselves are toxic (poisonous), can be harmful to the body. It is true that some of them will give symptoms of illness if administered orally (by mouth) or parenterally (by injection into the tissues). However, even though small amounts of some of them may be absorbed from the intestinal tract, they are completely detoxified before this occurs.

Thus, it is erroneous to say that symptoms of constipation, such as headache and dullness, are caused by *auto-intoxication*, or the absorption of toxic materials from the intestinal contents. Instead, these symptoms probably result from distension and irritation of the colon by the fecal mass, since packing the rectum with cotton will produce some of the same symptoms.

ABSORPTION

Absorption may be defined as the transport of material from the external to the internal environment. That is, the nutrients contained in ingested food are not really considered absorbed into the body until they get into the blood. The blood circulation then carries them to individual body organs and cells so they can be utilized.

Absorption in the Mouth. No absorption of food occurs in the mouth. Some drugs, however, are absorbed through the oral mucosa. These include nitroglycerine, a drug given to certain cardiac patients, and some of the hormones.

Absorption in the Stomach. Very little absorption takes place from the stomach. Absorption of certain drugs, alcohol, and glucose (if it is eaten as such) occurs to some extent. Even though the gastric mucosa is permeable to water, studies show that no water is actually absorbed through it.

Absorption in the Small Intestine. Almost all absorption of nutrients occurs through the walls of the small intestine. Because of the circular folds and villi, the small intestine has a large mucosal surface area. This area is estimated to be about ten square meters, or over ten square yards. Even if 80 to 90 per cent of the small intestine were removed, the remaining part would still provide for sufficient absorption.

Blood flow per gram of tissue is as high (0.5 to 1 cc. per minute per gram) in the small intestine as it is in the brain. Only the kidney and the thyroid gland have a higher flow rate. In addition, this part of the tract has good lymphatic drainage – remember that each of the millions of villi contains a lymphatic capillary. Even though the duodenum and jejunum contain relatively more villi than the ileum, absorption is equal in all three parts. *Motility*, or movement, is responsible for this since food moves faster through the proximal portion of the tube.

Absorption in the Large Intestine. Since chyme is exposed to the small intestine first, most of the absorbable materials are removed by the time it passes into the large intestine. The oral intake plus the secretions added by the digestive glands amounts to from 7 to 10 liters per day. Only 0.5 liter of this total passes the ileocecal valve, which means the remainder has been absorbed before it reaches the cecum. This does not mean that nothing is absorbed from the large intestine. Water and substances such as sodium and chloride (electrolytes) and certain vitamins are absorbed through the proximal half of the colon. In addition, glucose and some drugs can be absorbed if given by rectum.

Thus, this part of the tract performs important absorptive functions. The amount of water absorbed here determines the consistency of the feces. If the contents remain in the colon too long, more water than usual is absorbed from them, resulting in a constipated stool. Conversely, if the contents pass through too rapidly, not enough water is removed, resulting in diarrhea.

Mechanisms of Absorption. The two basic mechanisms involved in the absorption of materials from the intestinal tract are *diffusion* and *active transport.* Chapter 1 described diffusion as the movement of a substance through a membrane, with the movement occurring from the area of higher concentration to the area of lower concentration. This type of movement is called passive transport; that is, as long as the concentrations remain as stated, no energy need be applied to the system to keep diffusion going.

Active transport, on the other hand, means movement of a substance through a membrane *against* a concentration gradient. That is, if the substance to be absorbed is in higher concentration outside the intestine than it is inside, energy is required to move the substance out of the tract and into the blood. Active transport is an important mechanism because it permits the absorption of essential materials even when they are present in higher concentrations outside the tract.

Routes of Absorption. At the mucosal level, some substances, such as glucose, pass from the lumen (space) of the intestine *into* the epithelial cells of the mucosa. Other materials, such as water, pass *between* the epithelial cells.

Beyond the mucosal cells substances pass into either the blood capillaries or the lymphatic capillaries in the villi. The blood is collected into the portal system whereas the lymph passes up through lymphatic vessels into the thoracic duct. Since the thoracic duct opens into the left subclavian vein, all absorbed materials eventually enter the blood circulatory system.

Absorption of Nutrients. A sufficient amount of all nutrients needed to maintain normal body function must be absorbed from the digestive tract each day. These nutrients include food such as starches, fats, and proteins, which are used for energy and for building or repairing tissue. Water is especially important because it is a univeral solvent and composes a large part of the body substance. In addition, small but essential amounts of vitamins and minerals are needed for regulatory purposes.

CARBOHYDRATES. The simple sugars, products of carbohydrate digestion, are absorbed into the blood capillaries of the villi. Most of them are absorbed by active transport rather than by diffusion. The most rapidly transported sugar is galactose, then glucose, and finally fructose. Certain materials, such as cyanide, can block the absorption of monosaccharides. The average adult male can absorb about 70 grams of glucose per hour from the intestinal tract.

PROTEINS. Proteins are absorbed as amino acids by the active transport mechanism into the blood capillaries. As with glucose, the rate of absorption is decreased by the presence of metabolic poisons.

FATS. After digestion, fat is absorbed in one of two ways. By the action of *bile salts*, 60 to 70 per cent is absorbed as highly emulsified particles, each composed of one, two, or three fatty acids (mono-, di-, or triglycerides) and glycerol. These pass into the lacteals of the villi and by way of the thoracic duct eventually get into the blood circulation. They are carried by the blood to fat depots, or storage areas, in the connective tissue layers under the skin and between the muscles.

The remaining fats (30 to 40 per cent) are broken down by the *lipases* in the small intestine to fatty acids and glycerol. The action of bile then makes these compounds water-soluble so they can pass into the mucosal cells. From here they pass into the blood capillaries in the villi and are thus transported to the liver by way of the portal system (Fig. 8–17).

Absorption of Water and Electrolytes. Electrolytes—sodium, chloride, potassium, calcium, etc.—are absorbed through the intestinal wall by active transport. As they are absorbed, their concentration in the intestinal contents will decrease, and water will pass out of the intestine by simple osmosis.

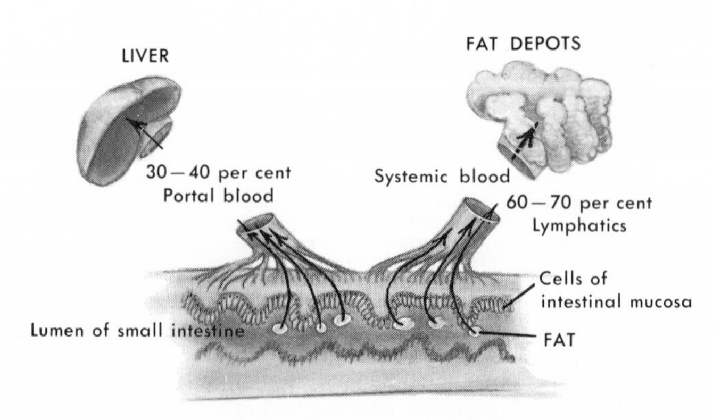

Figure 8–17. Fat absorption.

METABOLISM

Metabolism can be defined as the sum of all the chemical processes necessary to keep the body cells in the living state. It can be roughly divided into a building-up phase called *anabolism* and a breaking-down phase called *catabolism.*

Metabolism is a broad term, and the processes it includes are usually grouped under several subheadings. *Energy metabolism* refers to the mechanisms by which the body converts food into energy. *Carbohydrate metabolism* is essentially the metabolism of glucose and related substances. *Nitrogen metabolism* refers to the processes which handle amino acids and other protein compounds. *Fat metabolism* refers to the chemical steps involving fats and related substances. The metabolism of any substance handled by the body is a very complicated process and requires a knowledge of biochemistry for its understanding. Only a few basic metabolic principles will be considered here.

Energy Metabolism. The efficiency of the body in converting food into energy forms compares favorably with the efficiency of man-made devices such as the gasoline engine. The energy produced is used by the body to perform work. All functions, such as muscle contraction, gland secretion, growth and repair of body tissue, require energy.

CALORIES. The energy or caloric value of food is measured in terms of *large calories* (Cal.). A large calorie is the amount of heat needed to raise the temperature of 1000 grams of water one degree centigrade. Caloric values in round numbers for the three types of foodstuffs are: carbohydrates, 4 Cal. per gram; fats, 9 Cal. per gram; and proteins, 4 Cal. per gram.

Most foods are mixtures of carbohydrates, fats, and proteins. In order to determine the caloric value of a piece of bread, for example, one must know how many grams of each of the three materials it contains. Each weight is then multiplied by its caloric value and the three are added.

Example: Bread is approximately 52 per cent carbohydrate, 9 per cent protein, and 2 per cent fat. If one slice weighs 25 grams, then:

Carbohydrate:	13.0 grams × 4 Cal. per gram	52.0 Cal.
Protein:	2.3 grams × 4 Cal. per gram	9.2 Cal.
Fat:	0.5 gram × 9 Cal. per gram	4.5 Cal.
	Total	65.7 Cal.

It should be remembered that vitamins and minerals do not contain calories.

BASAL METABOLIC RATE. The basal metabolic rate (BMR) is the expression of the amount of energy used by the body when it is under basal conditions. Basal conditions are satisfied when the person has had no food for 12 hours, has lain quietly for 30 minutes before the measurement is taken, and has remained lying quietly but awake during the measurement. Therefore, the BMR is a measure of the calories needed by the body to perform only essential activities such as breathing, heart beat, and kidney function. Using this value, a person's total caloric needs can be determined by adding to the BMR requirement the calories he needs for his daily activities.

The BMR is related to certain physiological variables. It is usually higher in males than females and lower in overweight persons. It decreases with older age.

People living in tropical climates have lower metabolic rates than those living in cold regions. Certain diseases, especially those of the thyroid gland, can cause variations in the BMR.

Thus, total calorie needs vary from person to person. Children and adolescents require extra calories for growth needs. Pregnant and lactating women need additional calories, as do men engaged in strenuous physical work. Calories must be decreased if an obese person wishes to lose weight. Contrary to student opinion, mental activity does not use up calories.

Tissue Metabolism. Carbohydrates are the body's chief source of ready energy, and it uses them in preference to fat or protein. Glucose is the form in which sugar is utilized by the body cells. The transport of glucose through the body and its passage into the cells are facilitated by insulin. If the need arises, glucose can be formed in the body from available fat and protein, a process called *gluconeogenesis*.

After absorption, glucose can be used immediately by the cells for energy, or it may be stored in the liver as a complex compound called *glycogen*. Muscle cells can also store extra glucose in the form of glycogen. When needed, glycogen can be converted back into glucose by a process called *glycogenolysis*.

Fats may be considered the body's reserve supply of energy. Stored in fat depots, fat also serves as insulation and as protection for internal organs. The liver removes fat from the depots when needed and metabolizes it. Excess carbohydrate can be converted to fat in the body, probably by the liver, although it loses about 15 per cent of its original energy in the process.

Under certain conditions—uncontrolled diabetes, starvation, an all-fat diet—essentially no carbohydrates are metabolized. The body then turns to its fat stores to supply energy. This excess burning of fat results in the accumulation of fat breakdown products in the body, giving rise to a condition called *ketosis*. Ketosis in turn can cause severe *acidosis*, because the ketone bodies are strong acids and require a large part of the body's alkaline supply to neutralize them.This depletion of the body's alkaline materials disrupts the delicate *acid-base balance* and is one of the most serious complications in uncontrolled diabetes.

Although protein can be used for energy since it has 4 calories per gram, there are other and more important uses for it. No body tissue can grow or repair itself without protein. Proteins are composed of amino acids, several of which are essential as building blocks for tissue. Furthermore, the quality of protein varies from one kind of food to another. The quality is determined by the kind and amount of amino acids it contains and by how readily these acids are absorbed by the body. Judged by these criteria, the protein of eggs, milk, kidney, and liver are of the highest quality. Muscle meats, poultry, and fish are next in quality. Cereals, nuts, and vegetables are the poorest sources of high quality protein. Pure sugar, of course, contains no protein.

Minerals and Vitamins

MINERALS. Although minerals do not supply energy to the body or build tissue, they play vital roles in metabolism. Since their role is a regulatory one, only relatively small amounts of them are needed by the body. Without the proper level of sodium in the extracellular fluids, life cannot be sustained. Iron and copper are needed to make normal hemoglobin. The thyroid gland needs iodine to make its hormones. Calcium, stored in bones and teeth, is necessary for the proper discharge by nerve cells. Proper muscle contraction depends on several minerals, such as magnesium, potassium, and sodium. All these plus others such as chlorine, cobalt, and zinc are intimately concerned with the numerous metabolic processes which constantly take place in the body.

VITAMINS. Vitamins are organic compounds needed in small quantities by the body for normal procedure of metabolic processes. At present, some 15 vitamins are recognized as necessary for life, although the exact amounts needed are not known for all of them. A normal, well-balanced diet usually supplies sufficient quantities of these substances for a person in good health. Certain vitamins can be stored in the body, but others can not. For example, enough vitamin A can be stored in the liver to last for over a year without additional intake. On the other hand, a lack of some of the B vitamins will give rise to deficiency symptoms within a week.

Vitamin A is required for growth and the proper production of eye pigments. Many of the B vitamins such as thiamine, niacin, and riboflavin function as coenzymes in some of the important chemical steps concerned with metabolism. Lack of vitamin C results in poor wound healing and in the cessation of bone growth. Folic acid is necessary for growth and, along with vitamin B_{12}, is needed for the development of normal red blood cells. As in the case of the minerals, these are only a few examples of the importance in the diet of substances other than energy-giving materials.

OUTLINE SUMMARY

1. Introduction: The digestive system prepares materials essential for the nourishment of the body by breaking them down into particles small enough for absorption.
2. General Plan
    a. Tube from lips to anus – 27 feet long; four main layers
        1) Mucosa – lining layer containing smooth muscle, connective tissue, epithelium
        2) Submucosa – lies under mucosa; loose connective tissue with large blood vessels and nerve fibers
        3) Muscularis externa – usually two layers, inner circular and outer longitudinal
        4) Adventitia (above diaphragm) or serosa (below diaphragm)
    b. Peritoneum – double-layered serous membrane
        1) Parietal layer – lines abdominal cavity
        2) Visceral layer – covers organs
        3) Ligaments – double folds of peritoneum connecting organs
3. Anatomy of the Digestive System
    a. Mouth
        1) Vestibule – between the lips and cheeks and the teeth
        2) Cheeks – buccinator muscle
        3) Tongue – interlacing skeletal muscle bundles; papillae and taste buds
        4) Teeth – 20 deciduous, 32 permanent
        5) Palate – hard palate, soft palate
        6) Salivary glands – parotid, submandibular, sublingual
    b. Pharynx (see Chapter 7)

c. Esophagus – muscular tube lying posterior to trachea
d. Stomach – lies in upper part of abdominal cavity
  1) Entrance – cardiac orifice
  2) Fundus, body, pylorus
  3) Exit – pyloric orifice
  4) Greater and lesser curvatures
  5) Mucosa – gastric glands; rugae
e. Small intestine – 21 feet long, 1 inch diameter
  1) Extends from pyloric orifice to ileocecal valve
  2) Duodenum – first 12 inches
    a) **C**-shaped around head of pancreas
    b) Duodenal papilla – opening for pancreatic and common bile ducts
    c) Submucosal (Brunner's) glands
  3) Jejunum – next 8 feet; ileum – last 12 feet
    a) Intestinal glands (crypts) in mucosa
    b) Circular folds and villi
f. Large intestine – 4½ to 5 feet long, 2½ inches diameter
  1) Extends from ileocecal orifice to anus
  2) Three gross features – teniae coli, haustra, epiploic appendages
  3) Parts
    a) Cecum – blind pouch; appendix attached
    b) Ascending colon – passes upward to right colic flexure
    c) Transverse colon – crosses upper abdominal cavity
    d) Descending colon – descends downward from left colic flexure
    e) Sigmoid colon – **S**-shaped
    f) Rectum – from sigmoid colon to anus
g. Accessory organs
  1) Liver – largest gland in body
    a) Four lobes – right, left, quadrate, caudate
    b) Porta – hepatic artery, portal vein, bile ducts
    c) Liver lobule – functional liver unit; portal area, sinusoids, bile canaliculi, histiocytes
    d) Liver functions – regulatory, intermediary metabolism, detoxification, bile production
  2) Gallbladder
    a) Fundus, body, and neck
    b) Cystic duct with spiral valve
    c) Stores and concentrates bile; 35 to 50 cc. capacity
  3) Pancreas – 6 to 9 inches long
    a) Lies behind and under the stomach
    b) Head, body, and tail
    c) Main pancreatic duct carries pancreatic juice into duodenum
    d) Islet cells secrete hormones insulin and glucagon

4. Passage of Food Through the Digestive Tract
  a. Types of movement
    1) Peristalsis – basic propulsive movement
      a) Ring of contraction which moves forward, squeezing contents ahead of it
      b) Stimuli – presence of material in the tube and nerve supply
    2) Segmental movements – mixing movements
      a) Ring-like standing wave contractions

b) Aid digestion and absorption by mixing contents and digestive juices

b. Passage through pharynx and esophagus – by peristalsis

c. Passage through stomach
   1) Peristaltic waves – move toward pylorus
   2) Mixing movements
   3) First portion empties in 30 minutes; empties completely in two to four hours

d. Passage through small intestine
   1) Vigorous peristaltic waves progressively involve 5 to 15 cm. of the tube
   2) Segmental contractions mix contents
   3) Gastroileal reflex
   4) Transit time – three to six hours

e. Passage through large intestine
   1) Rhythmic segmental contractions only in ascending colon
   2) Peristaltic rushes three to four times a day over the remainder
   3) Material moved from cecum to distal colon in ten hours

f. Defecation – elimination of feces from rectum

5. Digestion – chemical breakdown of ingested food
   a. Enzymes – secreted by digestive glands (see Tables 8–1 and 8–2)
      1) Amylases – break down carbohydrates
      2) Proteinases and proteases – break down proteins
      3) Lipases – break down fats
   b. Action in the mouth
      1) Salivary amylase (ptyalin) – changes cooked starch to maltose (action limited)
      2) Mucus – lubricates food; assists swallowing; liberates taste
   c. Action in the stomach
      1) Mucus – protects gastric mucosa
      2) Hydrochloric acid (HCl) – gives acidity to gastric juice; antiseptic action; softens proteins
      3) Pepsin – powerful proteinase
      4) Gastric lipase – weak action on emulsified fats
      5) Hormone gastrin – stimulates secretion of gastric juice
   d. Action in the small intestine
      1) Hormones (Table 8–3)
         a) Secretin and pancreozymin – stimulate pancreas
         b) Enterogastrone – inhibits gastric juice
         c) Enterokinase – changes trypsinogen to trypsin
         d) Cholecystokinin – stimulates gallbladder
      2) Pancreatic juice – contains powerful enzymes
         a) Proteases – trypsin, chymotrypsin, carboxypeptidase
         b) Pancreatic amylase
         c) Pancreatic lipase
      3) Bile – secreted by liver; stored in gallbladder
         a) Brownish yellow, bitter, slimy fluid
         b) Contains water, bile pigments, bile salts, cholesterol
      4) Succus entericus (intestinal juice)
         a) Peptidases, disaccharidases, lipase
         b) Mucus by submucosal glands
   e. Action in the large intestine
      1) No digestion

2) Mucus – lubrication; neutralizes bacterial acids
3) Feces – bacteria, food residues, secretions, cell debris
4) Putrefaction – products of protein decomposition

6. Absorption – transport of material from external to internal environment
   a. Mouth – certain drugs absorbed
   b. Stomach – glucose, alcohol, certain drugs absorbed
   c. Small intestine – almost all food absorption occurs here
      1) Circular folds and villi greatly increase surface area
      2) High blood flow per gram of tissue
   d. Large intestine – absorbs water and electrolytes
   e. Mechanisms of absorption – diffusion and active transport
   f. Routes of absorption – from mucosal epithelial cells into villi
      1) Carbohydrates, protein, water, electrolytes – pass into blood capillaries of villi
      2) Fats – 60 to 70 per cent pass into lacteals; remainder passes into blood capillaries
7. Metabolism – sum of all chemical processes in the body
   a. Two phases
      1) Anabolism – building-up processes
      2) Catabolism – breaking-down processes
   b. Energy metabolism – conversion by body of food into energy
      1) Caloric values per gram – carbohydrate and protein 4, fat 9
      2) BMR – energy used by body under basal conditions
   c. Tissue metabolism
      1) Carbohydrates – body's chief energy source
         a) Gluconeogenesis – formation of glucose from fat or protein
         b) Glycogenolysis – conversion of glycogen into glucose
      2) Fats – body's reserve supply of energy
         a) Ketosis – overabundance of fat breakdown products accumulates in body
         b) Acidosis – depletion of body's alkaline reserve in attempt to neutralize acid ketone bodies
      3) Proteins – needed for tissue growth and repair
         a) Essential amino acids needed in diet to build tissue
         b) Quality of protein varies with food
      4) Minerals and vitamins – play regulatory role in metabolism

REVIEW QUESTIONS

1. Name and describe the layers of the digestive tube.
2. What is the mesentery?
3. Where are the parotid glands located?
4. What are rugae? How do they differ from circular folds?
5. How does the jejunum differ grossly from the ileum?
6. In what portions of the digestive tube are glands found in the submucosa?
7. Describe a villus.
8. What is the porta of the liver? Describe a liver lobule.
9. What does the gallbladder do?
10. What is peristalsis? How does it differ from segmental movements?
11. What is defecation? How does it occur?
12. What are the functions of mucus secretion in various parts of the digestive tube?
13. What are the functions of the bile salts?

14. What is succus entericus?
15. How much digestion takes place in the large intestine? Explain.
16. Where does almost all absorption of food take place? What features does this part of the tract have that enhance absorption?
17. How does active transport differ from diffusion?
18. In what state is each of the three foodstuffs absorbed?
19. Describe fat absorption.
20. If a certain food weighing 50 grams is 15 per cent protein, 40 per cent carbohydrate, and 5 per cent fat, what is the caloric value of the food?
21. Name some conditions that help determine a person's calorie needs.
22. Why is it harmful to the body when an excess of fat is burned?
23. True or false: Everybody should take supplementary vitamins every day. Explain.

ADDITIONAL READING

Green, D. E.: The metabolism of fats. Sci. Amer. (Jan.), 1954.

Leevy, C. M., Ed.: Evaluation of Liver Function. Indianapolis, Ind., The Lilly Research Laboratories, 1965.

Neurath, H.: Protein-digesting enzymes. Sci. Amer. (Dec.), 1964.

Chapter 9

THE URINARY SYSTEM AND WATER BALANCE

INTRODUCTION

In the nineteenth century, a famous French scientist named Claude Bernard introduced the concept of the constancy of the internal environment. This concept refers to the physiological mechanisms in the body that keep it operating in a state of *homeostasis* (see p. 12), which is not subject to changes from the external environment. An example would be the body temperature which is maintained at 98.6° F. even when the external temperature varies from below zero to over 100°.

Some of the most important of these mechanisms are those which maintain the constancy of the blood plasma within narrow limits. They operate to keep the relative concentrations of the various blood constituents at a stable level. It is also important that the acid-base balance of the blood remain constant, since even small changes will result in death. Finally, certain products of metabolism, especially *urea*, must be eliminated from the body as they are formed.

The organs which are responsible for maintaining the relative constancy of the blood plasma are the *kidneys*. They, together with the *ureters*, the *urinary bladder*, and the *urethra*, form the urinary system (Fig. 9–1). Since water balance is closely related to kidney function, it will be discussed in this chapter as will acid-base balance.

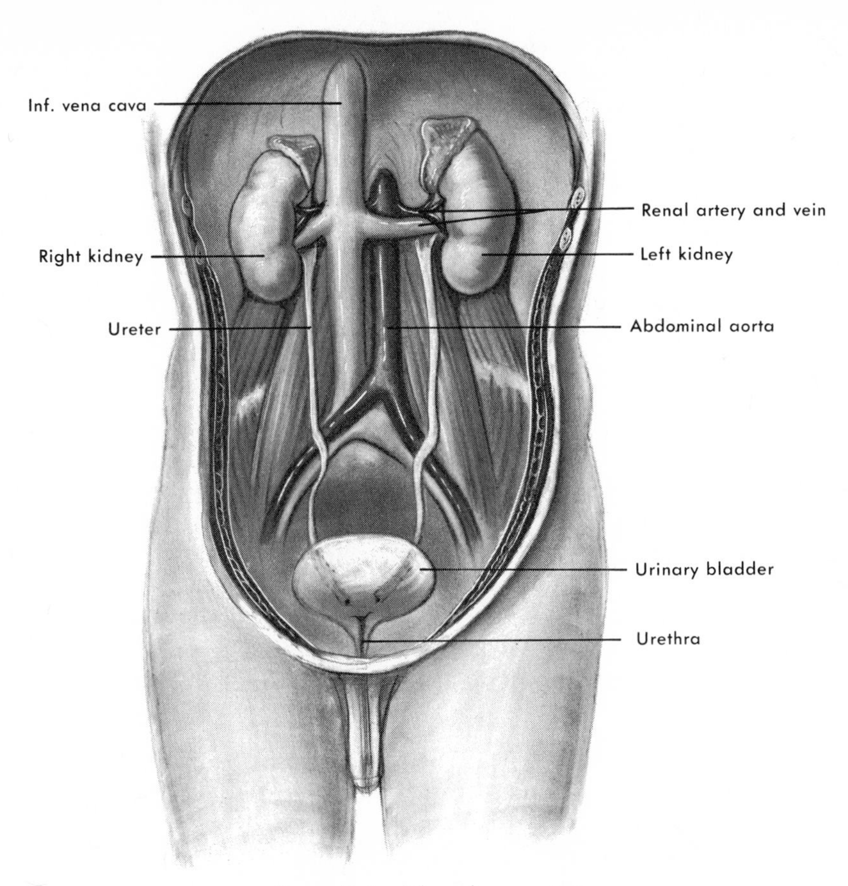

Figure 9–1. The urinary system.

THE KIDNEYS

Location. The kidneys are two bean-shaped organs lying retroperitoneally on either side of the vertebral column. They extend from the level of the twelfth thoracic vertebra down to the level of the third lumbar. Because the liver lies above it, the right kidney is usually slightly lower than the left.

About 11 cm. long and 5 cm. wide, the kidneys are embedded in fat and fibrous connective tissue, which help to keep them in place and serve as protection. Together the kidneys weigh less than 1 per cent of the total body weight. A connective tissue capsule encloses each organ except for a slit on the medial surface, which is called the *hilum.*

Gross Structure. If the kidney is examined in coronal section, it shows two rather distinct areas or zones (Fig. 9–2). The outer zone or *cortex* has a granular appearance, whereas the inner *medulla* has a striated (striped) look. The appearance of both areas is a result of their microscopic structures, which will be described below.

The upper portion of the ureter expands into the *renal pelvis.* This area is

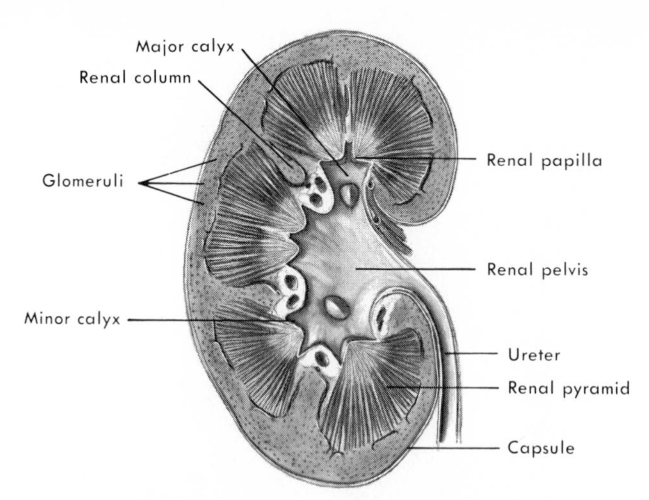

Figure 9–2. Coronal section of a kidney.

composed of three funnel-shaped ducts called the *major calyces.* Each major calyx in turn receives two or three smaller ducts, or *minor calyces.*

Microscopic Structure. The structural and functional unit of the kidney is the *nephron* (Fig. 9–3), and each kidney contains over a million nephrons. A nephron consists of a tuft of blood capillaries called a *glomerulus* plus a 1¼ inch long tubule. A cup-like *glomerular capsule* (Bowman's) surrounds the capillary tuft and is continuous with the *proximal convoluted* portion of the tubule. The tubule straightens out to form the *descending limb* of *Henle's loop,* then turns back as the *ascending limb.* As the loop approaches the proximal tubule, it again makes several twists and is called the *distal convoluted tubule.* After its last turn, the distal tubule empties into a *collecting duct* along with the distal tubules of several other nephrons.

In general, the cortex of the kidney contains glomeruli and proximal and distal convoluted tubules, whereas the medulla contains Henle's loops and the collecting ducts. This division is not clear-cut, however, since glomeruli also occur in the medulla and parts of the collecting ducts may be seen in the cortex.

A *lobe* of the kidney consists of a wedge-shaped section of medulla, called a *renal pyramid,* plus the cap of cortex above its base. The apex of the pyramid points toward the hilum, and its tip is the *renal papilla.* The collecting ducts run through the medullary area to finally open through the renal papillae and empty into the minor calyces. The *renal columns* are formed by cortical substance which passes down between the pyramids.

The human kidney contains eight to ten lobes. The dividing line between two lobes is marked by the presence of an *interlobar artery.* The outer surface of the adult kidney is smooth, giving no indication of the inner lobar structure. In the human fetus, however, the kidney surface is indented along the lines separating the lobes. Each lobe is further divided into *lobules* by interlobular arteries (Fig. 9–4).

Blood Supply. The function of the kidney is intimately related to its blood supply (Fig. 9–5). After entering the hilum, the renal artery divides into in-

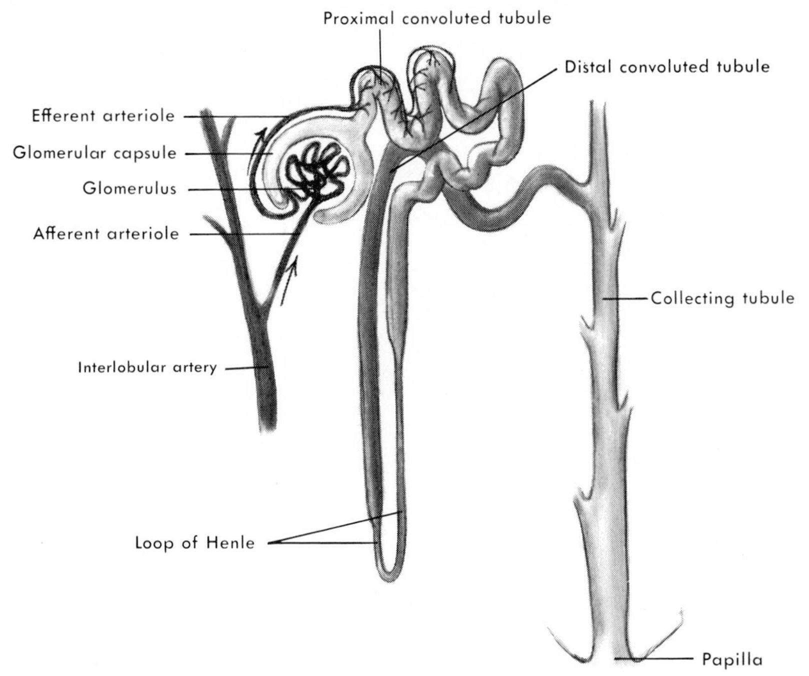

Figure 9–3. Microscopic view of a nephron.

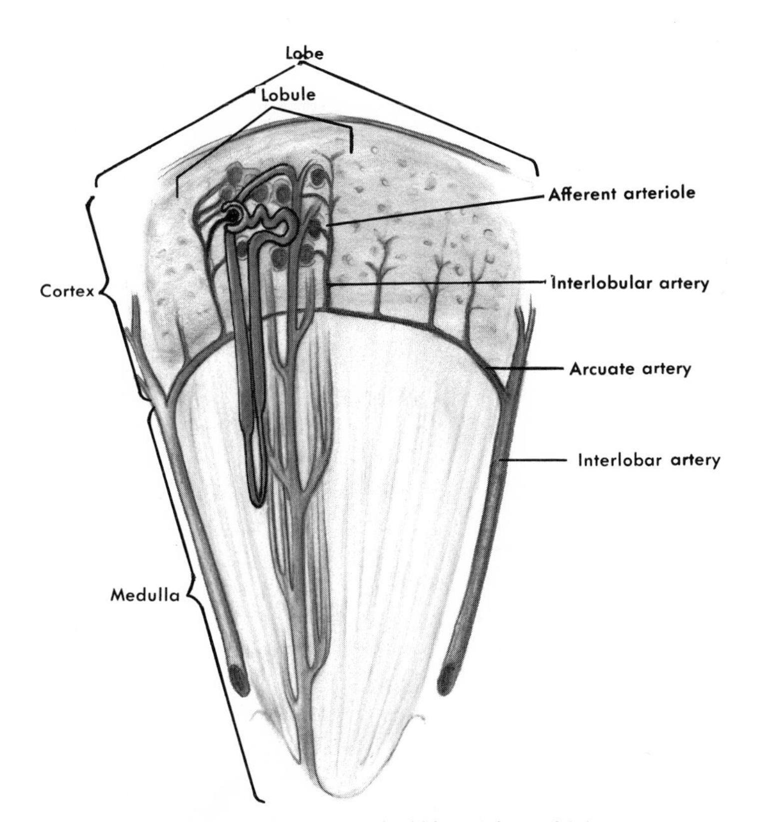

Figure 9–4. Schematic of a kidney lobe and lobule.

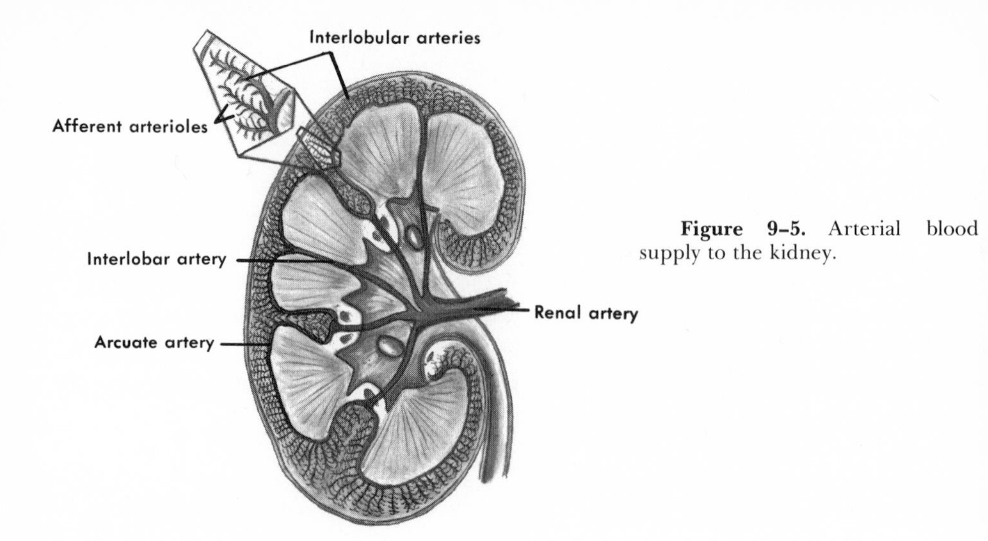

Figure 9-5. Arterial blood supply to the kidney.

terlobar arteries, which pass in a peripheral direction between the pyramids. When these arteries reach the boundary between the medulla and cortex, they branch off at right angles to form bow-shaped vessels called the *arcuate* (arched) *arteries.* From the arcuate arteries come numerous *interlobular arteries,* which pass between lobules.

Finally, each interlobular artery gives off many *afferent arterioles,* each of which forms a glomerulus. After circulating through the glomerulus, the arterial blood leaves by way of an *efferent arteriole.* This vessel delivers blood to a second capillary bed which surrounds the tubules. Venous blood is collected into veins, whose courses parallel those of the arteries, and is finally collected into the renal vein, which empties into the inferior vena cava.

FORMATION OF URINE

The function of the kidney is to filter out certain substances from the blood as it passes through the glomeruli. Certain of these substances in varying amounts are ultimately excreted from the body as *urine.* The secretion of urine by the kidney involves three main processes: filtration of blood through the glomeruli, reabsorption from the tubules, and tubular secretion.

Glomerular Filtration. As blood passes through the glomeruli, a certain portion of the plasma passes out of the capillaries and into the glomerular capsule by the process of filtration. Glomerular filtration proceeds because of hydrostatic pressure furnished by the blood. The *glomerular filtration rate,* or GFR, varies directly with the arterial blood pressure. That is, when the blood pressure goes up, the GFR increases and vice versa. The ratio of the amount filtered to the total blood plasma flow is called the *filtration fraction,* or FF.

The glomerular filtrate has almost the same composition as the tissue fluid which passes from the capillaries into the tissues in other parts of the body. In other words, it contains no red blood cells and practically no plasma proteins. Under normal conditions, the GFR averages about 125 cc./min.; that is, about 125 cc. is formed from the blood in the glomeruli and passes into the surrounding glomerular capsules. The 24 hour total is approximately 180 liters (1 liter = 1^+ quart). Thus, if the normal plasma flow through both kidneys is 650 cc./min. and the GFR is 125 cc./min., then the FF is 125/650 or nearly 20 per cent. It should be noted that these amounts can vary considerably under certain physiological conditions.

Tubular Reabsorption. The glomerular filtrate passes from the capsule into the proximal tubule, through Henle's loop, through the distal tubule, and finally into the collecting duct. During its passage, from 97 to 99 per cent of the water and dissolved substances are reabsorbed through the tubular cells into the interstitial fluid. The filtrate passes from the glomerular capsule to the collecting duct because of a pressure gradient. That is, the pressure in the capsule is 20 mm. Hg, whereas that at the distal end of the collecting duct is only 2 mm. Hg.

Reabsorption occurs by either diffusion or active transport. In the latter case, each substance so absorbed has its own "carrier" which transports it from the lumen of the tubule through the tubular cells and into the body fluids (Fig. 9–6). The carrier is apparently an enzyme which combines with the substance to carry it across the cell. After releasing the substance, the carrier returns to the lumen end of the cell to combine with more of the substance. That each material to be reabsorbed by active transport has its own exclusive carrier is shown by the fact that certain drugs will prevent the reabsorption of specific substances but not of others.

Eighty-seven per cent of the water is reabsorbed as the filtrate passes through the proximal tubule. In general, most of the substances important in metabolism, such as glucose and amino acids, are also reabsorbed in the proximal tubule. The reabsorption of these substances is almost total. Thus, the fluid leaving the proximal tubule contains considerably less water and no glucose, amino acids, or acetoacetic acid.

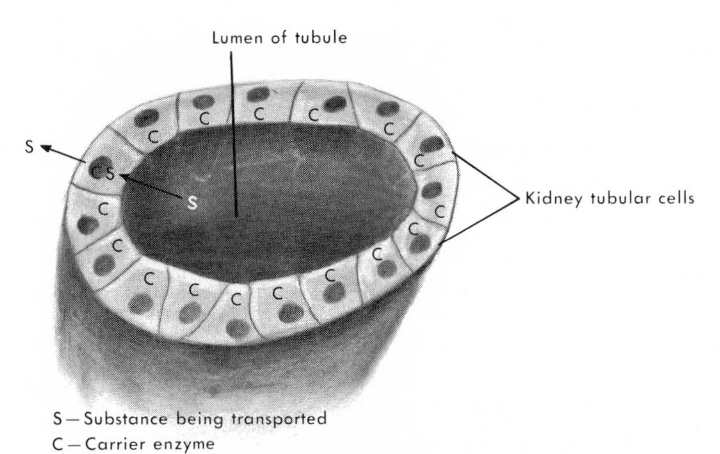

Figure 9–6. Tubular reabsorption by active transport.

Table 9-1. TUBULAR REABSORPTION BY THE KIDNEY

Substance in Filtrate	*Where Reabsorbed*	*Amount Reabsorbed (%)*
Water	Proximal (87%) and distal	up to 99
Glucose	Proximal	100
Amino acids	Proximal	98
Protein (if present)	Proximal	100
Sodium	All parts	up to 99
Potassium	Proximal	90 to 100
Chloride	?	99
Urea	Proximal	60
Uric acid	Proximal	90
Creatinine		0

Variable amounts of the electrolytes are reabsorbed in different parts of the tubule. Sodium, for example, can be reabsorbed in all portions of the tubule, but potassium reabsorption is apparently handled in the proximal portion only. By varying the amounts of electrolytes reabsorbed, the kidney regulates their concentrations in the body fluids. The most important of these substances are sodium, potassium, calcium, chloride, magnesium, sulfate, and phosphate.

Additional water is reabsorbed in the distal tubule – the amount varying with the body's state of hydration. If the body has sufficient water, the urine will be more dilute; that is, less water will be reabsorbed through the distal segment. Conversely, by concentrating the urine it secretes, the kidney conserves water when it is needed, thus promoting more water reabsorption and resulting in a scanty concentrated urine.

Other important substances handled by the kidney are the end products of metabolism, which are constantly being formed in the body. The kidney must remove these from the blood since they are of little or no use to the body. Substances such as *urea, creatinine,* and *uric acid* undergo little or no reabsorption in the tubules and pass through them to be excreted in the urine. The essence of renal function, then, is twofold. The kidney must actively reabsorb from the tubular filtrate those substances which are needed by the body, such as glucose and electrolytes, and it must refuse to reabsorb those which are harmful, such as urea in any significant amounts (Table 9-1).

Tubular Maximum (T_M). It has been stated that the kidney will totally reabsorb through the tubules certain substances such as glucose. There is, however, a definite limit to the amount of such a substance that can be actively reabsorbed per unit of time. The reason for this limit is that such reabsorption depends on the amount of the specific carrier that is available.

For example, if the glucose present in the filtrate does not exceed 320 milligrams per minute, all of it will be reabsorbed. Any excess over this amount, however, will not be reabsorbed but will pass into the urine. Therefore, the *tubular maximum,* or T_M, for glucose is 320 milligrams (mg.) per minute. Actually, there is a large safety margin in the case of glucose, since ordinarily the tubular load averages only 125 mg. per minute.

There is a tubular maximum for each of the amino acids, for uric acid, for protein, and for several other constituents. The reabsorption limits for these substances, as in the case of glucose, are usually not exceeded unless some disease process operating in the body results in an upset in their metabolism. In such cases, more of the particular substance than usual is present in the blood, thus increasing the amount presented to the tubules. (Of course, kidney disease itself may be responsible for the abnormal presence of substances in the urine.) Therefore, as long as the blood plasma concentration of any of these "threshold" substances remains at or below a certain level, essentially all that enters the tubules will be reabsorbed.

Tubular Secretion. In addition to actively or passively reabsorbing certain substances, the kidney tubular cells are capable of actually adding substances to the filtrate. The mechanism of transport is the same as in reabsorption, but the direction across the cell is reversed; that is, the substance is carried *from* the tubular cells *into* the lumen of the tubule.

Although potassium is normally reabsorbed from the proximal tubule, it also may be added to the filtrate by the distal tubular cells if its concentration in the body fluids becomes too high. Creatinine not only is not reabsorbed by the tubule but is actually secreted into the filtrate by the proximal tubule.

Whenever certain drugs are present in the body fluids, they are secreted into the filtrate by the proximal tubules. Penicillin is such a drug, and this is why it is difficult to keep an adequate level of the drug in the body. Another drug, Benemid, is often administered with penicillin because it poisons the carrier system for penicillin in the tubular cells and thus prevents them from secreting it into the filtrate and ultimately into the urine.

Rate of Urine Formation. If 125 cc. of glomerular filtrate is formed each minute, by the time it leaves the distal tubule and enters the collecting duct, only 1 cc. remains. This means that the other 124 cc. has been reabsorbed from the filtrate as it passed through the various segments of the tubule. Thus the rate of urine formation is 1 cc. per minute. Therefore, although approximately 180 quarts of glomerular filtrate is formed each day, only 2 quarts or less is excreted as urine.

Characteristics and Composition of Urine. Normal urine is a clear, amber-colored liquid. The eating of certain foods may cause a color change; beets, for example, impart a reddish tinge. The characteristic odor of voided urine changes upon standing to that of ammonia. Freshly voided urine is usually acid in reaction, but different types of diets may alter this. A high protein diet increases the acidity, whereas a vegetarian diet may result in an alkaline urine.

Water makes up 95 to 99 per cent of normal urine. The remainder is composed of metabolic waste products (urea, uric acid, etc.), mineral salts (sodium, potassium, etc.), and pigment (urochrome), which gives urine its color.

In certain diseases, urine will contain substances not normally found. The presence of blood (*hematuria*), sugar (*glycosuria*), or pus (*pyuria*), for example, are indications of some pathological condition in the body. Painful urination (*dysuria*), unusually large amounts of urine (*polyuria*), or very small amounts (*oliguria*) are also indicative of abnormal body function. *Anuria*, the complete absence of urine production, indicates kidney shutdown and is a very grave sign.

THE URETERS

The *ureters* are two tubes about 10 to 12 inches long and 1/5 inch in diameter. They lie behind the parietal peritoneum and convey urine from the hilum of the kidney down to the urinary bladder. The upper end of each ureter is expanded inside the kidney to form the renal pelvis. Each tube is lined with mucous membrane, which is thrown into longitudinal folds. The muscular coat consists of inner longitudinal and outer circular layers of smooth muscle. (Note that this arrangement is just the reverse of that in the intestinal tube.) The covering layer of the ureter is adventitia.

The ureters pierce the posterior surface of the bladder, entering at a slight angle. Small folds of mucosa form valves over the openings, preventing bladder contractions from forcing urine back into the ureters.

Urine is collected in the renal pelves of the kidneys and passes down the ureters and into the urinary bladder. Gravity alone is not responsible for this passage. Urine is forced down the ureters by peristaltic waves of contraction, which sweep over them at the rate of from one to five per minute.

THE URINARY BLADDER (Fig. 9–7)

The *urinary bladder* is a collapsible bag lying in the anterior half of the pelvis behind the symphysis pubis. Parietal peritoneum covers its superior surface only, the remainder of the bladder lying below the peritoneum. When filled, the bladder contains about 500 cc. of urine and rises up into the abdominal cavity.

The bladder wall is similar to that of the ureter but is much thicker. The epithelium of the mucosa is *transitional* in type; it is about five cells thick when the bladder is empty but only two or three cells thick when it is filled. Folds in the

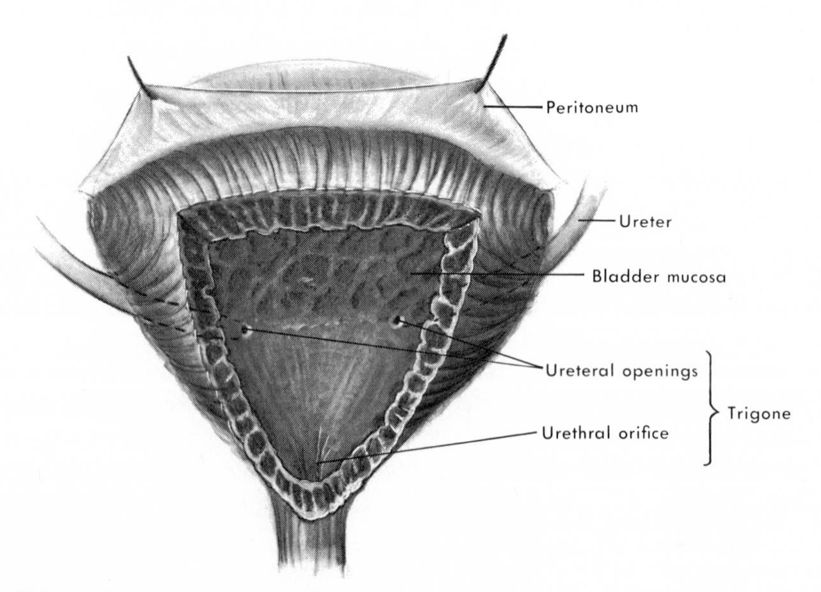

Figure 9–7. The urinary bladder as viewed from the front (cut to show openings).

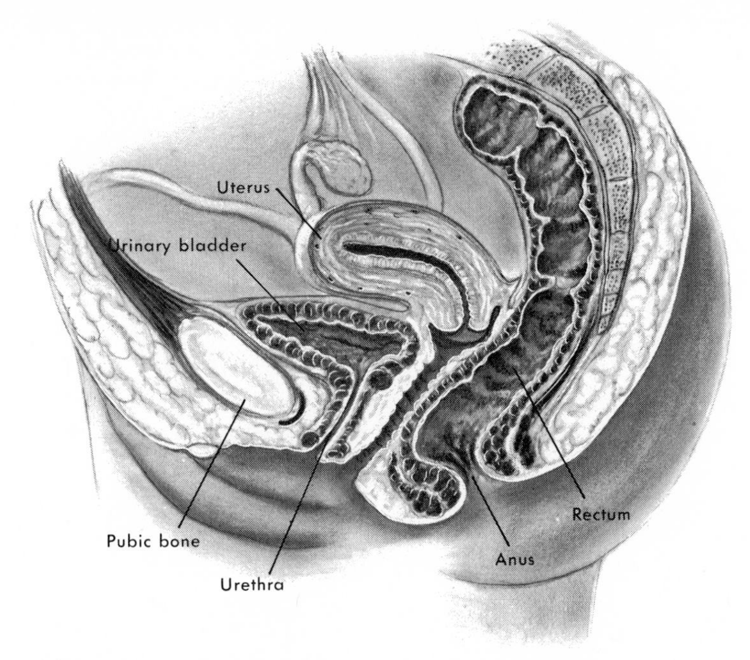

Figure 9–8. The female bladder and urethra – midsagittal view.

mucosa are present when the bladder is empty. The middle circular coat of the smooth muscle layers is the most prominent one. The *trigone* of the bladder is a triangular area. The base points of this area are formed by the two ureteral openings, and the lower apex is formed by the urethral opening.

THE URETHRA

The *urethra* is a tube which conveys urine from the bladder to the outside of the body. It is different in the two sexes since the male urethra is also part of the reproductive system. In addition to carrying urine from the bladder, it also allows for passage of the male reproductive fluid. For this reason, the male urethra will be discussed in Chapter 11.

The female urethra is much shorter than that of the male, averaging 3 to 4 cm. (about 1½ inches) in length (Fig. 9–8). It is fairly straight and has a mucous membrane lining. There are two layers of smooth muscle in its wall. Its termination is the *urinary meatus*, which opens anterior to the opening of the vagina. Urethral glands open at the sides of the orifice, which is surrounded by a voluntary muscle sphincter.

MICTURITION

The process of emptying the bladder is called *micturition, urination,* or *voiding*. In babies, it is an involuntary first-level reflex. In the voluntary learned action, it is a third-level reflex.

As the bladder fills with urine, its smooth muscle layers stretch to accommodate the fluid. It is thought that parasympathetic nerves maintain the tone of the main (*detrusor*) muscle of the bladder wall during filling. When the bladder is moderately distended by a volume of about 350 cc., sensory stretch receptors in the muscle are stimulated and they convey to the central nervous system the desire to void. Efferent (parasympathetic) impulses from the sacral segments of the spinal cord then produce contraction of the detrusor muscle and relaxation of the internal (trigonal) urethral sphincter (Fig. 9–9).

Thus, the *micturition reflex* is a completely automatic cord reflex, but it can be influenced by centers in the brain. It can be facilitated, that is, made more powerful, by impulses from the pons and hypothalamus. Conversely, cortical centers inhibit the reflex by decreasing contractions of the bladder muscle. When the time is appropriate, the cortical centers stimulate the sacral cord centers to begin a micturition reflex and also to inhibit the voluntary external urethral sphincter so that urination can occur.

When control over the micturition reflex is learned, it is possible to voluntarily postpone the act of voiding. This voluntary action apparently abolishes for a time the afferent stretch receptor impulses. It is likewise possible for the individ-

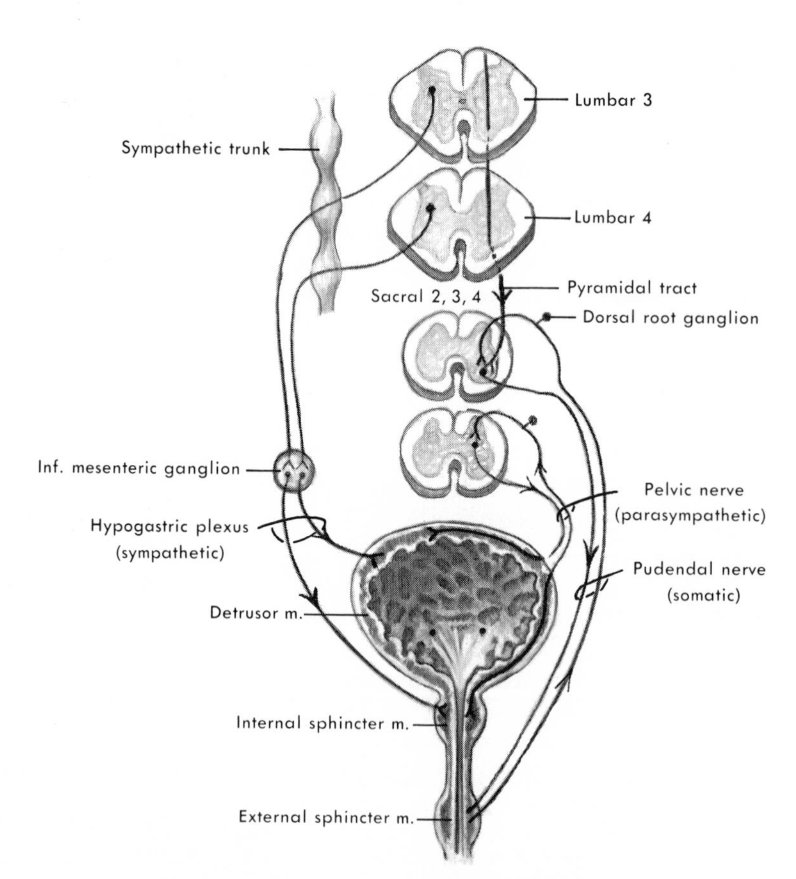

Figure 9–9. Nerve supply to the bladder.

ual to voluntarily void even though the desire to do so is absent. This voluntary control over a reflex activity points out the influence of the voluntary centers of the central nervous system over the autonomic system.

The sympathetic system appears to be relatively unimportant, since the bladder continues to function normally after its sympathetic supply (hypogastric nerve) is cut. Cutting the pelvic (parasympathetics and afferents) nerve, however, results in an *autonomous bladder*. In this case there is no sensory or motor function, and the bladder must be drained.

WATER BALANCE

Importance of Water. The most abundant constituent in the body is water. The body of a 70 kg. (154 lbs.) man contains about 46 liters or two thirds of his body weight. Water is necessary to all cells and performs several important functions in metabolism. It has the highest heat capacity of any liquid and thus acts as an effective buffer to keep body temperature from rising to high levels. Since, except for metals, water is the best conductor of heat, the body can dissipate large amounts of heat by losing some of its water.

Water will dissolve more chemicals than any other known solvent and, because it reacts with so few chemicals, it is a very stable compound. Water transports metabolites from one site in the body to another, and is the means by which waste products are eliminated in the form of urine, perspiration, and feces. As a component of saliva, it assists in swallowing, and acts as a lubricant in the elimination of fecal matter.

Distribution of Body Water. Total body water is distributed among three compartments (Fig. 9–10). The blood plasma, or fluid portion of the blood, makes up 5 per cent of the body weight. Fifteen per cent of the body weight is made up of the *interstitial fluid*, or the fluid between the body cells. Finally, 50 per cent of the body weight is composed of intracellular fluid, or fluid within the cells.

The muscles contain the highest percentage of water by weight, while the skeleton contains the lowest. The skin is the largest reservoir of extracellular fluid (plasma plus interstitial fluid), whereas muscle has the largest amount of intracellular fluid. Interstitial fluid acts as the "middle man." Through it, all nutrients pass from the blood plasma into the intracellular fluid. Waste products, on the other hand, pass in the reverse direction – from the cells to the blood plasma.

Absorption and Excretion of Water. Total water requirement ranges from 2000 to 3500 cc. per day. The bulk of water ingested in liquids and foods is absorbed through the mucosa of the intestines. There is a steady loss of fluid by insensible perspiration (via the skin and lungs), sensible perspiration (sweating), and via the kidneys. Strenuous exercise in a humid climate can result in a fluid loss of up to 4 liters an hour by sweating. Of course, when copious sweating occurs, urine production by the kidneys is decreased. The volume of urine formed by the kidneys can vary from about 1/3 ml. per minute to 15 ml. per minute. Thus the kidneys contribute greatly to the maintenance of normal fluid volume.

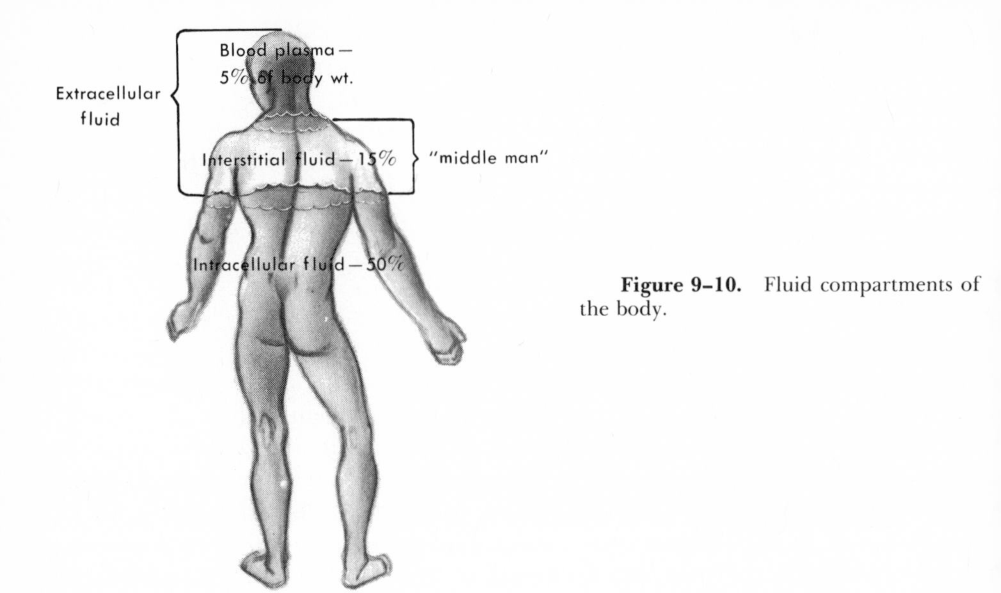

Figure 9–10. Fluid compartments of the body.

Edema and Dehydration. It is impossible to dilute the blood, except temporarily, by drinking water; any excess of water is stored in the interstitial fluid. *Edema*, then, is the expansion of the interstitial fluid volume. There are many causes of edema, in which increased permeability of the capillaries allows excess fluid to leak out into the blood. Some of these conditions are severe burns, bacterial poisons, or lack of certain vitamins (especially vitamins B and C). Kidney disease and heart disease can cause excessive retention of salt and water, leading to an edematous condition.

Dehydration refers to the reduction of the interstitial fluid volume. (If the fluid loss is gradual, however, all three water compartments suffer equally.) When extreme fluid loss occurs, water from the interstitial compartment helps to keep both the blood plasma fluid and the intracellular fluid constant. If the loss continues, water is drawn from the plasma to keep up the intracellular fluid level. In severe and prolonged dehydration, the intracellular compartment itself finally suffers water loss.

Dehydration involves not only a change in water balance but also a change in electrolyte balance. The most important of these electrolytes are sodium, potassium, chloride, and bicarbonate. Some of the pathological conditions which cause dehydration are habitual vomiting, pyloric obstruction, severe diarrhea, diabetes, and kidney failure.

Control of Fluid Intake. The body has two main mechanisms which control fluid volume. One of these determines the concentration of the fluid and involves *osmoreceptors* which are nervous components located in the hypothalamus of the brain. When the body becomes dehydrated, these structures respond by sending messages to the posterior lobe of the pituitary gland (see Chapter 10), calling for the release of a hormone, the *antidiuretic hormone* (ADH). This hormone goes via the blood to the kidneys where it inhibits urine production by these organs. The result, of course, is the conservation of body water. The stimulus which sets this mechanism in motion is the increased concentration of the blood (osmotic pressure) because of dehydration, a situation which is picked up or "sensed" by the osmoreceptors.

A second mechanism involves *volume* receptors. These are nerve endings located in the walls of the great veins and the atria of the heart. When a drop in blood volume occurs, as in severe hemorrhage, these receptors inform the posterior pituitary, which increases ADH secretion, thus decreasing urine production. Changes in blood volume also apparently affect the secretion of another hormone, *aldosterone*, by the adrenal cortex. This hormone acts to decrease sodium excretion by the kidneys. Since sodium secretion parallels water secretion, if sodium secretion decreases, water secretion likewise decreases and urine volume is diminished.

Thirst. Thirst is the mechanism that governs the ingestion of water. It is difficult to define thirst, but animal studies have provided some clues. Stimulation of the hypothalamus in goats, for example, can cause them to drink up to 40 per cent of their body weight. If small quantities of a hypertonic salt solution (1 to 2 per cent) are injected into an animal, it will seek out any available water. Thus, there apparently is a "thirst center" located most likely in the hypothalamus. Cells here are similar to those which regulate ADH secretion; that is, they monitor the concentration of the blood.

ACID-BASE BALANCE

Acids, Bases, and Buffers. The balance between acidic and basic substances in the body fluids is a critical one, since necessary metabolic activities can proceed only if this aspect of body chemistry is kept within proper limits. The mechanisms used by the body to insure maintenance of this balance, as well as the conditions which alter it, will be described briefly.

An *acid* is any substance which, in solution, will give up or lose its hydrogen ions. (An *ion* is a chemical atom that has an electrical charge.) For example, when hydrochloric acid (HCl) is added to water, the following *ionization* occurs:

$$HCl \rightleftharpoons H^+ + Cl^-$$

The arrows going in both directions mean that, initially, the reaction proceeds in both directions, coming to a precise balance point called the equilibrium point. If the concentrations of any of the three substances (HCl, H Cl) should change for any reason, then the reaction will again proceed in both directions until a new balance or equilibrium point is reached. Acids whose hydrogen ions break away from the parent molecule when the acid is placed in solution are said to be *dissociated*. The more highly dissociated the acid, the stronger it is. Water, under ordinary circumstances, does not undergo dissociation; therefore, it is considered to be electrically neutral.

Acidity, then, is a measure of the number of free hydrogen ions that are present in a chemical solution; it has nothing to do with whether or not the solution "tastes acidic." This measurement is referred to as the *pH* of the substance. The pH scale ranges from 1 to 14; water has a pH of 7, the midpoint in the pH scale. All acids have a pH of less than 7, and the stronger the acid, the lower the number.

A *base* is a compound which takes up or *accepts* hydrogen ions when in solution. All bases have a pH greater than 7; the stronger the base, the higher the pH number.

A *buffer* is any solution which can withstand large changes in pH when an acid or base is added. Buffer systems are usually a mixture of a weak acid and its salt or a weak base and its salt. When a weak acid (such as acetic acid) and its salt (sodium acetate) are added to a strong acid (such as HCl) or to a strong base (such as sodium hydroxide), the following reactions occur:

1. Weak acid and its salt plus a strong acid:

$$\begin{matrix} CH_3COOH \text{ (acetic acid)} \\ + \\ CH_3COONa \text{ (sodium acetate)} \end{matrix} + HCl \longrightarrow \begin{matrix} CH_3COOH \\ + \\ CH_3COONa \end{matrix} + NaCl \text{ (sodium chloride)}$$

Thus, the strong acid (HCl) is converted to a weak acid (acetic) plus another salt (NaCl). In contrast with the HCl, which exists in solution almost entirely as H and Cl, the acetic acid exists primarily as CH_3COOH, and very little H is now free in solution. Thus, the pH of the buffer solution changes very little when a small amount of strong acid is added to it.

2. Weak acid and its salt plus a strong base:

$$\begin{matrix} CH_3COOH \\ + \\ CH_3COONa \end{matrix} + NaOH \longrightarrow 2CH_3COONa + H_2O$$

The strong base (sodium hydroxide) is converted to a salt and water.

Body Fluids. When food is burned in an incinerator, the by-products will be acidic, basic, or neutral in reaction. So too when food is burned by the body; its final products after metabolism are similar in this respect. Acid-forming foods include meats, eggs, fish, and seafood. Most fruits and vegetables are basic substances, while butter and milk are neutral, or nearly so. In normal metabolism, a general mixed diet will result in the production of a large excess of acid residue, primarily in the form of carbon dioxide (CO_2). CO_2, when combined with water, generates H and bicarbonate (HCO_3) and is the most common source of H or acid in the body. For example, a 3000 calorie diet of an average adult will produce about 480 liters (about 500 quarts) of CO_2 a day. This is equivalent to approximately 2 liters of concentrated HCl.

Therefore, the body is called on to excrete large amounts of acid and, at the same time, to conserve base in order to maintain the proper balance of pH. Most of this acid is excreted as CO_2 by the lungs and as inorganic and organic acids by the kidneys.

The pH of body fluids is held within narrow limits by the presence of buffer systems which operate according to the principles already described. Most body fluids are slightly alkaline, as shown in the following chart:

FLUID	pH
Whole blood	7.25–7.45
Blood plasma	7.25–7.45
RBC'S	7.1 –7.3
Lymph	7.4
Cerebrospinal fluid	7.4
Exceptions:	
Gastric juice	1.0 –5.0
Urine (sometimes)	2.0

Alterations in Acid-Base Balance. If the pH of body fluids moves outside of the normal limits as listed, the resulting condition will be either *acidosis* (pH below the lower limit) or alkalosis (pH above the upper limit). Either condition poses a serious threat to life unless it is remedied without delay.

Acidosis can be the result of (1) the body's failure to excrete sufficient acid, (2) the production of more acid than the body can excrete, or (3) a loss of the body's normal alkaline reserve. Alkalosis can result from (1) hyperventilation of the lungs (excessive loss of CO_2), (2) severe gastric vomiting (loss of HCl), or (3) addition of large amounts of alkali, such as in the administration of bicarbonate compounds. Table 9-2 lists some conditions in which the acid-base balance may be altered.

FUNCTIONS OF BUFFERS, LUNGS, AND KIDNEYS

Buffers. Body buffers are present both extracellularly (mostly in the blood plasma) and intracellularly. Blood buffers are bicarbonate, hemoglobin, and plasma proteins. Buffering action by the blood is rapid and efficient in normal body metabolism. Almost all buffers are used to neutralize the CO_2 produced by the body cells. If the capacity of the body's buffer systems is exceeded for any reason, then acidosis or alkalosis will result. The acid-base status of an individual can be determined by measuring the hydrogen ion activity and either the CO_2 or the bicarbonate concentration of the blood or plasma.

Lungs: Respiratory Compensation. Respiratory adjustment of pH occurs rapidly. This is accomplished by variation in the ventilation rate and thus in CO_2

Table 9-2. CONDITIONS LEADING TO ACID-BASE IMBALANCE

Acidosis (Blood pH Below 7.25)	*Alkalosis (Blood pH Above 7.45)*
Uncontrolled diabetes	Hyperventilation of lungs
Cardiac failure	Excessive antacid ingestion
Renal failure	Severe gastric vomiting
Severe diarrhea	
Impaired lung ventilation	
Hypoxia	
Barbiturate poisoning	
Excessive acid ingestion	

concentration in the lung alveoli. In alkalosis, there is a depression of ventilation which has the effect of "saving" CO_2. A twofold increase in the ventilation rate will cut the alveolar CO_2 concentration nearly in half. Thus, in a matter of minutes, the body can rid itself of large amounts of acid in the form of CO_2. The lungs, however, cannot directly restore any loss in the bicarbonate reserves.

Kidneys: Renal Compensation. The kidneys can change the pH of the urine back and forth between acid and alkaline, depending upon the need. These organs have the ability to excrete large amounts of hydrogen ion by utilizing several mechanisms. One way is by the formation of ammonia (NH_4) which can then be excreted in exchange for sodium. This mechanism not only gets rid of hydrogen ions, but also conserves sodium. Thus, the kidneys perform two important functions: excretion of acid and conservation of base. Renal compensation, however, is slow, as it takes several hours or even days.

OUTLINE SUMMARY

1. Urinary System – kidney helps maintain constancy of internal environment
   a. Regulates concentration of blood constituents
   b. Regulates acid-base balance of body
   c. Removes end products of metabolism from body
2. Kidneys – secrete urine
   a. Paired, bean-shaped organs on either side of vertebral column
   b. Outer cortex and inner medulla
   c. Three funnel-shaped major calyces; several minor calyces
   d. Nephron – structural and functional unit of kidney
      1) Glomerulus – tuft of blood capillaries
      2) Tubule – proximal, Henle's loop, distal; leads to collecting duct
   e. Kidney lobe – renal pyramid
   f. Arterial blood supply:
      renal artery → interlobar arteries → arcuate arteries → interlobular arteries → afferent arterioles → glomeruli → efferent arterioles → tubular capillaries
3. Formation of Urine
   a. Glomerular filtration
      1) GFR – varies directly with blood pressure; averages 125 cc./min.
      2) FF – ratio of amount filtered to total plasma flow; averages 20 per cent
   b. Tubular reabsorption – selective process by tubular cells by diffusion and active transport
      1) In proximal tubule – 87 per cent of the water; all glucose and amino acids; electrolytes, urea
      2) In distal tubule – water (up to 12 per cent); some electrolytes
      3) Tubular maximum – limit of substance reabsorbed per unit of time; depends on carrier
   c. Tubular secretion – tubular cells add substances to filtrate
   d. Rate of urine formation – 1 cc./min.; 1250 to 3000 cc./day

   e. Urine composition – clear, amber-colored, acid reaction
      1) 95 to 99 per cent water
      2) Metabolic waste products
      3) Mineral salts
   f. Abnormal substances in urine – blood, bacteria, pus
4. Ureters – carry urine from kidneys to bladder
   a. Two tubes, 10 to 12 inches long
   b. Renal pelvis – upper expanded portion
   c. Folds in mucosa; two smooth muscle layers
   d. Peristaltic waves convey urine down ureters
5. Urinary bladder – collapsible bag lying in pelvis; stores urine
   a. Mucosa has transitional epithelium
   b. Three smooth muscle layers (detrusor muscle)
   c. Trigone – three openings for ureters and urethra
6. Urethra – carries urine from bladder to outside body
   a. Male – longer than in female; also part of reproductive system
   b. Female – 3 to 4 cm. long
   c. Urinary meatus – opening surrounded by external sphincter
7. Micturition (Urination, Voiding) – emptying of bladder
   a. Micturition reflex – set up by moderately distended bladder
      1) Stretch afferents in bladder wall – carry desire to void in central nervous system
      2) Parasympathetic efferents from sacral cord – contraction of detrusor muscle and relaxation of internal sphincter
   b. Reflex facilitated by impulses from pons and hypothalamus
   c. Reflex inhibited by cortical impulses
   d. Control of external urethral sphincter is learned action
   e. Cutting pelvic nerves results in autonomous bladder
8. Water balance
   a. Water is universal solvent for all chemical reactions in body
      1) Elimination of waste products
      2) Aids in swallowing
      3) Lubricates intestinal tract
      4) Helps maintain body temperature
   b. Distribution of body water – three compartments
      1) Blood plasma – 5 per cent of body weight
      2) Interstitial fluid ("middle man") – 15 per cent of body weight
      3) Intracellular fluid – 50 per cent of body weight
   c. Absorption and excretion of water
      1) Absorbed through intestinal wall
      2) Excreted through lungs, skin, intestine, kidneys
   d. Edema – expansion of interstitial fluid volume
   e. Dehydration – reduction of interstitial fluid volume
      1) Intracellular compartment last to suffer loss
      2) Involves upset in electrolyte balance
   f. Control of fluid intake
      1) Osmoreceptors in hypothalamus; "thirst center"
      2) ADH from posterior pituitary
      3) Volume receptors in veins and heart; aldosterone secretion
9. Acid-Base Balance
   a. Acid – substance which donates hydrogen ions
   b. Base – substance which accepts hydrogens

c. Buffer – can withstand pH changes; mostly in blood
d. Acid-base balance changes
    1) Acidosis – pH below 7.25
    2) Alkalosis – pH above 7.45
    3) Lungs and kidneys – compensatory mechanisms for pH alterations

REVIEW QUESTIONS

1. Explain what is meant by the "constancy of the internal environment." List some of the body processes involved in its maintenance.
2. What structures give the kidney cortex its granular appearance?
3. Trace the course of a drop of fluid from the glomerulus to the urinary bladder.
4. What is the difference between filtrate and urine?
5. Describe a kidney lobe.
6. Would you expect a decrease in urine formation if the arterial blood pressure drops? Why?
7. If the renal plasma flow were 700 cc./min. and the GFR 140 cc./min., what would be the FF?
8. In what ways does the composition of the filtrate change as it passes from the glomerular capsule to the collecting duct?
9. Explain what is meant by the T_M of a substance.
10. Why must doses of penicillin be given at regular intervals to maintain a certain level in the body?
11. Explain the micturition reflex. Can it be voluntarily controlled?
12. What is the functional relationship of the three body water compartments to each other?
13. In a normal person, what is the best guide to follow to insure sufficient water intake?
14. Explain the role of the hypothalamus in the control of fluid intake.
15. What is the meaning of *dissociation* in relation to acids?
16. How do buffers operate to maintain constant pH?
17. Describe the differences between renal and respiratory compensation in conditions of acid-base imbalance.

ADDITIONAL READING

Reidenberg, M. W.: Renal Function and Drug Action. Philadelphia, W. B. Saunders Company, 1971.

Smith, H. W.: The kidney. Sci. Amer. (Jan.), 1953.

Chapter 10

THE ENDOCRINE GLANDS

INTRODUCTION

As already noted in previous chapters, there are many structures in the body called *glands*, which produce secretions such as saliva, mucus, and pancreatic juice. These secretions are carried from the glands to their respective destinations by tubes or channels called *ducts*. In addition to these duct or *exocrine* glands, there are discrete masses of secreting cells in different parts of the body which do not have ducts. These ductless or *endocrine* glands pour their secretions, called *hormones*, directly into the blood which flows through them. The circulation then carries the hormones to specific structures, or target organs, where they perform their functions.

For ease of study, the endocrine glands are usually considered as a system, even though they are located in different parts of the body (Fig. 10–1). Although their functions differ widely in some cases, there are many interrelationships among them, some of which will be discussed. The hormones produced by the gastrointestinal glands were considered in Chapter 8, since their functions related specifically to the digestive system. The organs of the endocrine system are the pituitary (hypophysis), the thyroid, the parathyroids, the pancreatic islet cells, the adrenals, and the gonads (sex glands). A note on the thymus gland will be included.

The endocrine system is the second most important system in assuring unity of function in the body, being superseded only by the nervous system. It takes longer for hormones to produce changes in body processes than it does for nerve impulses, but the effects last longer.

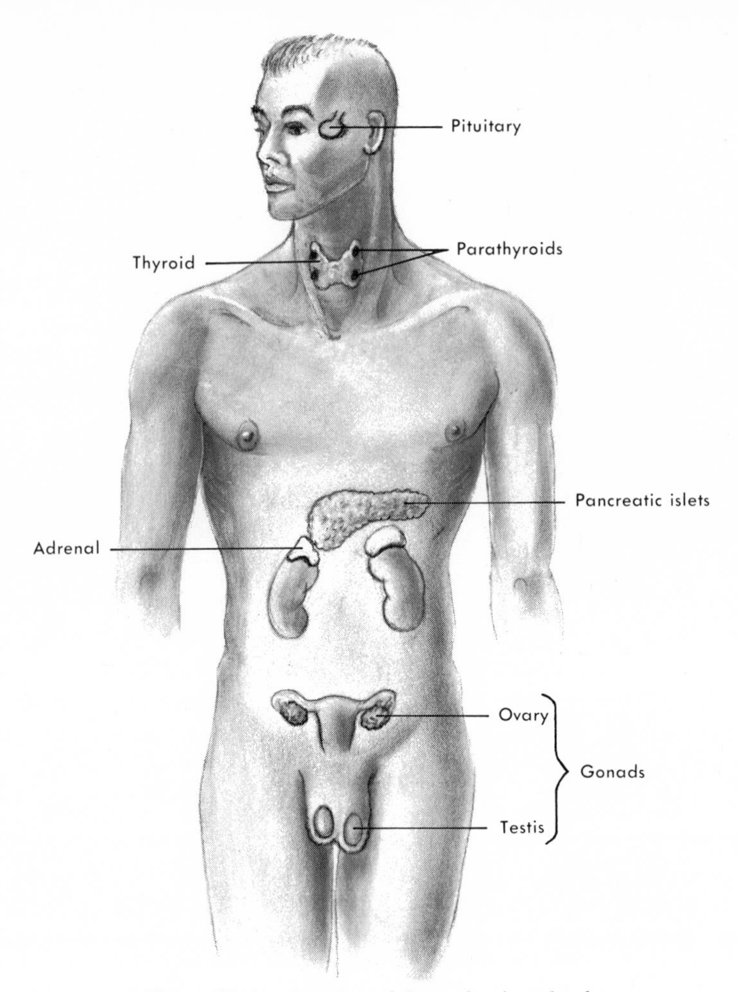

Figure 10–1. Location of the endocrine glands.

PROPERTIES OF HORMONES

Hormones are chemical substances produced by specific types of cells. Most of them are either *steroids*, which are compounds derived from fats, or proteins. They are secreted into and distributed by the blood to all parts of the body. Very powerful in small amounts, they either excite or inhibit physiological processes in the various body tissues. Indeed, some hormones are indispensable for life itself.

Hormones are not a product of, nor directly involved in metabolism. Thus by definition, such substances as carbon dioxide, glucose, amino acids, and minerals are not hormones. Moreover, since they are produced within the body *(endogenous)* and are not supplied by ingested food *(exogenous)*, hormones are distinct from vitamins. Furthermore, hormones will not function outside of the living, intact body cells.

Many hormones have widespread activity, influencing functions that are a part of body metabolism as a whole. For example, the metabolism of carbohydrate, calcium, phosphorus, salt, and water are all subject to regulation by certain hormones. Some hormones regulate overall growth of body tissues, whereas others are more specific in promoting local growth and development of certain organs. Hormonal deficiencies cause metabolic disturbances such as diabetes, rather than toxic disturbances such as those caused by bacterial poisons.

TYPES OF ENDOCRINE GLANDS

Pure endocrine glands are those whose only function is to produce hormones. *Mixed* glands (such as the pancreas) are those which produce both endocrine and exocrine secretions. Some books also list as endocrine nonglandular structures which produce substances with hormone-like activities. An example would be the sympathetic and parasympathetic nerves, which do release potent chemical substances from their endings.

Structurally, an endocrine gland may be one of two types, solid or storage. A solid cellular gland will have its secretions present in its interstitial spaces. A storage gland, on the other hand, has a space inside it in which to store its hormones.

Secretion of their hormones by the endocrine glands may be stimulated in various ways. Some glands pour forth their secretions in response to nervous stimulation from certain parts of the brain. These, then, are said to be under *nervous control.* The activity of other glands is regulated by the level of some substance in the blood, and this type of regulation is called *humoral control.* Finally, the increased activity of one gland may directly cause the increased activity of another gland. Such a relationship is called *reciprocal release* or *feedback*, since the increased level of hormones produced by the second gland inhibits or decreases the activity of the initial gland.

Hyperfunction and *hypofunction* refer to endocrine function above or below the range accepted as normal. Either condition usually gives rise to symptoms which enable the investigator to understand what effects a hormone has on body metabolism. Indeed, much of our knowledge of hormonal activity has been derived from such observations.

GLANDS OF THE ENDOCRINE SYSTEM

THE PITUITARY GLAND (THE HYPOPHYSIS)

Description. The pituitary, or hypophysis, is sometimes referred to as the master gland, and, in a functional sense, it is really two glands. The *anterior pituitary* is an outgrowth of the roof of the pharynx. The *posterior pituitary* develops from the third ventricle of the brain. During fetal life, the anterior lobe severs its connection with the pharynx and joins with the posterior lobe. Thus the entire gland is attached to the underside of the brain by a stalk which is called the *infundibulum.* The *intermediate lobe* is a subdivision of the anterior lobe and lies between it and the posterior lobe. It has no proved functional significance in man.

The entire pituitary gland is about the size of a small pea. It lies in the sella turcica of the sphenoid bone and is roofed by the meningeal dura mater except where its stalk attaches to the brain (Fig. 10–2). Periosteal dura lines the sella. Eighteen to 20 small arteries from the arterial circle (of Willis) and the internal carotid give the gland a rich blood supply.

Anterior Lobe Hormones. The anterior lobe hormones perform a variety of functions, both metabolic and stimulating. They regulate the general growth and development of the body, the metabolism of carbohydrate, fat, and protein, the development of sex characteristics, and the functions of other endocrine glands (Fig. 10–3). There are six known anterior lobe hormones, four of which have been chemically purified. All but one of the six are tropic hormones; that is, they have a direct influence on other endocrine glands.

GROWTH HORMONE (GH). This hormone increases the rate of growth of all body cells. It does so by promoting the deposition of protein and the mobilization of fat from the tissues. This growth is not just a simple increase in size as in obesity; the need is an increase in mass of symmetrical composition. In addition, GH has a *diabetogenic* or anti-insulin effect: It acts upon the blood sugar by increasing its level, whereas the hormone insulin decreases the blood sugar level.

Although the anterior lobe obviously produces more GH during the years of rapid body growth, it continues to secrete some of the hormone even after growth is completed. It may be that GH is necessary for protein synthesis even in the nongrowing animal. It has not yet been explained how the anterior pituitary knows the optimal size of a particular individual and thus when to decrease the secretion of growth hormone at a specific point in development.

Hypofunction. An insufficient amount of GH in the young individual results in *dwarfism.* Dwarfism involves a symmetrical cessation of growth, especially of the epiphyseal cartilage cells in the long bones. Calcification of the bones, howev-

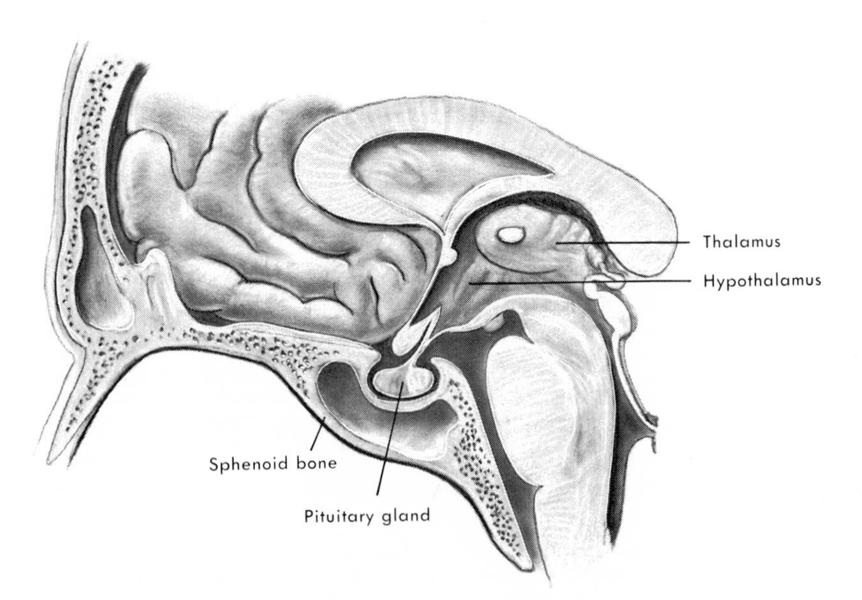

Figure 10–2. Location of the pituitary gland within the skull.

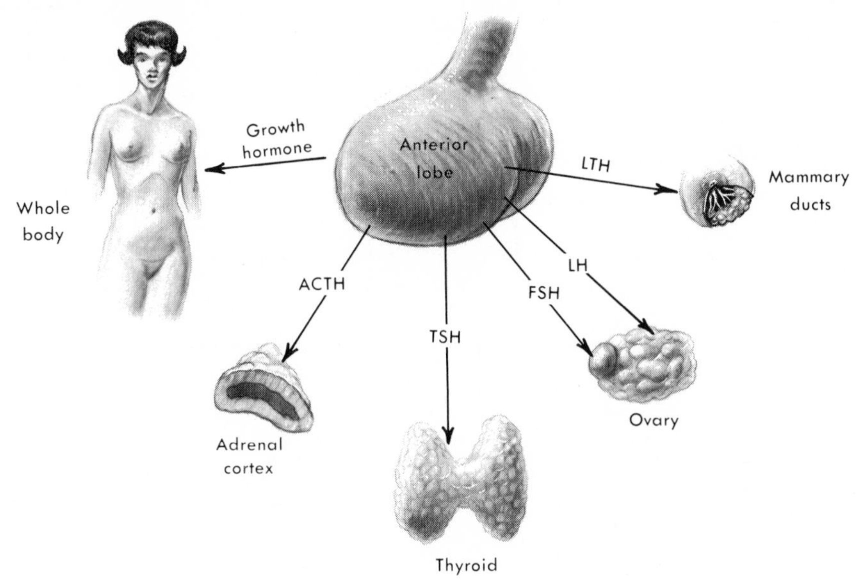

Figure 10–3. Hormones of the anterior lobe and their target organs.

er, proceeds normally. If GH production ceases in an adult, *Simmonds' disease* or *hypophyseal cachexia* occurs. Its signs are extreme weakness, with loss of weight and wasting of body tissues.

Hyperfunction. An excess of growth hormone in the young causes *gigantism.* There is an increase in the size of all organs as well as a symmetrical increase in the growth of the skeleton. In the adult, since growth has already been completed, overproduction of GH cannot cause gigantism. Instead, it results in a condition called *acromegaly.* In acromegaly skeletal changes occur predominantly in the face, hands, and feet, producing prominent cheek bones, a protruding jaw, and spade-like fingers. The viscera also increase in size, and diabetes may develop.

Adrenocorticotropic Hormone (ACTH). ACTH is responsible for the function and maintenance of part of another endocrine gland, the *adrenal cortex.* ACTH increases the rate at which adrenal cortical hormones are secreted, and it also increases the growth of the cortical cells. Like GH, ACTH has a diabetogenic effect.

Hypofunction. Decreased secretion of ACTH by the anterior lobe will lead to a decrease in the output of adrenal cortical hormones by that gland. In the absence of ACTH, the adrenal cortex will *atrophy* or waste away.

Hyperfunction. Overproduction of ACTH was formerly thought to be the cause of *Cushing's syndrome.* Now Cushing's syndrome is considered basically a disease of the adrenal cortex and will be discussed with that gland. Pituitary changes in Cushing's syndrome are a secondary effect.

Thyrotropic Hormone (TSH). The normal activity of the thyroid gland is directly regulated by TSH from the anterior lobe. It affects the uptake of iodine by the thyroid gland. TSH causes the thyroid to release its hormone into the blood. It also increases the number of cells in the thyroid and stimulates them to produce more thyroid hormone. A cold environment will also stimulate the production of TSH.

Hypofunction. Total loss of TSH results in depressed thyroid function and over a period of time causes thyroid atrophy.

Hyperfunction. Overproduction of TSH is given as the cause of a condition called *Graves' disease* or *exophthalmic goiter*. This disease will be discussed with the thyroid gland.

FOLLICLE-STIMULATING HORMONE (FSH). In the female, FSH causes an increase in the growth of the follicle cells that surround the egg cell, or ovum, in the ovary. The follicular cells in turn secrete *estrogens*, or female hormones. In the male, this anterior lobe hormone is called *germinal-stimulating hormone* (GSH). It initially stimulates the cells in the male gonad, or *testis*, which produce sperm.

LUTEINIZING HORMONE (LH). In the female, LH is responsible for growth and maintenance of the ovarian cells that produce both female hormones, the estrogens and *progesterone*. Luteinizing hormone is responsible for *ovulation*, the release of a mature egg cell from the ovary. The comparable hormone in the male is *interstitial cell-stimulating hormone* (ICSH). It stimulates the cells in the testis that produce the male hormone *testosterone*.

LUTEOTROPIC HORMONE (LTH OR PROLACTIN). In the pregnant female, LTH is responsible for the growth of the duct system in the mammary gland. It also continues the ovary's secretion of the female hormones, which was initiated by LH. There is apparently no distinct counterpart for LTH in the male.

Panhypopituitarism. Perhaps the best way to summarize the functions of the anterior pituitary hormones is to describe the effects in the adult resulting from an insufficient supply of all these hormones. Such a condition is called *panhypopituitarism*, which means decreased secretion of all anterior lobe hormones. This decreased secretion may be slight or marked, and it may occur suddenly or slowly over a period of time.

In general, the effects of panhypopituitarism are (1) decreased functioning of the thyroid gland (hypothyroidism), (2) decreased production of hormones by the adrenal cortex, (3) decreased production of gonadotropic hormones with loss of sexual function, and (4) decreased blood sugar. When severe loss of appetite resulting in emaciation accompanies the condition, the term *Simmonds' disease* is added.

Posterior Lobe Hormones. Two distinct hormones are secreted by the posterior pituitary.

VASOPRESSIN (PITRESSIN). This hormone has two separate actions. When injected into the body in large amounts, it causes a rise in the arterial blood pressure. In quantities normally secreted by the gland, however, there is no significant effect on the blood pressure. In addition, removal of the posterior lobe does not markedly alter the arterial blood pressure level.

Antidiuretic hormone (ADH) is the more important fraction of vasopressin. It promotes reabsorption of water through the distal tubules of the kidney. In the absence of ADH, the body loses large amounts of water through the kidneys. Thus the individual excretes large quantities of very dilute urine, a condition known as *diabetes insipidus*. (This diabetes is called "insipidus" to differentiate it from diabetes mellitus [sweet], a condition in which sugar is present in the urine.) Such an individual may excrete as much as 10 to 15 liters of urine per day, depending upon the extent of his thirst for water. The condition can be alleviated by administering as little as 0.1 microgram (1,000,000 micrograms = 1 gram) of powdered posterior pituitary gland. Changes in blood volume affect ADH production; an increase in blood volume leads to increased water excretion and decreased ADH secretion. The reverse is true with a decrease in blood volume. (See Chapter 9.)

OXYTOCIN (PITOCIN). This hormone causes powerful contractions of the pregnant uterus (womb). Its action increases as the pregnancy progresses toward the time for delivery (parturition). Oxytocin acts directly on the smooth muscle in the wall of the uterus and is considered to be at least partially responsible for initiating labor. Therapeutically it has its greatest use in controlling hemorrhage that may occur after delivery.

Oxytocin also causes milk present in the alveoli of the mammary gland to be expressed into the ducts of the gland so that the baby can suckle. The initial suckling by the baby is the stimulus for the release of this hormone. Oxytocin is one of the most powerful chemical substances known, and very minute amounts can cause very strong effects.

Both posterior pituitary hormones have been chemically isolated and identified. They are both proteins, and each is composed of eight different amino acids. The two hormones are identical in structure except for two of these amino acids.

Control of Pituitary Secretions. The mechanism for the secretion of pituitary hormones is rather complex. In the anterior lobe, the tropic hormones are secreted partly in response to the blood level of the hormones produced by the glands they stimulate (Fig. 10–4). For example, when there is a body need for thyroid hormone, the pituitary secretes TSH, which stimulates the thyroid gland to secrete more of its hormone. Then, when the thyroid hormone rises to its proper level, it inhibits the further secretion of TSH by the pituitary. This interrelationship is an example of the reciprocal release mechanism mentioned earlier:

Low blood thyroid hormone → ↑ TSH secretion by pituitary
Pituitary TSH → ↑ hormone secretion by thyroid gland
Proper blood level of thyroid hormone → ↓ TSH secretion

The hypothalamus in the brain also plays a role in regulating anterior lobe secretion. Cells in the medial basal part of the hypothalamus secrete *releasing factors* (RF) which are sent by way of nerve fibers to the portal blood vessels of the anterior lobe. Releasing factors for LH, GH, FSH, TSH, and ACTH stimulate the appropriate cells in the anterior pituitary to release these hormones into the blood. In the case of prolactin (LTH), however, its hypothalamic RF seems to *inhibit* its release. Experimental evidence for the actions of releasing factors shows that if the anterior lobe is removed from its normal location and implanted elsewhere in an animal's body, the only hormone it continues to release is prolactin.

The posterior lobe of the pituitary has an even closer relationship to the hypothalamus (Fig. 10–5). The posterior lobe cells bear such a resemblance to nerve cells that this lobe is sometimes called the *neurohypophysis*. There is good evidence to show that the cells of the neurohypophysis do not in fact actually secrete the posterior lobe hormones but simply store them until it is time for their release into the blood. Nuclei of nerve cell bodies in the hypothalamus actu-

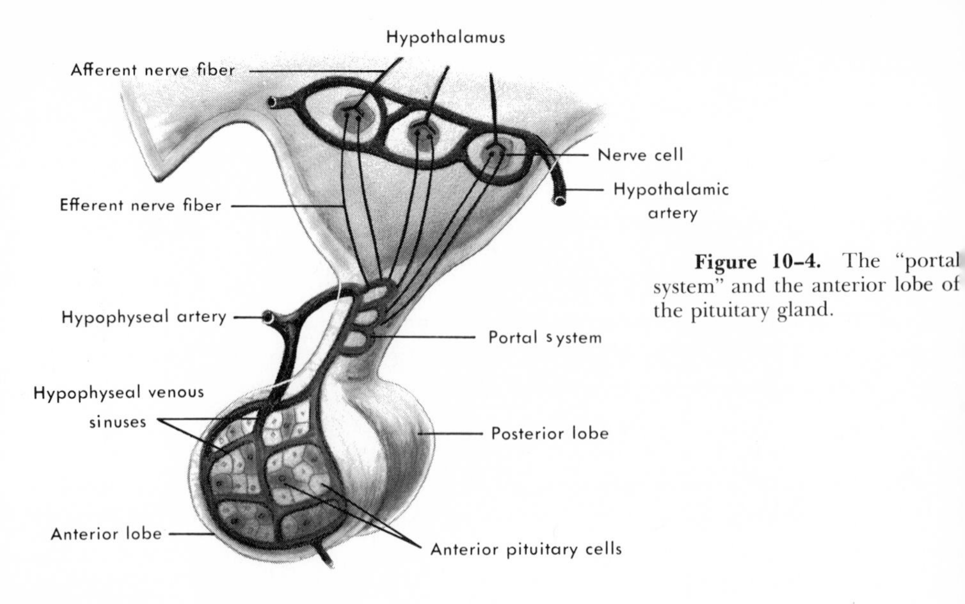

Figure 10–4. The "portal system" and the anterior lobe of the pituitary gland.

ally produce the hormones, which are then transported down nerve fibers and into the posterior lobe. One of the proofs for this explanation is that if the pituitary stalk is severed but the hypothalamus is left intact, the posterior lobe hormones continue to be formed.

THE THYROID GLAND

Description. The thyroid gland is located in the anterior aspect of the neck immediately below the thyroid cartilage of the larynx (Fig. 10–6). It consists

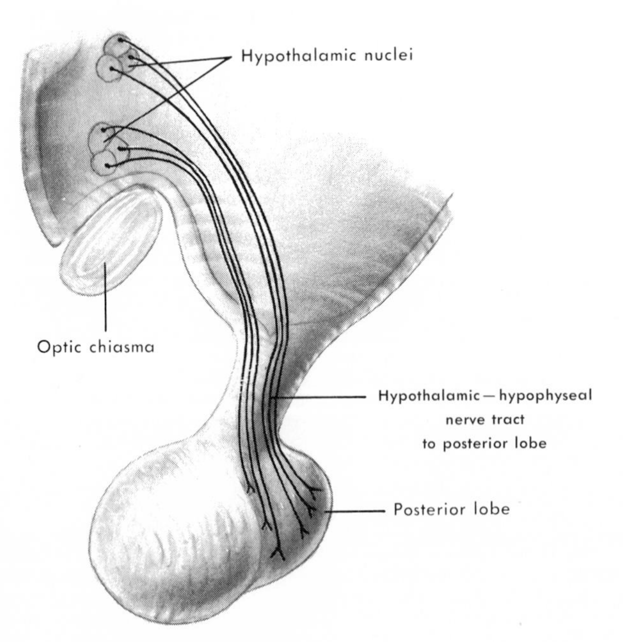

Figure 10–5. The hypothalamus and the posterior lobe of the pituitary gland.

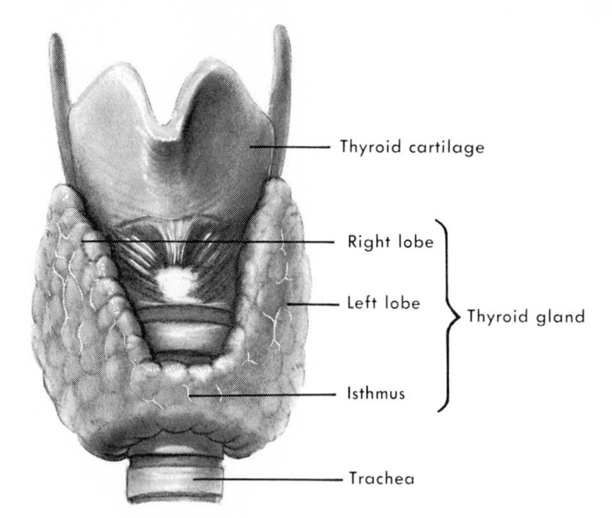

Figure 10–6. The thyroid gland—anterior view.

of two lateral *lobes* connected by a narrow central portion called the *isthmus*. The gland is composed of *follicles*, each of which is a cyst-like structure surrounded by epithelial cells. These cells secrete the thyroid hormones, which are then stored as *colloid* inside the follicles. The height of the follicular cells reflects changes in the gland's functional state. Gland size depends upon its activity; it is larger in the female and in the young. It contains 25 per cent of the body's iodine.

Functions. The functions of the thyroid gland are to increase the metabolism of all body cells and, in the growing animal, to promote ossification of the bones. Thyroid hormones have a calorigenic effect; that is, they increase oxygen consumption and thus energy production by the tissues. They also increase glucose absorption from the intestine and its utilization by the body cells. Therefore, thyroid gland activity acts to increase the basal metabolic rate. The thyroid was the first of the endocrines to be studied extensively, and the general principles of hormone function are a result of this.

Hormones. Three hormones are produced by the thyroid, but only one, *thyroxine*, is found in any quantity in the blood. The thyroid forms its hormones by taking *iodine* from the blood and combining it with the amino acid *tyrosine*. Since 65 per cent of thyroxine is iodine, the thyroid gland must be able to remove iodine from the blood rapidly and store it.

Hypothyroidism. The most characteristic sign of an underfunctioning thyroid is a low basal metabolic rate. After total removal of the gland, the basal metabolic rate gradually falls to 45 or 50 per cent of its normal value.

Extreme hypothyroidism in infancy and childhood is called *cretinism*. It results from either a congenital lack of a thyroid gland or insufficient iodine in the diet. The latter condition leads to an underproduction of thyroxine. A cretin will not mature physically or mentally. Skeletal growth is particularly retarded, but this condition can be reversed if thyroxine is supplied to the child. However, unless treatment is started within a few months after birth, the mental retardation will be permanent.

Myxedema refers to hypothyroidism in the adult human; it is not seen in animals. Such a person will have a puffy face because of fluid collection under the skin. He will be physically sluggish and will also show signs of slowed mental activity. Treatment is effective and consists of giving either thyroxine or extracts of the thyroid gland itself.

Hyperthyroidism. The common term for overactivity of the thyroid gland is *Graves' disease*. It is characterized by weight loss, increased sweating, intolerance to heat, muscular weakness, nervousness, and an increased BMR. In most cases, some protrusion of the eyeballs, or *exophthalmos*, is present. Treatment by removal of part of the thyroid usually alleviates most of the symptoms with the exception of exophthalmos, which is irreversible.

Goiter. The term *goiter* means any enlargement of the thyroid gland, and it may result from either under- or overfunctioning of the gland (Fig. 10–7). During puberty and pregnancy a *parenchymatous* or *simple goiter* may develop. This type of goiter is usually the result of an iodine deficiency in the diet. Without sufficient iodine, the gland cannot make enough hormone to keep up the normal blood level. The reciprocal release mechanism is then set in motion, and the anterior pituitary puts out increased amounts of TSH. In response to TSH, follicle cells (parenchyma) of the thyroid gland increase in number and height, thus increasing the size of the gland as a whole but reducing the amount of stored colloid within the follicles.

A *colloid goiter* usually follows the development of a parenchymatous goiter. It is assumed that after the demand for more thyroxine, as in the case of pregnancy, has passed, or if sufficient iodine is given, the gland is able to raise the hormone level in the blood. Subsequently, there is a decrease in the secretion of TSH, and the follicular cells of the thyroid become low cuboidal again. In addi-

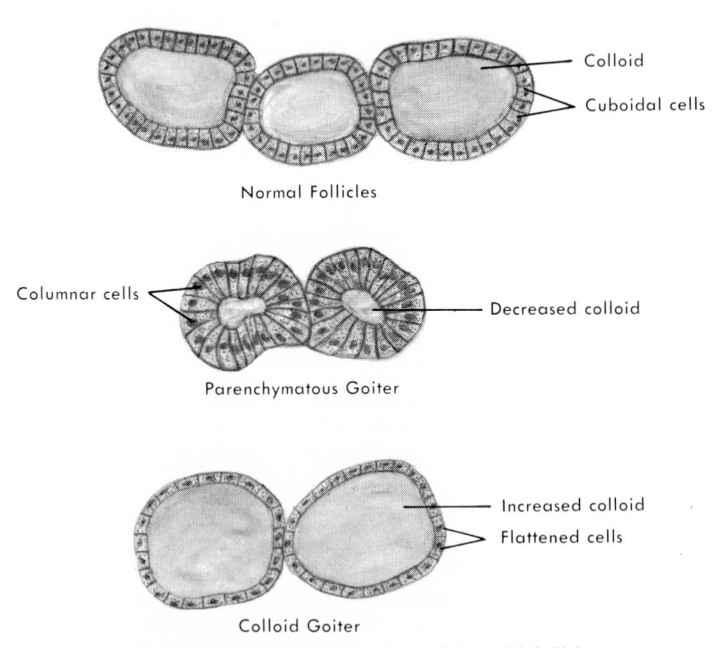

Figure 10–7. Microscopic view of thyroid follicles.

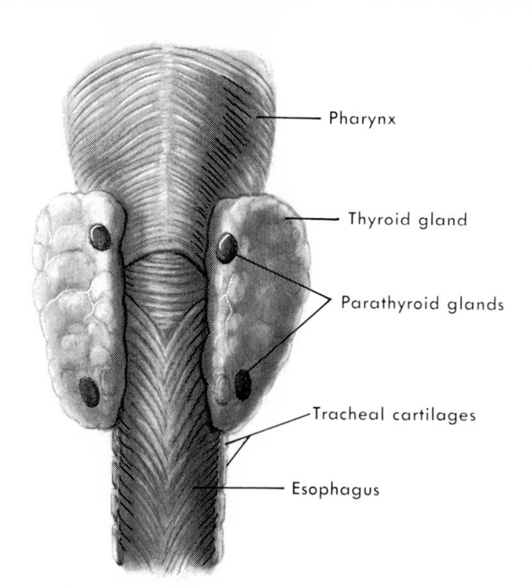

Figure 10–8. The parathyroid glands – posterior view.

tion, the follicles will store more hormone in the form of colloid, thus increasing the gland size.

Exophthalmic goiter is so named because it is often associated with Graves' disease and exophthalmos. Microscopically, the gland resembles a parenchymatous goiter. The exophthalmic goiter produces an excess of thyroid hormone. For some reason, this excess of thyroxine in the blood does *not* lead to decreased TSH production by the anterior pituitary as one would expect. Treatment consists of giving one of several compounds called *goitrogens*, which depress the production of thyroxine by the thyroid gland.

THE PARATHYROID GLANDS

Description. The parathyroid glands, usually four in number, are small flattened or oval bodies. They lie on or just under the posterior surface of the lateral lobes of the thyroid gland – hence the name *parathyroids* (Fig. 10–8). The glands originate from outpouchings of the pharynx and are solid glands, composed of masses and cords of epithelial cells. In man, their total weight is about 100 mg.

Function. The chief function of the one known parathyroid hormone is to regulate calcium and phosphorus metabolism in the body by maintaining serum calcium ion within narrow limits. Primary action is on bone, from which calcium and phosphorus are mobilized to maintain the proper blood level of these minerals. The hormone also controls renal excretion of phosphate, either by increased glomerular filtration of phosphate or, more probably, by decreasing its reabsorption by the kidney tubules.

Hypoparathyroidism. If all parathyroid tissue is removed from the body, the level of blood calcium falls (hypocalcemia), leading eventually to a condition called *tetany*. Lack of sufficient calcium causes the nervous system to become more and more excitable, and the nerve fibers begin to fire spontaneously.

These impulses pass to skeletal muscles, causing them to go into tetanic spasms. If the condition is not corrected, death results from suffocation because of spasticity of the respiratory muscles.

Along with a drop in the blood calcium in tetany, there is an increase in the blood phosphate level. This increase takes place because there is no parathyroid hormone to promote phosphate excretion by the kidney.

Hyperparathyroidism. Overproduction of the parathyroid hormone in man is usually caused by a tumor of the glands. The excess hormone causes increased removal of calcium and phosphate from the bones into the blood, leading to decalcification and weakening of the bones. In addition, since the kidney cannot handle the increased minerals, calcium phosphate becomes deposited in the soft tissues of the body. It is usually deposited first in the kidneys, but such deposits are also found in the liver, heart, arterial walls, and other organs.

The preceding is a description of *primary hyperparathyroidism.* Conditions which may lead to a low blood calcium level, such as a low calcium diet, pregnancy, lactation, or rickets, lead to *secondary hyperparathyroidism.* In such cases, the parathyroid glands enlarge (hypertrophy) in response to the low blood calcium level in an effort to produce more hormone. The result is a mobilization of large amounts of calcium and phosphate from the bones, greatly weakening them.

Recent but unconfirmed evidence has led to the description of a new parathyroid hormone called *calcitonin.* It is a hypocalcemic factor; that is, it lowers blood calcium. A slight increase above the normal serum calcium level apparently causes the secretion of calcitonin. This in turn leads to a prompt but transient drop in the plasma calcium level.

Control of Parathyroid Secretion. Unlike the thyroid gland, the parathyroids are not under the control of any anterior pituitary hormone. In other words, there is no parathyrotropic hormone. Rather, parathyroid secretion is controlled by the level of calcium in the blood. In this way, an increase in blood calcium decreases parathyroid activity, whereas a drop in blood calcium stimulates the glands (Fig. 10–9). Vitamin D is not considered to affect parathyroid activity; it does, however, increase reabsorption of calcium from the intestinal tract.

THE PANCREATIC ISLET CELLS

Description. The islet cells were described when the pancreas was discussed with the digestive organs. Two major types of cells, *alpha* and *beta,* make up the islets (Fig. 10–10). These masses of tissue are scattered throughout the pancreatic acinar cells, which secrete the pancreatic digestive juice. Each type of islet cells makes one of the two hormones secreted by the islets.

Hormones

INSULIN. Insulin is the hormone produced by the beta cells. It is a protein whose chemical formula is known. Insulin has three basic effects on carbohydrate metabolism: (1) It increases the rate of glucose metabolism. (2) It decreases the amount of glucose in the blood. (3) It increases the amount of glycogen stored in the tissues.

Although it is true that glucose can be metabolized and glycogen can be stored without insulin, these two processes are severely hampered by insulin

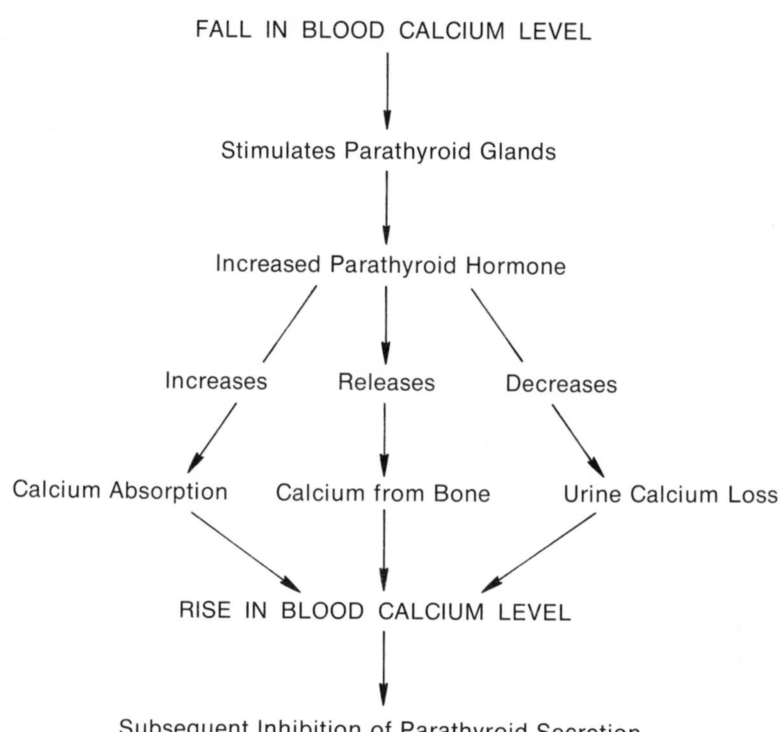

Figure 10–9. Parathyroid control.

deficiency. For example, insulin enables the body to change glucose to fat so that the fat can be stored. It also enables the body to burn available glucose for energy instead of burning protein or fat; this function is the so-called "sparing" action of insulin. Finally, without insulin, glucose goes into liver cells so slowly that it is metabolized instead of being converted to glycogen and stored. The mechanism of oral hypoglycemic agents such as orinase is not known.

Hypoinsulinism gives rise to the disease known as *diabetes mellitus*, which is the most common endocrine disorder known. Its severity depends upon the degree of insulin deficiency. It is a metabolic disease which affects many body functions. One sign of diabetes mellitus is an abnormally high blood sugar level, or *hyperglycemia.* Hyperglycemia in turn causes sugar to spill into the urine, a condition called *glycosuria.* Because it is unable to burn enough sugar to satisfy energy needs, the body turns to burning fats and protein. Excessive fat burning leads to *ketosis* (see Chapter 8), or the piling up of fat breakdown products in the body. Other body functions are also affected in diabetes. Diabetics are very susceptible to infections and the early onset of blood vessel disease.

Hyperinsulinism, or the secretion of too much insulin by the beta cells, is usually caused by a tumor of the islet cells. In such cases, the blood sugar is depressed and may drop low enough to cause fainting, coma, and convulsions.

GLUCAGON. The alpha cells of the islets secrete glucagon. The effect of this hormone is antagonistic to that produced by insulin; namely, glucagon causes an increase in the blood sugar level. (This relationship is similar to that

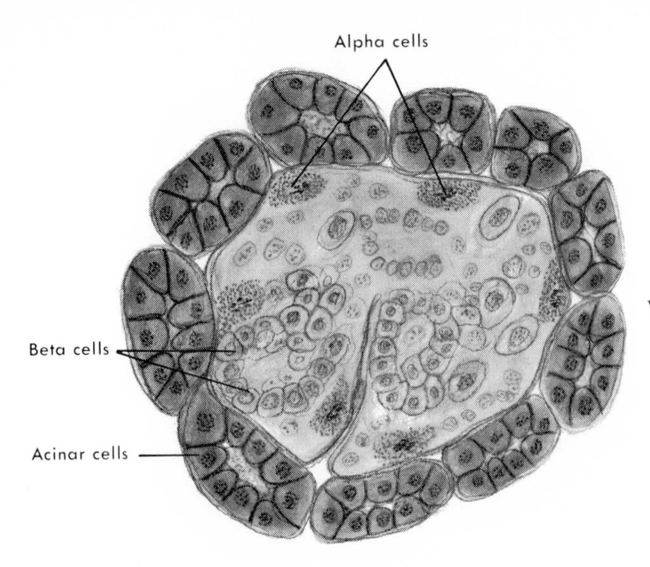

Figure 10–10. Microscopic view of pancreatic cells.

between parathyroid hormone and calcitonin). It does so by breaking down liver glycogen into glucose (glycogenolysis). In this way, glucagon prevents the blood sugar from dropping too low.

Control of Islet Cell Secretions. Insulin secretion is not stimulated by the direct action of the pituitary gland. The beta cells respond to the blood sugar level, increasing their secretion when this level rises above normal limits. In addition, hormones secreted by other endocrine glands have effects on the blood glucose level, which in turn affects the rate of insulin secretion. It has already been noted that some of the anterior pituitary hormones, such as GH, LTH, ACTH, and TSH, are diabetogenic in action. If animals are injected with sufficient quantities of these hormones over a period of time, they will develop true diabetes. The secretion of glucagon, on the other hand, is apparently under the control of anterior pituitary growth hormone.

THE ADRENAL GLANDS

Description. The adrenal glands are paired organs lying on the superior poles of the kidneys and are often referred to as the *suprarenal glands*. Roughly triangular in shape, each has an outer *cortex* and an inner *medulla* which function as separate glands (much like the two lobes of the pituitary) (Fig. 10–11).

The Medulla and Its Hormones. The adrenal medulla develops from the same part of the neural tube as the sympathetic nervous system. Therefore, it is not surprising that its hormones "mimic" sympathetic stimulation and thus are said to be *sympathomimetic* in action. Unlike some of the endocrines, the adrenal medulla is not necessary for life, since most of its functions can be performed by the sympathetic nervous system.

The medulla secretes two hormones, epinephrine and norepinephrine. *Epinephrine* is the major hormone, making up 80 to 90 per cent of the gland's secretions. Functions of both hormones are listed as being either *alpha* or *beta*.

Alpha Functions	Beta Functions
Constriction of skin blood vessels	Dilatation of skeletal muscle blood vessels
Pupil dilatation	Bronchodilatation
Piloerection (hairs "standing on end")	Heart muscle stimulation
Intestinal muscle relaxation	

Epinephrine has both alpha and beta functions. Norepinephrine has all the alpha functions plus one beta function, that of heart muscle stimulation. Consequently, epinephrine is the drug of choice for treatment of asthmatic attacks, since it is a bronchodilator and norepinephrine is not. Both are valuable for increasing the cardiac output and heart rate, since both are myocardial stimulants.

Both hormones have other important uses. Epinephrine is often mixed with anesthetics, such as Novocaine, to prolong their activity. It is also used as a pupil dilator for eye examinations and as a treatment for sensitivity or allergic reactions to drugs. Because it is a powerful vasoconstrictor, norepinephrine is useful for nasal congestion and circulatory shock.

In addition to mimicking the actions of the sympathetic nervous system, the medullary hormones increase the activity of all body cells including those not supplied with sympathetic nerve fibers. Moreover, although the effect of sympathetic stimulation appear more rapidly, the effects of these hormones last longer once they appear in quantity in the blood.

Secretion of medullary hormones is stimulated by the sympathetic nerve supply to the medulla. This stimulation can be caused by trauma, pain, exercise, hypoglycemia, and by many drugs, but not by alcohol or barbiturates.

The Cortex and Its Hormones. The adrenal cortex, unlike the medulla, is essential for life. It is composed of three concentric cellular zones, each producing different hormones. Cortical hormone functions bear no direct relationship to the medullary hormones.

Cortical hormones are all steroid in nature, and as a group are called *corticosteroids*. Although more than 30 such compounds have been isolated from cortical tissue, only three are recognized as the gland's principal hormones. These three are *cortisol, corticosterone,* and *aldosterone.* In addition, small amounts of both male and female sex hormones are produced by the innermost zonal cells.

Glucocorticoids. Cortisol and corticosterone are primarily concerned with carbohydrate metabolism and thus are called *glucocorticoids* (Fig. 10–12). They stimulate gluconeogenesis by the liver, causing a six- to ten-fold increase in the conversion of amino acids to glucose. In addition, they delay the rate of glucose utilization by the body cells. These two effects cause an increase in the

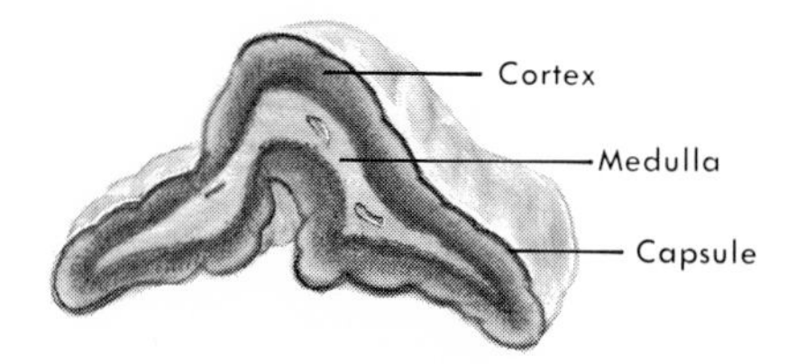

Figure 10–11. Cross section of the adrenal gland.

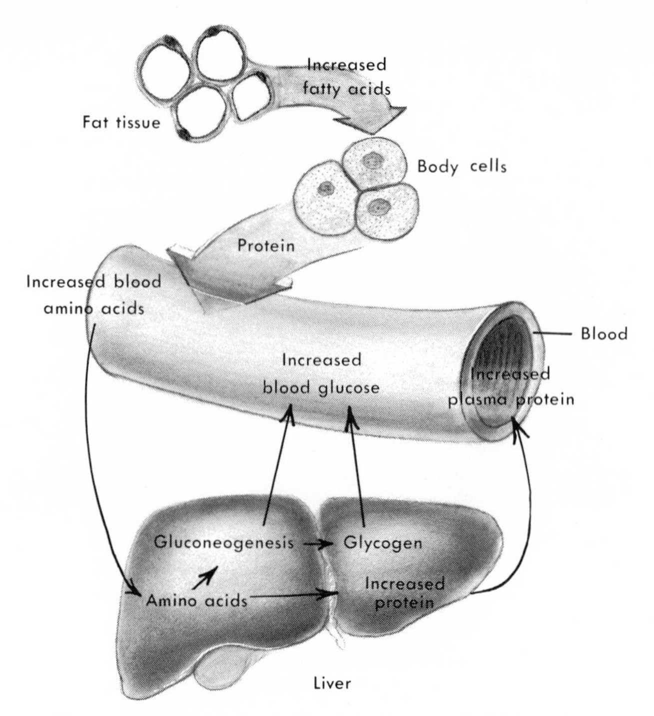

Figure 10–12. Diagram showing glucocorticoid functions.

blood sugar. Thus, if these actions are prolonged by giving excessive amounts of glucocorticoids, *adrenal diabetes* may result.

Glucocorticoids also decrease the protein stores in the body cells, thus increasing amino acids in the blood. The amino acids in turn are used by the liver to make plasma proteins. These hormones also affect fat metabolism by accelerating the processes by which fats are both stored and removed from storage. Thus, if glucocorticoids are given in excess to persons on an adequate diet, these individuals tend to become overweight. In the body, secretion of glucocorticoids follows when RF from the hypothalamus stimulate ACTH release by the anterior pituitary. ACTH then stimulates the adrenal cortex to produce glucocorticoids.

There has been a great deal of interest in the relationship between glucocorticoid secretion and response to stress. When the body is subjected to such stressful conditions as surgery, intense heat or cold, severe burns, or pregnancy, it has been determined that the adrenal cortex increases its release of glucocorticoids. The total significance of this reaction is not clear, but it helps to explain why persons with poorly functioning adrenal glands are helped through periods of stress by injections of these hormones.

Glucocorticoids are also useful in the treatment of inflammatory diseases such as rheumatic fever, arthritis, and acute kidney disease. Here again, the mechanism of action is not clear, but it may be similar to that which is involved in the stress reaction.

MINERALOCORTICOIDS. The principal hormone of the adrenal cortex that is involved in mineral balance and water metabolism is *aldosterone*. The main actions of aldosterone are to increase reabsorption of sodium by the kidney tubules, to decrease tubular reabsorption of potassium, and to increase the entrance of sodium into cells and simultaneously transfer potassium out. Since water is reabsorbed along with the sodium, aldosterone also helps control water balance in the body.

The secretion of aldosterone is not regulated by ACTH. In the normal person, enough aldosterone is secreted to maintain the proper level of sodium and potassium in the extracellular body fluids and to maintain normal extracellular fluid volume (Fig. 10–13). Thus evidence indicates that aldosterone secretion is controlled by the blood volume. When the extracellular sodium level and the fluid volume both drop below normal and when the potassium level is too high, nerve centers in the brain are stimulated. These nerve centers in turn relay impulses to the adrenal cortex, causing it to release additional aldosterone to the blood. This response is rapid, occurring within minutes of its initiation.

The structures thought responsible for detecting a fall in fluid volume are certain cells lining the blood vessels in the kidneys. These detector cells respond by secreting a substance called *renin* into the blood. Renin then reacts with a blood substance (angiotensinogen) to form *angiotensin*. The blood carries angiotensin to the adrenal cortex, stimulating it to secrete aldosterone. The hormone, in turn, acts upon the kidney tubules to decrease the excretion of sodium and water, thus increasing the extracellular fluid volume.

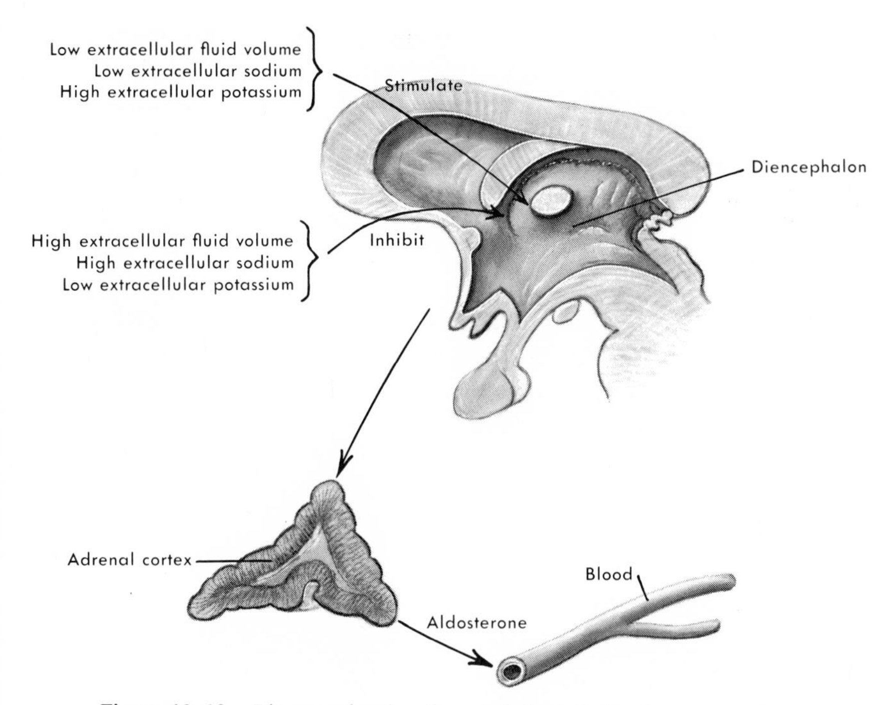

Figure 10–13. Diagram showing the regulation of aldosterone secretion.

Sex Hormones. The normal adrenal cortex secretes certain amounts of the male sex hormones called *androgens*. In addition, it also produces very small amounts of the female sex hormones. Sex hormones probably exert very little influence upon the physiology of the normal person. Difficulties can arise, however, if the cortex begins to secrete large amounts of these hormones.

The Relationship of the Cortex to Other Endocrines

Relationship to the Thyroid. Thyrotropic hormone and ACTH are secreted by the same cells in the anterior pituitary. If a glucocorticoid is given to an individual, it causes a decrease in the amount of ACTH secreted by the pituitary. This decrease results in increased TSH secretion by these pituitary cells, which in turn causes increased secretion of thyroxine by the thyroid gland. On the other hand, giving thyroxine will decrease pituitary TSH secretion while increasing ACTH output. The effect in this case is an increase in glucocorticoid secretion by the adrenal cortex.

Relationship to the Gonads (Sex Glands). Complete absence of adrenal cortical secretion will cause atrophy (reduction in size) of both male and female gonads. The atrophy can be explained by the fact that the anterior pituitary cells which secrete ACTH and TSH also produce FSH and LH. When the adrenal cortex is not functioning, ACTH secretion is greatly increased, and thus FSH and LH secretions are decreased. Without these gonadotropic hormones, the sex glands atrophy.

Abnormal Adrenal Cortex Function

Hypoadrenalism. Hypoadrenalism results from the failure of the cortex to produce corticoid hormones and gives rise to the condition known as *Addison's disease*. The classic picture of this disorder was described by Thomas Addison, a nineteenth century English physician. At that time the disease was commonly caused by tuberculosis of the adrenal glands, resulting in their bilateral destruction. The person with Addison's disease is anemic and very weak, has bronzing of the skin, and is highly susceptible to infections. Severe shock and death follow if the proper hormones are not administered. Today, the most frequent causes of Addison's disease are *idiopathic*—that is, of unknown origin. In a certain number of cases, cancer is responsible, especially cancer which spreads from the bronchi of the lung.

Hyperadrenalism. Hyperfunction of the adrenal cortex results in Cushing's disease and is generally caused by enlargement of both adrenal glands, most often because of a tumor. The individual with Cushing's disease demonstrates the effects of increased secretion of glucocorticoids, mineralocorticoids, and sex hormones. The disease occurs most often in the adult female. The upset in protein metabolism leads to breakdown of the body tissues and weakening of the bones. Increased secretion of glucocorticoids causes an increased blood sugar leading to adrenal diabetes, which may become permanent diabetes if it continues for a period of time.

The increase in male sex hormone secretion results in masculinizing effects in the female, with the growth of a beard, deepening of the voice, and even baldness. Electrolyte and water imbalance lead to high blood pressure and puffiness around the face. Treatment for Cushing's disease consists of the surgical removal of excess adrenal cortical tissue.

THE GONADS

The male and female sex glands produce hormones and so are included in the endocrine system. However, since they are such integral parts of the reproductive systems, they will be discussed in the following chapter.

THE THYMUS

The thymus is a flat, two-lobed organ lying in the chest cavity behind the sternum. It is relatively large in childhood but stops growing when adolescence begins. It then progressively atrophies with age, all but disappearing in the adult. Because it obviously fulfills its duties early in life, its exact function has been a mystery until recent years. Since it was previously considered to have some endocrine function, it was traditional to list the thymus with the endocrine glands.

It is now known that the thymus has the job of producing lymphocytes early in life. Moreover, most authorities agree that the thymus is the primary source of these cells, which are then sent by the gland into the blood and eventually to the spleen and lymph nodes. From these primary cells come the cells responsible for the body's development of immunity to disease. That is, they produce *antibodies*, substances which help fight infections from bacteria and viruses.

In defense of the hormone theory, however, recent experimental evidence has shown that the mouse thymus produces a hormone-like factor. This factor is sent from the thymus to the spleen and lymph nodes of the animal, stimulating these organs to produce lymphocytes from their own cells. Thus, it may be that the thymus plays a dual role, acting both as lymphoid tissue and as an endocrine organ.

OUTLINE SUMMARY

1. Introduction
    a. Endocrines are ductless glands; exocrine glands have ducts
    b. Endocrines pour their hormones directly into the blood
    c. Endocrines help regulate body functions
2. Properties of Hormones
    a. Most hormones are either proteins or steroids
    b. Hormones are endogenous (not supplied by food)
    c. Hormone deficiencies cause metabolic diseases
3. Types of Endocrine Glands
    a. Pure endocrine glands produce only hormones
    b. Mixed glands produce both hormones and exocrine secretions
    c. Solid endocrine glands store their secretions in spaces between the cells
    d. Storage glands have spaces surrounded by cells for hormone storage
    e. Endocrine stimulation is under:
        1) Nervous control
        2) Humoral control
        3) Reciprocal release or feedback control
4. The Pituitary Gland (The Hypophysis)
    a. Located in sella turcica of sphenoid bone
    b. About the size of a small pea; has anterior, posterior, and intermediate lobes

c. Anterior lobe hormones
  1) Growth hormone (GH) – increases growth rate of all body cells
    a) Hypofunction – dwarfism in the young; cachexia in the adult
    b) Hyperfunction – gigantism in the young; acromegaly in the adult
  2) Adrenocorticotropic hormone (ACTH) – function and maintenance of adrenal cortex
    a) Hypofunction – atrophy of the adrenal cortex
    b) Hyperfunction – secondary effects of Cushing's disease
  3) Thyrotropic hormone (TSH) – normal activity of thyroid gland
    a) Hypofunction – depressed thyroid function and atrophy
    b) Hyperfunction – Graves' disease (exophthalmic goiter)
  4) Follicle-stimulating hormone (FSH) – increased growth of ovarian follicular cells; germinal-stimulating hormone (GSH) – stimulates sperm-producing cells in testis
  5) Luteinizing hormone (LH) – growth of ovarian cells producing progesterone; interstitial cell-stimulating hormone (ICSH) – stimulates cells producing testosterone
  6) Luteotropic hormone (LTH) – growth of mammary duct system in pregnant female

d. Panhypopituitarism – decreased secretion of all anterior lobe hormones

e. Posterior lobe hormones
  1) Vasopressin (Pitressin) – antidiuretic hormone (ADH) promotes reabsorption of water through kidneys; lack of ADH causes diabetes insipidus
  2) Oxytocin (Pitocin) – causes contraction of the pregnant uterus

f. Control of pituitary secretions
  1) Anterior lobe – reciprocal release mechanism; hypothalamic releasing factors
  2) Posterior lobe – hypothalamus actually produces hormones which are stored in the posterior lobe

5. The Thyroid Gland
  a. Located in anterior aspect of neck, below the thyroid cartilage
  b. Formed by two lateral lobes and an isthmus
  c. Composed of follicles which store colloid
  d. Main hormone is thyroxine – increases metabolism of all body cells
    1) Hypothyroidism – cretinism in child; myxedema in adult
    2) Hyperthyroidism – Graves' disease with exophthalmos
  e. Goiter – any enlargement of the thyroid
    1) Parenchymatous type – result of iodine deficiency
    2) Colloid type – result of decreased TSH secretion
    3) Exophthalmic type – results in excess thyroxine production

6. The Parathyroid Glands
  a. Usually four in number; lie on the thyroid gland
  b. Solid masses of epithelial cells
  c. Hormone – regulates calcium and phosphorus metabolism
    1) Hypoparathyroidism – decreased blood calcium causes tetany
    2) Hyperparathyroidism
      a) Primary – tumor causes overproduction of hormone which removes calcium and phosphorus from the bones
      b) Secondary – hypertrophy of the glands caused by low blood calcium

   d. Calcitonin – hypocalcemic factor
   e. Control of parathyroid secretion – by blood calcium level
7. The Pancreatic Islet Cells
   a. Alpha and beta cells in the pancreatic islet tissue
   b. Two hormones
      1) Insulin – produced by the beta cells
         a) Regulates carbohydrate metabolism
         b) Hypoinsulinism – causes diabetes mellitus, with hyperglycemia, glycosuria, ketosis
         c) Hyperinsulinism – depressed blood sugar leads to fainting and coma
      2) Glucagon – produced by alpha cells; causes increase in blood sugar level
   c. Control of islet cell secretions
      1) Insulin – secreted in response to increased blood sugar
      2) Glucagon – controlled by growth hormone
8. The Adrenal Glands
   a. Paired organs lying on the superior poles of the kidneys
   b. Composed of an inner medulla and an outer cortex
   c. Medulla and its hormones
      1) Epinephrine (major hormone) – has all alpha and beta functions (see p. 225)
      2) Norepinephrine – has all alpha functions and one beta function
      3) Both hormones mimic the sympathetic nervous system
   d. Cortex and its hormones (corticosteroids)
      1) Glucocorticoids – cortisol and corticosterone
         a) Stimulate gluconeogenesis and rate of glucose utilization
         b) Secretion stimulated by ACTH via the hypothalamus
         c) Uses – to treat stress reactions and inflammatory diseases
      2) Mineralocorticoids – aldosterone
         a) Involved in mineral balance and water metabolism
         b) Secretion regulated by extracellular sodium and fluid volume which stimulate brain centers
      3) Sex hormones – small amounts of male and female hormones normally are produced
   e. Abnormal cortical functions
      1) Hypoadrenalism – causes Addison's disease
      2) Hyperadrenalism – causes Cushing's disease
9. The Gonads – discussed in Chapter 11
10. The Thymus
   a. Organ in the chest cavity which is large in children but atrophies with age
   b. Function is to produce lymphocytes early in life
   c. Possible hormone activity

REVIEW QUESTIONS

1. How do hormones differ from vitamins? From glucose?
2. Name (1) a mixed gland, (2) a storage type endocrine gland, (3) a solid type endocrine gland.
3. What is meant by reciprocal release? Give an example.

4. What is meant by the diabetogenic effect of a hormone? List some hormones having this effect.
5. What are the effects of panhypopituitarism?
6. What is diabetes insipidus? What causes it?
7. What role does the hypothalamus play in pituitary secretion?
8. How does the thyroid gland make its hormone?
9. Describe the three types of goiter.
10. What endocrine disorder is caused by a low calcium diet?
11. Describe the effects of insulin on carbohydrate metabolism.
12. What controls pancreatic islet secretions?
13. What is the drug of choice for treatment of an acute asthmatic attack? Why?
14. How can the administration of cortisol cause overweight?
15. What are glucocorticoids used for?
16. Explain the regulation of aldosterone secretion.
17. What is Addison's disease? Cushing's disease?
18. Explain why the thymus probably should not be grouped exclusively with the endocrine glands.
19. Explain what is meant by the antagonistic activity of parathyroid hormone and calcitonin.

ADDITIONAL READING

Burnet, Sir M.: The thymus gland. Sci. Amer. (Nov.), 1962.
Rasmussen, H.: The parathyroid hormone. Sci. Amer. (April), 1961.
Rawson, R. W.: The thyroid gland. Clinical Symposia, Ciba Pharmaceutical Company, Vol. 17, No. 2., 1965.

Chapter 11

THE REPRODUCTIVE SYSTEMS

INTRODUCTION

A chief characteristic of all living things is the ability to reproduce themselves. Simpler forms of life reproduce by a direct division of their substance into two equal portions, each of which becomes a new individual. Reproduction in higher forms involves more complex mechanisms in order to produce new members of the species.

Sexual reproduction is the method by which man and the higher animals perpetuate themselves. Each of the two sexes is equipped with a *gonad*, which provides a reproductive cell, and a set of ducts and other *accessory organs*. Each structure is responsible for a particular function, the sum total of which is the production of a live offspring.

MALE REPRODUCTIVE SYSTEM (Fig. 11–1)

MALE GONAD

The *testis* is the male gonad. The paired testes are two oval bodies which begin their development in the embryo's abdominal cavity. Toward the end of fetal life they begin to descend from their original position and finally take their place in the *scrotum*. The functions of the testes are to produce the *spermatozoa*, or male sex cells, and the male hormones. Each testis is composed of a large number of tightly coiled seminiferous tubules, which are separated into 250 lobules by connective tissue (Fig. 11–2). The epithelial cells of the walls of these tubules are the *germinal cells*, which produce the spermatozoa. Between the

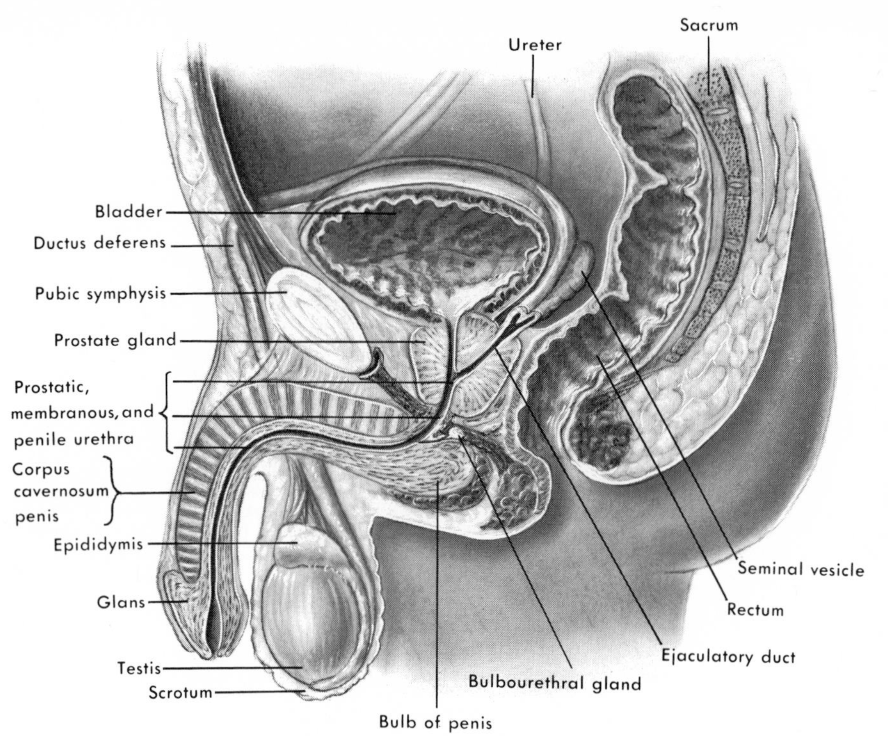

Figure 11–1. The male reproductive system.

seminiferous tubules lie the *interstitial cells*, which produce the *androgens*, or male hormones.

Testosterone. The most important of the androgens is *testosterone*. It is responsible for the normal growth, development, and function of the male reproductive organs. It is necessary for the development and maintenance of secondary sex characteristics in the male. As the young male enters puberty (adolescence), increasing amounts of testosterone are produced by the testicular cells. This hormone causes the appearance of such secondary sex characteristics as voice change, beard growth, increased strength and tone of muscles, increased bone growth, increased physical endurance, and increased cardiac output and hemoglobin level. In addition, there is an increase in the growth of the accessory sex organs.

If the testes are removed (castration) from an immature male, these changes do not occur, and the individual is said to suffer from *infantilism*. Castration of the mature male results in involution (regression) of the sex organs with possible loss of the sex drive.

Production of Sperm. Each seminiferous tubule measures 1 to 3 feet in length, and both testes together contain 275 yards of tubules. The lining cells of the tubules are involved in *spermatogenesis*, or the production of spermatozoa (Fig. 11–3). The initial germ cells, or *spermatogonia*, multiply and produce a new generation of cells, called *primary spermatocytes*. Each of these divides to form two smaller *secondary spermatocytes*. Each secondary spermatocyte then divides, giving

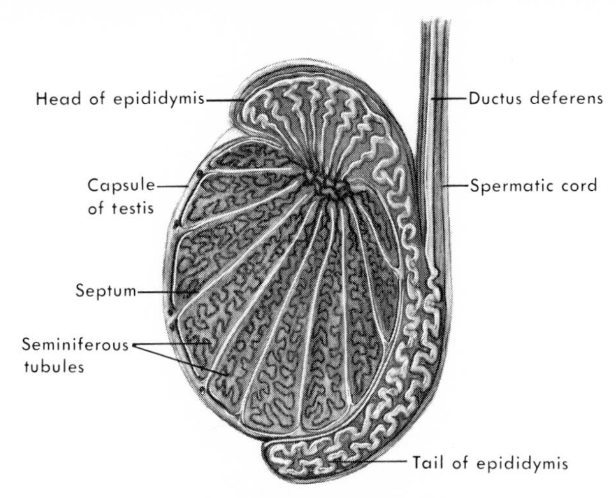

Figure 11-2. Testis and epididymis.

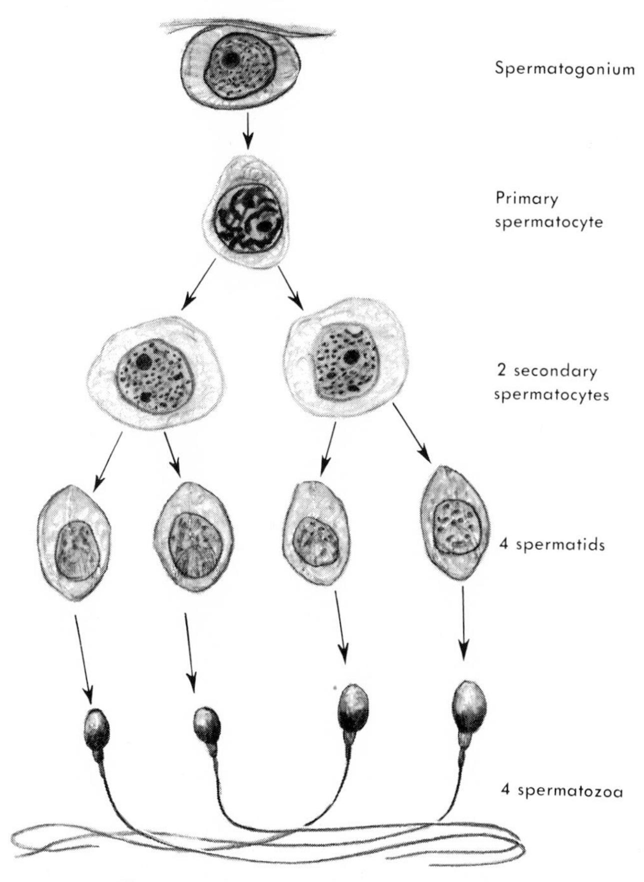

Figure 11-3. Formation of spermatozoa.

rise to two *spermatids*. Spermatids do not divide but become transformed into spermatozoa. Thus, each primary spermatocyte gives rise to four spermatozoa, each of which is capable of fertilizing a mature egg cell.

ACCESSORY ORGANS

In the male, the accessory organs of reproduction include a series of ducts plus several exocrine glands.

Ducts. The *epididymis* is a 20-foot-long coiled tube formed by the fusion of several tubules into which the seminiferous tubules empty (Fig. 11–2). It fits over the superior end and posterolateral border of the testis, its tail nearly reaching the inferior end of the gonad. The tail straightens out to form the ductus deferens, or sperm duct.

The *ductus deferens*, the direct continuation of the duct of the epididymis, ascends from the testis, passes out of the scrotum, and traverses the abdominal wall in a region called the *inguinal canal*. As it does so, it is joined by blood vessels, nerves, lymphatics, and tissue wrappings from the abdominal layers—all of which are collectively called the *spermatic cord* (Fig. 11–4).

After entering the abdominal cavity, the ductus deferens loses its wrappings and passes to the back of the urinary bladder (Fig. 11–5). As it swings anteriorly around the bladder, the ductus deferens is joined by the *ejaculatory duct*. The two ejaculatory ducts are formed by the union of the ducts of the seminal vesicles with the narrowed ends of the ductus deferentes. This union takes place within the prostate gland. Here each duct opens into the first part of the urethra. From this point on, the urinary and reproductive systems in the male share the urethra as a common passageway.

The Urethra. The male urethra is about 20 cm. long and serves for the passage of both urine and reproductive fluid. Many small urethral glands add their alkaline secretions to the contents of the tube. Pursuing a twisting course after it leaves the urinary bladder, the urethra may be divided into three regions or portions:

The *prostatic urethra* is the first portion. It descends through the prostate gland in a slightly forward curve. It receives the prostatic secretions plus the contents of the ejaculatory ducts.

The *membranous urethra* is the second portion. It pierces the urogenital diaphragm behind the symphysis pubis. The fibers of the external sphincter of the urinary bladder surround it.

The *penile urethra* is the third and longest of the three regions. It will be discussed with the penis.

Glands. The reproductive fluid of the male is called *semen*. It is composed of the spermatozoa from the testes plus fluid which is added by glands. The *seminal vesicles* are two blind pouches lying on either side of the urinary bladder. They secrete an alkaline fluid which passes out of them through short ducts. These ducts join with the ductus deferentes to form the ejaculatory ducts as mentioned previously.

The partly muscular, partly glandular *prostate gland* lies under the urinary bladder. It is roughly cone-shaped with its base adjoining the inferior aspect of the bladder. The urethra enters the prostate gland near the middle of its base, and there the urethra receives the contents of the ejaculatory ducts. Glandular

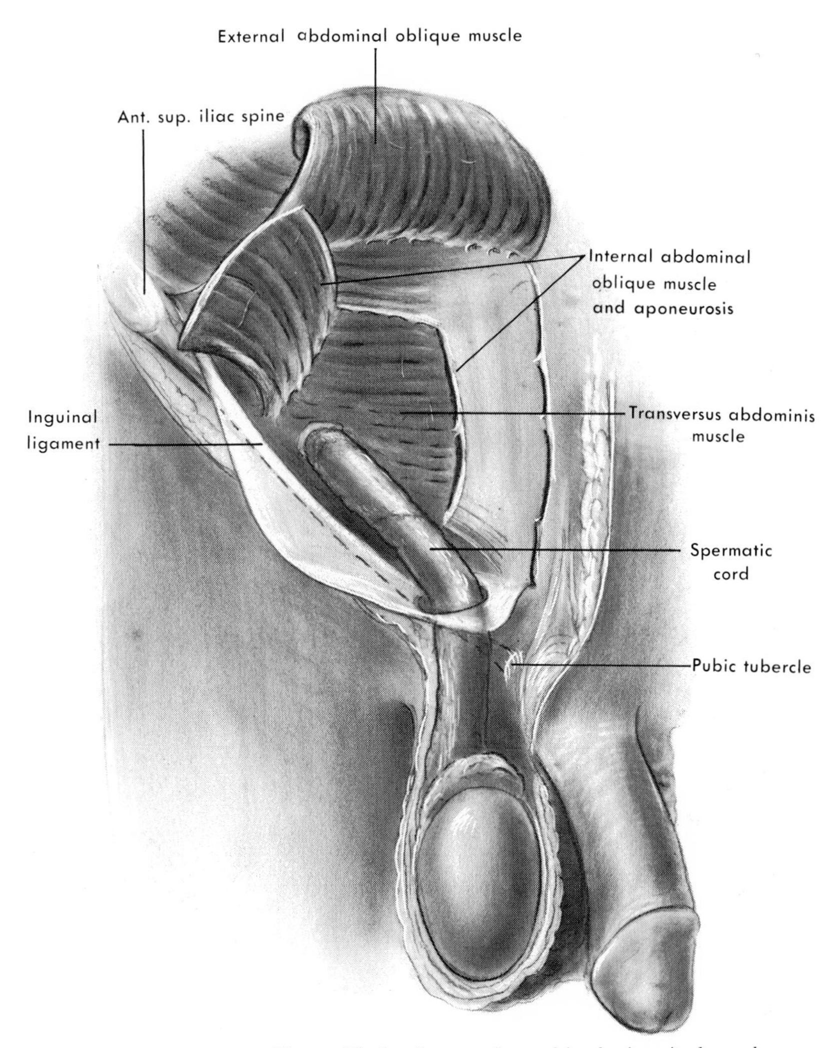

Figure 11–4. Spermatic cord in the inguinal canal.

cells of the prostate gland produce a sticky secretion which is added to the semen.

The paired *bulbourethral glands* (Cowper's) are pea-sized organs lying on either side of the membranous urethra with their ducts opening into it. These glands produce an alkaline fluid which, along with that from the seminal vesicles, helps to neutralize the acidity of the urethra.

EXTERNAL GENITALIA

The *scrotum* is the sac which contains the testes. It is developed from the layers of the abdominal wall. By a midline septum it is divided into two chambers, one for each testis. The scrotum contains muscle fibers which contract during exercise or exposure to cold temperatures, causing it to become smaller and wrinkled.

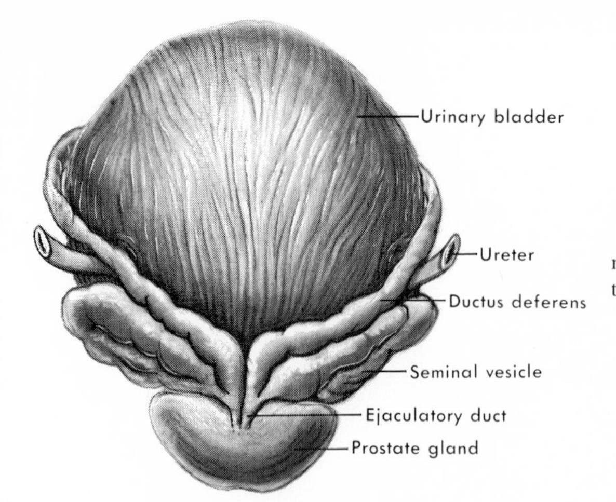

Figure 11-5. The bladder, seminal vesicles, and prostate gland—posterior view.

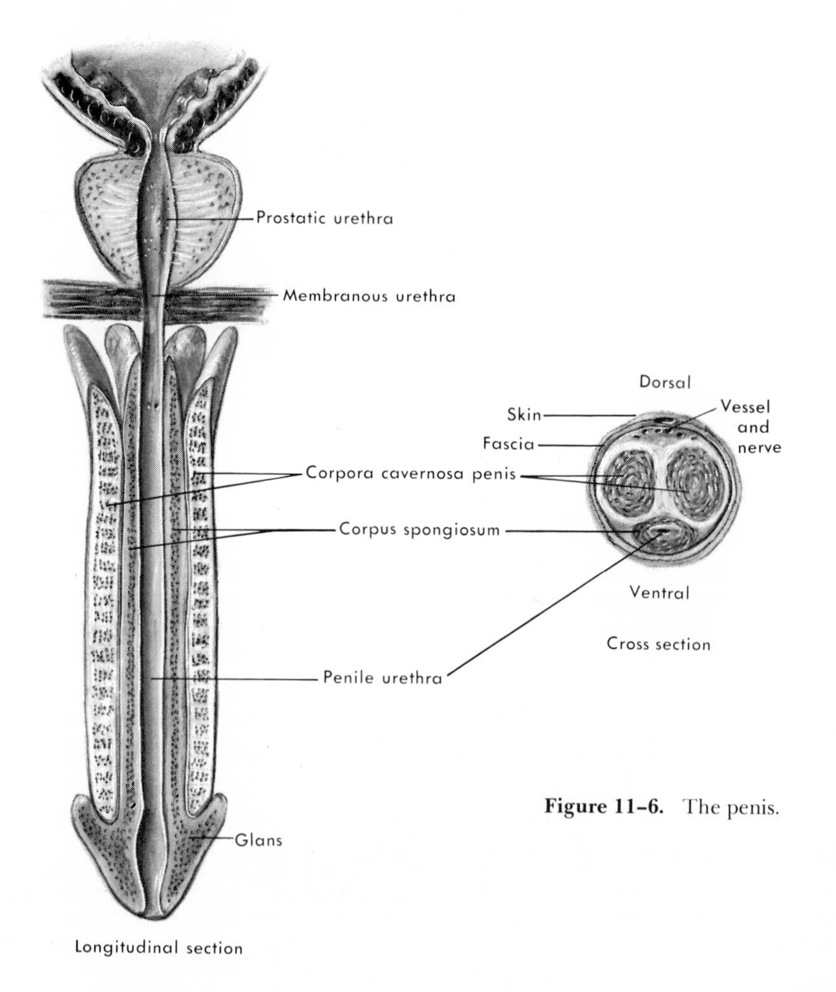

Figure 11-6. The penis.

The *penis* is the intromittent organ of the male reproductive system. That is, it conveys the semen from the male into the vagina of the female. It is composed of *cavernous* or *erectile* tissue and is traversed by the spongy portion of the urethra (penile urethra) (Fig. 11–6). The three portions of the penis are the *shaft*, the *glans*, and the *prepuce*.

In cross section the shaft shows three cylinders of erectile tissue. The two dorsolateral *corpora cavernosa penis* and the ventral *corpus spongiosum* contain a large number of small spaces which are connected to the vascular system. Under nervous stimulation the veins which drain these cavernous spaces become compressed, and they remain filled with venous blood, causing the organ to become enlarged and rigid.

The penile urethra traverses the corpus spongiosum. The distal cone-shaped glans is formed by an expansion of the corpus spongiosum. The prepuce is a free fold of skin along the base of the glans. It is the tissue removed in circumcision.

In summary, the spermatozoa proceed along the following course (Fig. 11–7):

Seminiferous tubules (of testes) → epididymides → ductus deferentes → ejaculatory ducts → prostatic urethra → membranous urethra → penile urethra → penile meatus.

Ejaculation, or the actual expulsion of semen from the urethra, is caused partly by contraction of the smooth muscle of the prostate and seminal vesicles. It also is due in part to peristaltic contractions in the testes, epididymides, and ductus deferentes. Though spermatozoa can live for several weeks in the reproductive ducts of the male, they have a life span of only 24 to 72 hours in ejaculated semen.

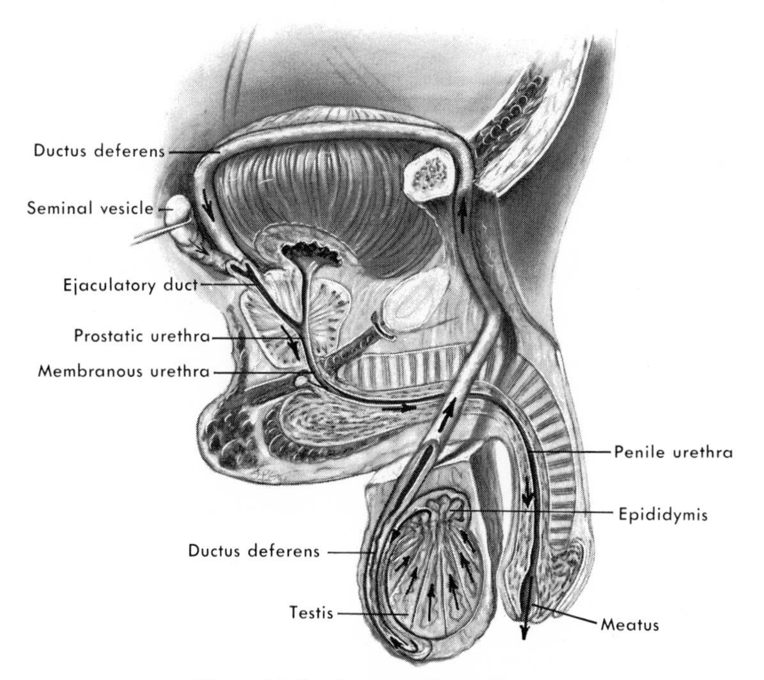

Figure 11–7. Passage of spermatozoa.

FEMALE REPRODUCTIVE SYSTEM (Fig. 11–8)

FEMALE GONAD

The paired *ovaries* are flattened oval bodies lying against the lateral walls of the pelvis. Enclosed in an extension of the *broad ligament* of the uterus, each ovary lies in a shallow depression of parietal peritoneum. The *suspensory ligament* conducts blood vessels and nerves to the gonad.

Oögenesis is the course of differentiation of the female sex cell. It takes place in three stages which are as follows (Fig. 11–9):

Proliferation. Germinal epithelium covers the surface of the ovary, with connective tissue lying immediately below it. The germinal cells begin proliferating during fetal life to produce egg cells, or *ova*. These cells sink into the cortex of the ovary and continue to multiply. Later in fetal life a layer of other epithelial cells encloses the egg cells to form *primary follicles.*

Shortly after birth, the formation of ova stops—the total at this time numbering from 40,000 to 300,000. With occasional exceptions, there is no advanced follicular development beyond the primary stage until the onset of puberty. From that time until some 30 years later, larger follicles in various growth stages are always present in the ovary. Several years after the end of the child-bearing span, follicles are no longer seen.

Growth. During the growth stage of oögenesis, a crop of enlarging follicles is produced during each menstrual cycle. Most of these follicles fail to reach ma-

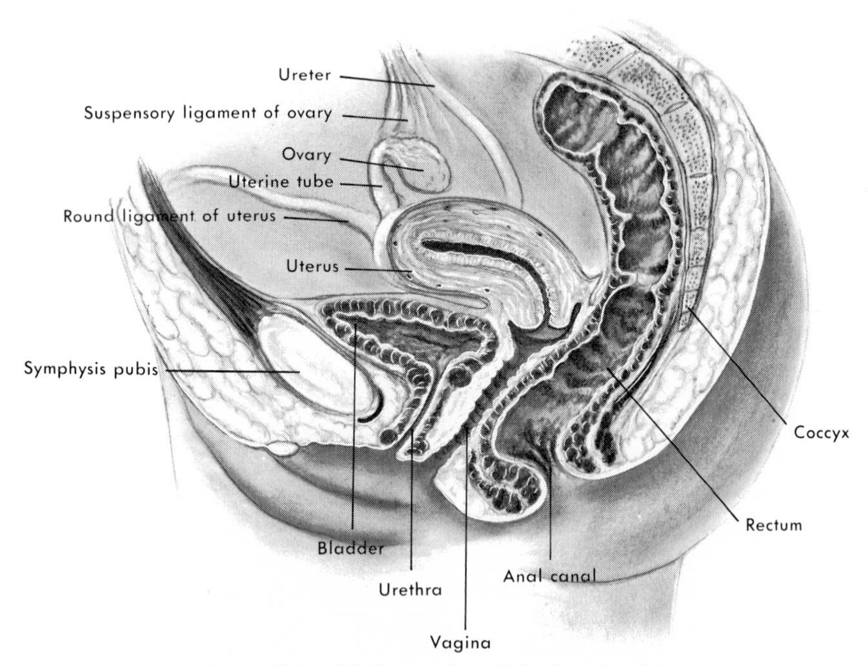

Figure 11–8. Median section of the female pelvis.

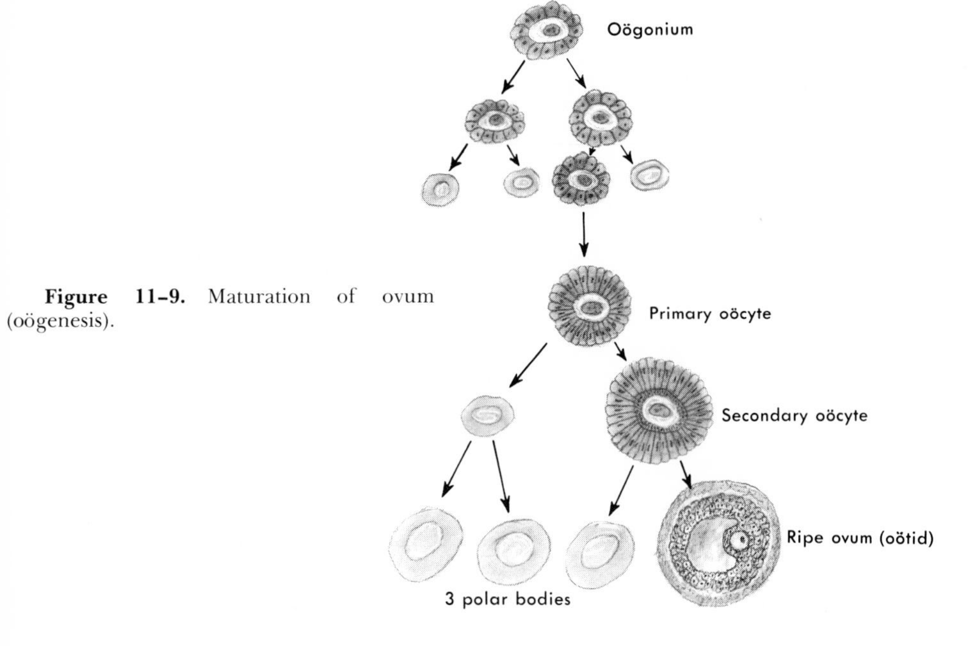

Figure 11–9. Maturation of ovum (oögenesis).

turity and regress by a process called *atresia*. Of the original number of ova in the ovary, only some 300 to 400 will mature during the female reproductive years. The rest of the ova suffer atresia, and thus follicles in various stages of regression are present in the ovary during these years.

At the beginning of growth, the human ovum measures 0.02 mm. in diameter and has one layer of follicular cells surrounding it. When the egg cell grows to a diameter of 0.14 mm., it is called the *primary oöcyte.* As growth continues, follicular cells proliferate and form a layer several cells thick around the ovum. Fluid spaces begin to appear between the follicular cells. These unite to form a fluid-filled cavity in the follicle called the *antrum.* As the follicle grows, the egg cell moves from its center and becomes eccentrically buried in a mound of follicular cells.

The follicle finally is converted into a *mature* (graafian) *follicle*, which is a sac filled with follicular fluid and surrounded by a membrane called the *theca.* During the final stages of growth, the follicle approaches the surface of the ovary and eventually raises a blister-like elevation on it. A full grown human follicle has a diameter of 9 to 12 mm. and usually contains only one ovum.

Maturation. During two specialized maturation divisions, each of the chromosomes splits only once, so that each of the four cells formed has a reduced but complete single set of chromosomes replacing the double set of the primary oöcyte. Although the nuclei of the cells are equal, the cytoplasm divides unequally. The result is one large, ripe ovum and three rudimentary ova called polar bodies.

HORMONES

The primary female hormones are the *estrogens*, which are produced by the thecal cells surrounding the follicle. Estrogens, of which the most important is *estradiol*, have several functions. They inhibit follicle-stimulating hormone (FSH) secretion and stimulate luteinizing hormone (LH) secretion by the anterior pituitary. They induce ovulation and cause growth of the uterus, uterine tubes, and ducts of the mammary glands. Estrogens are responsible for the development of secondary sex characteristics in the female. Growth in the female usually ceases several years earlier than in the male because estrogens cause earlier closing of the epiphyseal plates in the long bones. These hormones cause increased deposition of fat in the subcutaneous tissues, especially in the breasts, thighs, and buttocks. They are also responsible for the growth of pubic hair and the development of a smooth, soft-textured skin.

Progesterone is the female hormone secreted by the *corpus luteum*, the name given to the mature follicle after ovulation occurs. The most important function of progesterone is to prepare the lining of the uterus for implantation of a fertilized ovum. The alveoli of the mammary glands enlarge and become secretory under the influence of progesterone during pregnancy; thus, it is sometimes referred to as the "pregnancy hormone." During pregnancy, this hormone also inhibits ovulation by the ovary.

ACCESSORY ORGANS

The female accessory organs of reproduction are the uterus, the uterine tubes, the vagina (Fig. 11–10), the external genitalia, and the mammary glands.

Uterus. The uterus is a hollow, pear-shaped organ lying in the pelvis between the urinary bladder and the rectum. About 7.5 cm. long and 5.0 cm. broad, it lies horizontal in orientation. Its undersurface rests on the urinary bladder, and its rounded intestinal surface faces the cavity of the pelvis. There is an angulation between the body and the cervix of the organ and between the cervix

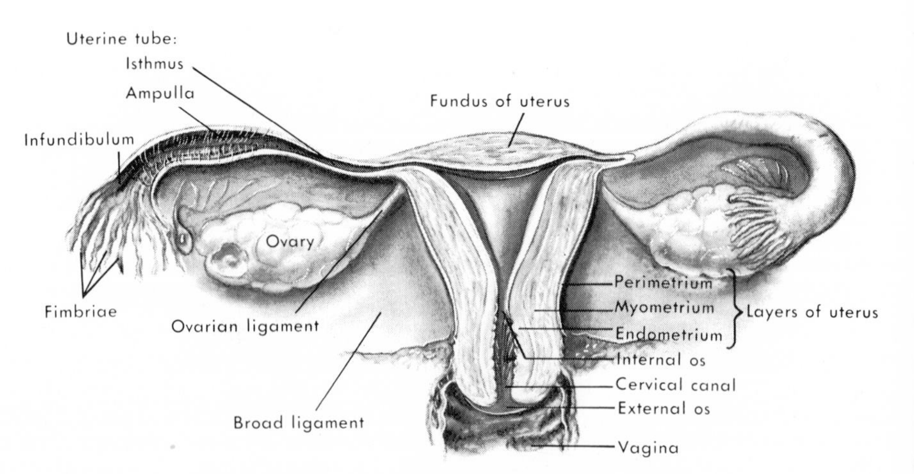

Figure 11–10. Internal female reproductive organs.

and the vagina. As urine accumulates in the bladder, the uterus is readily displaced upward.

The *fundus* is the blunt, rounded free extremity of the uterus. It is the part lying above a line joining the entrances of the two uterine tubes. The *body*, or main portion, narrows as it proceeds from the fundus to the *isthmus*, or entrance. The *broad ligaments* of the uterus sweep lateralward from the body. The tapered vaginal end is the *cervix*, or neck, and constitutes the lower 2 cm. of the organ. The vagina is attached around the circumference of the cervix.

The wall of the uterus consists of three layers. The outer layer, or *perimetrium*, is a thin serous coat formed by the peritoneum. The middle *myometrium*, or muscle layer, is very thick and is formed by three poorly defined layers of smooth muscle. The middle circular layer is the thickest of the three layers and contains the large blood vessels of the uterine wall. During pregnancy the smooth muscle fibers hypertrophy and become up to ten times as long and several times as thick as in the nonpregnant organ.

The *endometrium* is the inner mucous membrane coat of the uterine wall, and it lines the body and fundus. Its connective tissue layer has simple tubular glands that open through the overlying epithelium into the lumen of the uterus. An unusual feature of the endometrium is that is consists of two chief layers. These layers are the thick, superficial *functionalis*, which is shed during menstruation, and the thinner deep *basalis*, which regenerates the functionalis after each menstrual flow.

Uterine Tubes. The two muscular uterine tubes lie on either side of the uterus, joining the body just below the fundus. They extend laterally toward the ovaries. The narrow *isthmus* of each tube opens into the uterine cavity. The middle wide portion, or *ampulla*, arches over the ovary. The *infundibulum* is the funnel-like open end of the tube and is formed by a number of fringed processes called *fimbriae*. The fimbriae are spread over the medial surface of the ovary.

Vagina. The vagina is a muscular tube about 7.5 cm. long which passes from the cervix of the uterus to the outside of the body. It has an inner mucosa, a middle smooth muscle layer, and an outer connective tissue layer. The mucosa does not contain glands, so the vagina is lubricated by secretions from the mucous glands in the cervix. The vagina is the receptive organ for the penis and also provides for the excretion of the menstrual flow. During parturition, it acts as the terminal portion of the birth canal. Its opening lies posterior to that of the urethra and anterior to the anus.

EXTERNAL GENITALIA (VULVA) (Fig. 11–11)

The *labia majora* ("major lips") are two prominent longitudinal folds of skin containing fat and some smooth muscle. They contain large sebaceous glands, and after puberty, hairs are present in the skin of the outer surface. The labia meet at their anterior ends to form the *mons pubis*, a rounded area anterior to the pubic symphysis. Posteriorly, the lips do not meet, but end in a mass of fibrofatty tissue called the *perineal body*.

Two smaller folds of skin, the *labia minora*, lie between the labia majora. The labia minora contain sebaceous glands but no fat or hairs. The *clitoris* is a small rounded mass of erectile tissue covered by sensitive epithelium. It is partially hooded by a *prepuce*, which is formed by the meeting of the labia minora.

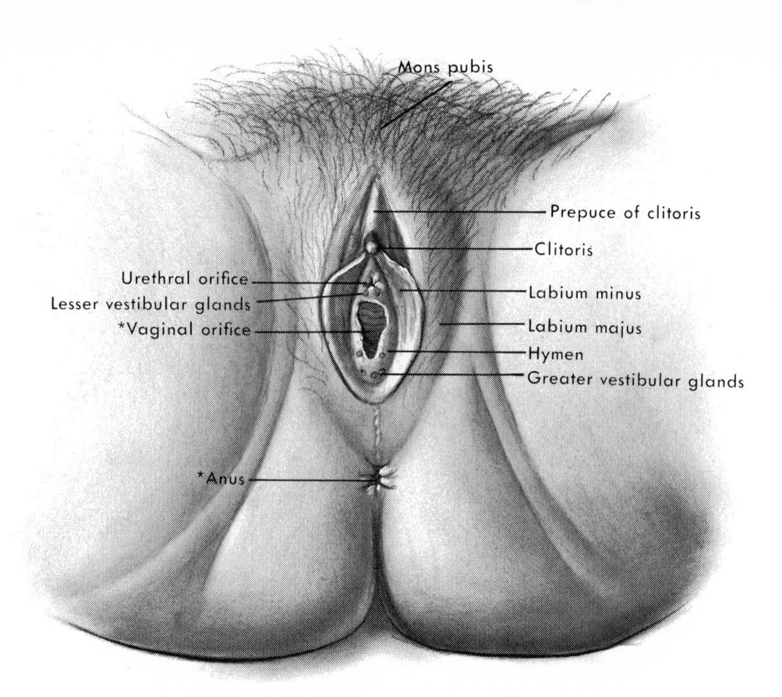

*The clinical perineum lies between these two openings.

Figure 11–11. External female genitalia.

Between the labia minora lies an area called the *vestibule.* Its floor contains the openings for the urethra, vagina, and *greater vestibular glands.* The *hymen* is a fold of mucous membrane which partially covers the posterior portion of the vaginal entrance. This membrane varies in size and shape in different individuals, and when ruptured, it enlarges the entrance to the vagina. Many small *lesser vestibular glands* open by way of ducts between the urethral and vaginal orifices.

THE MAMMARY GLANDS

The breasts, or mammary glands, are modified sweat glands which secrete milk. They lie over the chest muscles, each extending from the border of the sternum to the middle of the axilla, or armpit (Fig. 11–12). Their shape varies with age and functional activity; after repeated pregnancies, they become more elongated. Each gland is composed of 15 to 20 lobes, each of which opens through its duct at the tip of the nipple. The formation and secretion of milk by the glands is regulated by hormones during pregnancy. The *areola* is a pigmented area surrounding the nipple, which darkens during pregnancy.

THE MENSTRUAL CYCLE

The menstrual cycle in the female reflects the monthly rhythmical changes in the secretion of hormones and in the reproductive organs themselves. These cyclical changes take place during the entire reproductive period of the female, from the onset of puberty to the menopause. Each cycle is usually completed in 28 days, but the number of days varies greatly, even in normal healthy women.

Two main events occur during the menstrual cycle. First, a single mature ovum is released from the ovaries each month. Second, the lining of the uterus is prepared for implantation of the egg cell should the cell become fertilized. Each cycle may be divided into three stages or phases of unequal length. The changes taking place in each phase will be described.

THE PROLIFERATIVE PHASE (POSTMENSTRUAL, REPAIR, OR ESTROGENIC PHASE). This period, lasting about ten days, involves the development of the mature ovum and its surrounding follicle, which has already been described. Two anterior pituitary hormones, FSH and LH, stimulate the growth and development of the mature follicle.

Under the influence of FSH, several of the primary follicles in the ovary begin to grow; only one of these will reach the final stages of development. LH acts with FSH to promote the continued growth of the ovum and follicle with enlargement of the antrum.

The estrogens secreted by the thecal cells act upon the endometrium of the uterus causing its epithelial cells to proliferate. This thickening of the endometrium is accompanied by an increase in the number of its capillaries, thus increasing the vascularity of the lining. The endometrial glands begin secreting a thin, watery mucus. With the rise in the estrogen level, FSH from the pituitary gland is inhibited.

THE PROGESTATIONAL PHASE (PREMENSTRUAL, PROGRAVID, OR SECRETORY PHASE). *Ovulation* occurs on the first day of this 12- to 14-day period. The mature follicle on the surface of the ovary ruptures, expelling the ovum into the abdominal cavity. LH is particularly responsible for the last stages of follicular growth and for ovulation. The ruptured follicle, under the influence of LH, undergoes a series of changes in which its cells take on a distinct yellow color. It is then called the *corpus luteum*, or "yellow body," and begins to secrete progesterone.

Progesterone has its primary effect on the continued growth of the endometrium. Under the influence of progesterone, the lining doubles in thickness. The

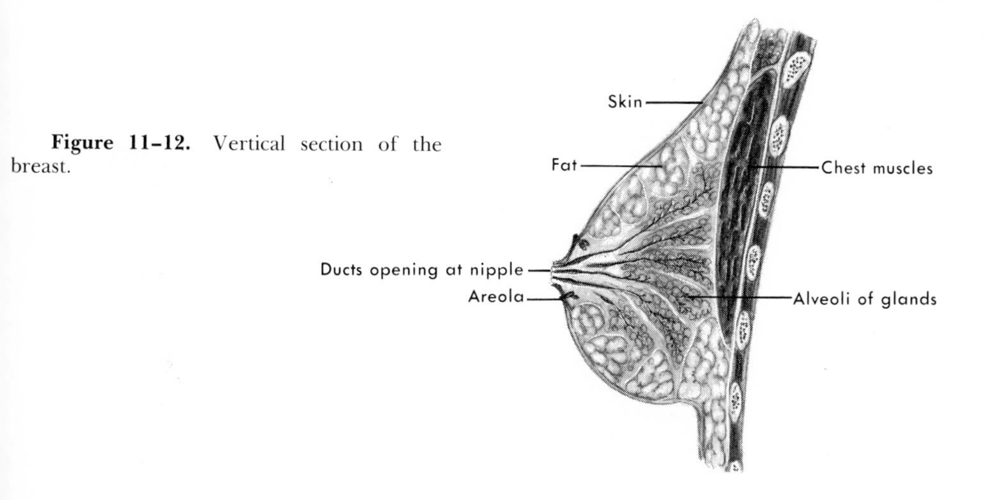

Figure 11–12. Vertical section of the breast.

glands, which previously were straight, become tortuous and begin to secrete large quantities of thick mucus. Progesterone also prevents contraction of the myometrium of the uterus, so that if implantation occurs, the fertilized ovum will not be expelled.

If the ovum is not fertilized during its passage through the uterine tube, the corpus luteum begins to degenerate. In about 12 days it loses its characteristic yellow color and becomes the *corpus albicans*, or "white body." Further degeneration follows, and the corpus albicans becomes replaced by connective tissue, which forms a scar on the surface of the ovary.

On the last day of this phase, the coiled arteries in the endometrium constrict, causing *ischemia* (lack of blood) and *necrosis* (death) of the tissue forming the functionalis layer. Constriction of the coiled arteries is probably caused by a sudden fall in the blood level of estrogen, since this hormone is a dilator of endometrial blood vessels. At this time, the progesterone level in the blood also declines.

THE MENSTRUAL PHASE. The menstrual flow occurs during the last five days of the cycle. The constricted coiled arteries become dilated, and pools of blood accumulate below the endometrium (Fig. 11–13). The necrotic function-

PHASE:	PROLIFERATIVE	SECRETORY	ISCHEMIC	MENSTRUAL
MAIN FEATURE	Mitosis	Hypertrophy	Ischemia	Denuded surface
GLANDS	Straight and narrow	Corkscrew	Very contorted	FUNCTI
STROMA	Abundant	Incr. edema Incr. vascularity	Incr. white blood cells	ONALIS
COILED ARTERIES	Not present in superficial 1/3	Reach superficial 1/3	Very contorted	LOST
THICKNESS	2–3 mm.	6 mm.	3–4 mm.	1/2–1 mm.
	6 to 14 days	15 to 26 days	27 to 28 days	1 to 5 days

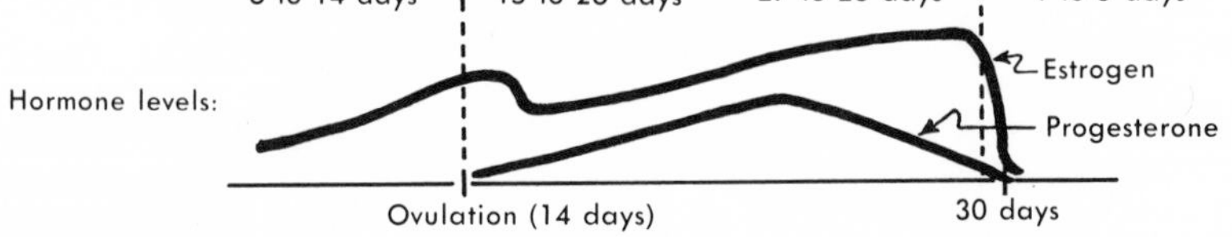

Hormone levels:
Estrogen
Progesterone
Ovulation (14 days)
30 days

Figure 11–13. Summary of endometrial changes during the menstrual cycle.

alis tissue begins to separate from the uterine wall, and together with the accumulated blood, it forms the menstrual flow. Myometrial contractions help to expel the material from the uterine cavity. The total menstrual flow amounts to about 70 cc., half of which is blood and half is serous fluid. Because a *fibrinolysin*, or dissolver of fibrin, is secreted along with the flow, the blood normally does not clot.

Thus the functionalis layer of the endometrium is lost during menstruation. Before this phase ends, however, the blood level of FSH has begun to rise. Thus, a new follicle is ripening in the ovary, and the denuded uterine lining is already being repaired in anticipation of a new cycle.

FERTILIZATION AND IMPLANTATION

If the ovum liberated at ovulation is fertilized by a sperm cell from the male, it becomes implanted in the endometrial wall. The corpus luteum, which begins to degenerate during the progestational phase of the menstrual cycle, must continue to function if the pregnancy is to be maintained. If the corpus luteum is removed before the eleventh week of pregnancy, spontaneous abortion of the fetus is likely to occur.

Fertilization. After its release from the ovary at ovulation, the ovum makes its way into the oviduct. The exact mechanism as to how it accomplishes this feat is not known. It may be either that the fimbriated end of the tube plasters itself on the surface of the ovary and thus guides the egg on its way, or that the tube literally sucks the egg into itself from the abdominal cavity.

After spermatozoa or *sperm* have been deposited in the vagina, they must migrate from there by their own power. They are aided in this struggle by the presence of a tail or *flagellum* which enables them to become highly motile. Approximately one to two hours elapse before the sperm appear at the outer end of the oviduct. Spermatozoa can live in the female reproductive tract for at least 48 hours, but in their passage they must swim against the beat of cilia and tubal contractions. It may be that a sort of chemical "radar" helps them to zero in on the ovum.

Fertilization usually takes place in the middle or outer one-third of the oviduct, and occurs 12 to 24 hours after ovulation. Although the ovum is surrounded by sperm, only one actually manages to penetrate the egg; it is as though this penetration is a signal to the ovum to reject all other sperm. After fertilization, the ovum is called a *zygote*. It undergoes several mitotic divisions as it passes down the oviduct and enters the uterus as the *morula* (16-cell stage) on about the third day.

Implantation. On the fourth day after fertilization, the morula becomes the *blastocyst*. The blastocyst lies free in the uterine cavity for about two days; then it begins to implant itself. This usually occurs in the posterior wall of the body of the uterus. Implantation of the blastocyst in any other site, such as in the abdominal cavity, the oviduct, or the ovary, will result in an *ectopic* pregnancy.

The outer cell mass of the blastocyst, collectively called the *trophoblast*, actually digests the uterine tissue. During the second week of development, the blastocyst becomes embedded in the uterine mucosa and endometrial cells cover the defect in the wall.

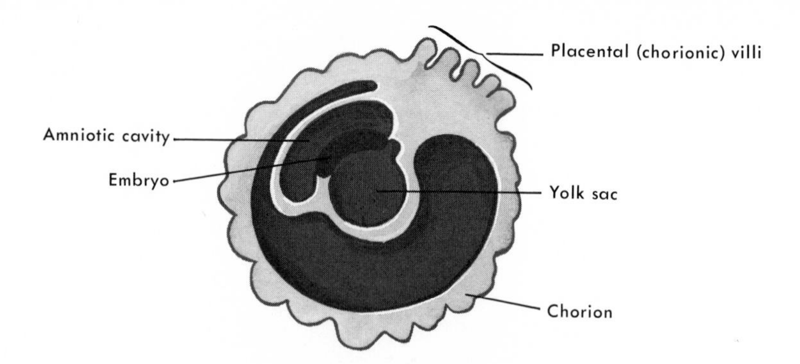

A. Early membrane formation

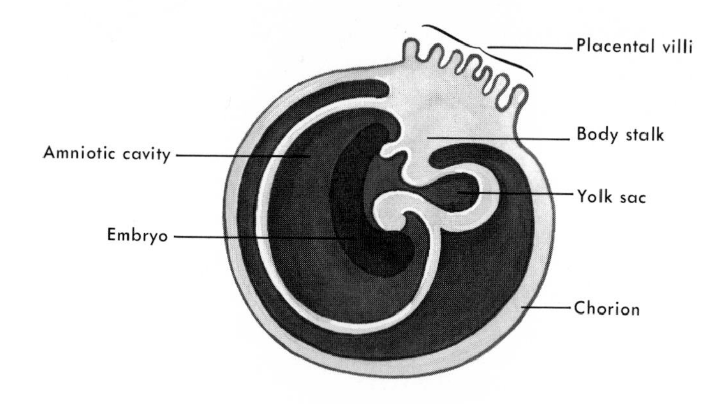

B. Expansion of amnion

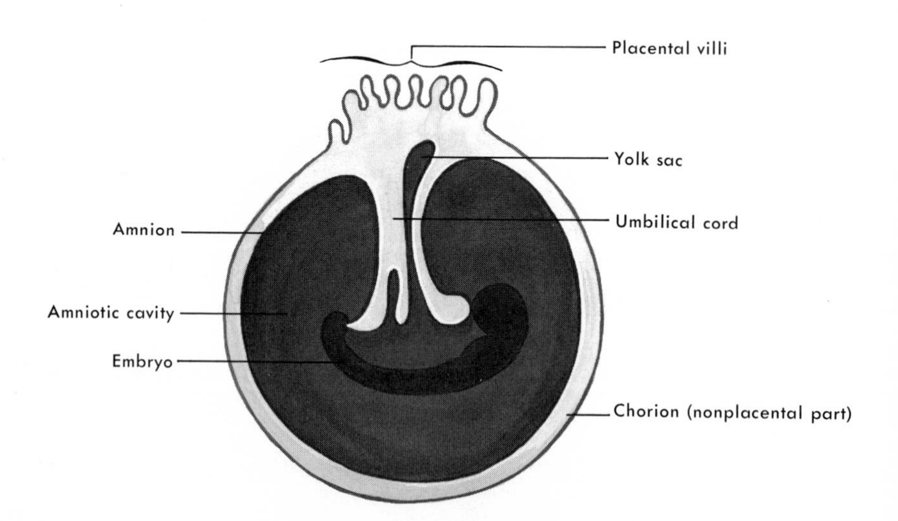

C. Appearance of umbilical cord

Figure 11-14. Fetal membrane development.

GERM LAYERS AND FETAL MEMBRANES (Fig. 11–14)

The inner cell mass develops into two groups of cells, an outer germ layer or *ectoderm,* and an inner germ layer or *entoderm.* A cavity forms within each of these two cell groups; cells surrounding the ectoderm or amniotic cavity become the *amnion,* while entodermal cells form the *yolk sac.* Scattered cells between the ectoderm and entoderm form the third germ cell layer, the *mesoderm.* (Each germ layer gives rise to sheets of epithelium; ectoderm also produces nervous tissue while mesoderm forms muscle and connective tissue. See Table 11–1.)

The trophoblast later combines with the mesoderm to form the *chorion.* This extraembryonic membrane continues to grow throughout pregnancy along with the fetus, serving as the outer barrier between it and the uterus. The amnion expands rapidly, finally coming into contact with the chorion. Its cavity, expanded by the accumulation of amniotic fluid, eventually completely surrounds the embryo and its umbilical cord. This fluid allows the fetus to move and offers it protection from injury.

PREGNANCY AND THE PLACENTA

Development of the Placenta. The newly implanted embryo is nourished by secretions from its own covering cells, the trophoblast. However, to provide for continuance of the intrauterine life of the new individual, a highly specialized structure called the *placenta* develops in the uterus (Fig. 11–15). Shortly after implantation, the trophoblastic cells of the fetus spread progressively deeper into the mother's tissues. In one week, connections are formed between the trophoblasts and the blood vessels of the uterus. By the end of the third week, the necessary relationships which permit the required exchanges between mother and fetus have developed. Thus the placenta is formed by the trophoblastic cells of the fetus plus the uterine endometrial cells of the mother. In general, the placenta may be likened to two parallel tissue layers with a maternal blood space between them. The *umbilical cord* is the structure which mediates the exchange of materials; it is the "life line" of the fetus.

Functions of the Placenta. It has been said that "as the placenta goes, so goes the pregnancy." That is to say, many changes in pregnancy can be explained by placental functions. Two hundred and fifty cubic centimeters of fetal blood passes through the placenta every minute. This volume represents 57 per

Table 11-1. Some Derivatives of Primary Germ Layers

Ectoderm (Outer Layer)	*Mesoderm (Middle Layer)*	*Entoderm (Inner Layer)*
1. Skin (includes hair, nails)	Muscle (all types)	Larynx, trachea, lungs
2. Epithelium of mouth, nasal cavity, sense organs	Epithelium of blood vessels, kidney, gonads, body cavities	Epithelium of pharynx, thyroid, parathyroid
3. Nervous tissue, adrenal medulla	Connective tissue (includes blood and lymphoid tissue	Digestive tube, bladder, urethra, vagina

cent of the fetal cardiac output. As for the mother, the uterine blood flow increases as the weight of the fetus increases.

The general function of the placenta is to act as a transfer mechanism. The fetus must be provided with a continuous supply of oxygen and nutrients. Thus oxygen, glucose, amino acids, vitamins, and minerals are transferred across the placental barrier by both diffusion and active transport. Protective substances, such as antibodies to certain diseases, also cross the barrier.

The placenta also acts as a storage mechanism, particularly during the first trimester of pregnancy. Because it grows so rapidly during this time, the placenta is able to store quantities of nutrients such as calcium, iron, and proteins, which are used for later fetal growth.

Carbon dioxide and waste materials must be excreted by the fetus. Thus, the placenta has an excretory function in that these substances diffuse from fetal to maternal blood and are excreted with the mother's own waste products.

Hormones in Pregnancy (Fig. 11–16). Certain hormones are required to insure the continuance of pregnancy. It is obvious that if the endometrial lining of the uterus should be cast off as it is during menstruation, the newly implanted embryo would be destroyed. Detachment of the endometrial lining is prevented by a substance called *chorionic gonadotropin,* which is secreted by certain of the placental cells.

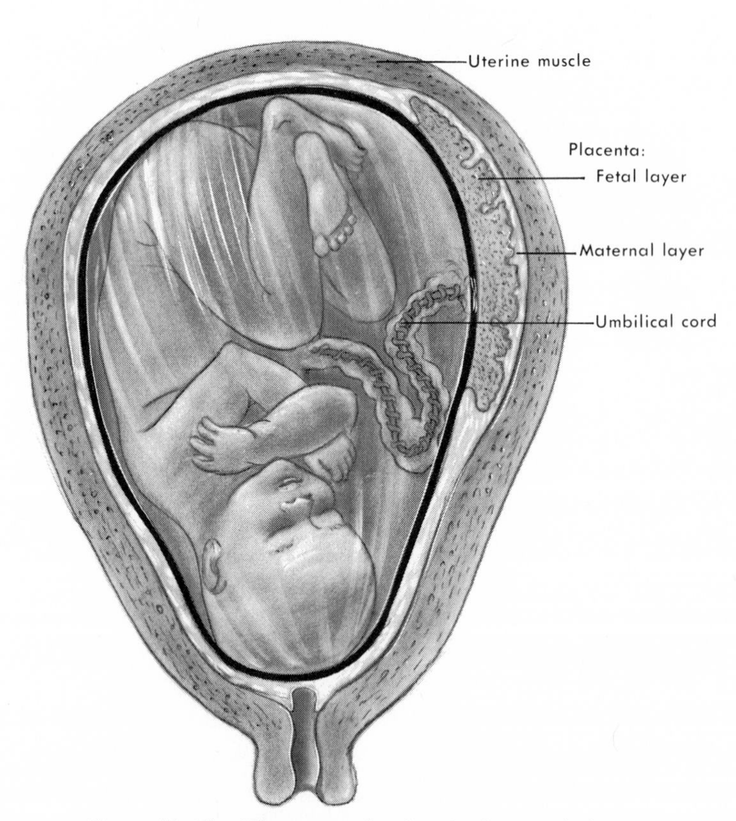

Figure 11–15. The uterus, showing the fetus and placenta.

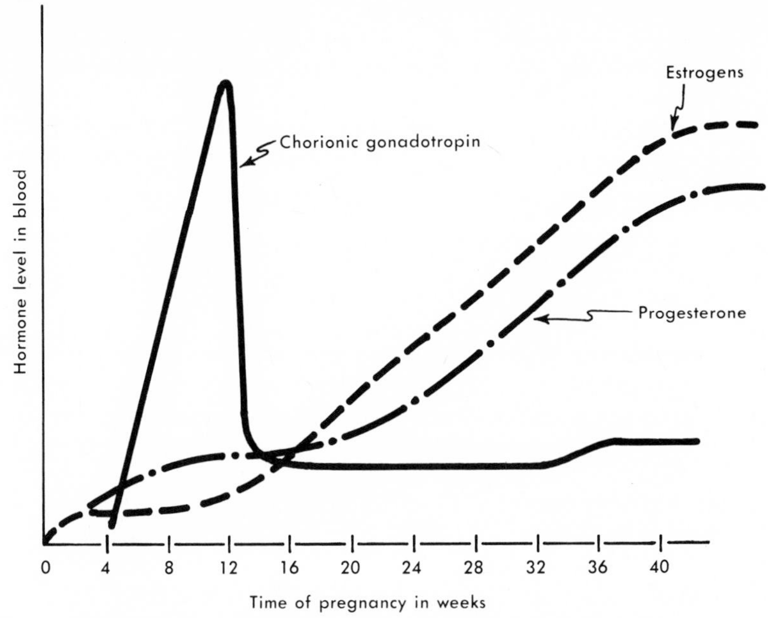

Figure 11–16. Graph showing hormone levels during pregnancy.

Chorionic gonadotropin is very rich in hormones. It performs many of the same functions as the anterior pituitary gonadotropins. For example, it maintains the corpus luteum so that instead of degenerating, the corpus luteum continues to produce estrogens and progesterone. These hormones in turn insure the continued growth of the endometrium, especially during placental development. Chorionic gonadotropin also has an interstitial cell-stimulating hormone (ICSH) effect on the testes of male fetuses. ICSH stimulates the testes in the fetus to produce testosterone, which in turn causes these organs to descend into the scrotum as the pregnancy nears completion.

As the period of pregnancy advances, the placenta itself secretes estrogens and progesterone. The estrogens cause growth of the mammary gland tissue and enlargement of the uterus and external genitalia. The relaxation of the pelvic ligaments, which occurs prior to the passage of the fetus through the birth canal, is also attributed to estrogen. Progesterone is of particular importance in preventing contractions of the uterine muscle during pregnancy, thus preventing a premature expulsion of the fetus. Progesterone also helps to prepare the breasts for milk secretion.

OUTLINE SUMMARY

1. Male Reproductive System
    a. Testis – male gonad
        1) Paired organs located in scrotum
        2) Interstitial cells produce androgens, mainly testosterone
        3) Germinal cells produce spermatozoa

b. Ducts
   1) Epididymis – 20-feet coiled tube from collecting tubules of testis
   2) Ductus deferens – direct continuation of epididymis; contained in spermatic cord
   3) Ejaculatory duct – union of ductus deferens and seminal vesicle ducts
   4) Urethra – passage for both urine and semen
c. Glands – add alkaline secretions to semen
   1) Seminal vesicles – two blind pouches on either side of urinary bladder
   2) Prostate – lies under bladder, surrounding urethra
   3) Bulbourethral glands – two small glands on either side of membranous urethra
d. External genitalia
   1) Scrotum – bipartite sac housing testes
   2) Penis – intromittent organ of erectile tissue; traversed by penile urethra
e. Passage of spermatozoa – see p. 239

2. Female Reproductive System
   a. Ovary – female gonad
      1) Paired oval bodies lying in pelvis, supported by ligaments
      2) Oögenesis – germinal cells produce ova; growth and maturation of follicles
      3) Hormones
         a) Estrogens – produced by follicular cells
         b) Progesterone – pregnancy hormone; secreted by mature follicle
   b. Accessory organs
      1) Uterus – hollow, pear-shaped organ between bladder and rectum; three layers
      2) Uterine tubes (oviducts) – on either side of uterus; convey ovum from ovary
      3) Vagina – muscular tube from uterus to outside of body
   c. External genitalia (vulva)
      1) Labia majora – large, prominent folds of skin
      2) Labia minora – smaller folds of skin
      3) Clitoris – small mass of erectile tissue
      4) Vestibule – area between labia minora; urethral and vaginal orifices; glands
   d. Mammary glands (breasts)
      1) Modified sweat glands that produce milk
      2) Composed of 15 to 20 lobes
3. Menstrual Cycle – monthly rhythmical changes in female
   a. Proliferative phase – 10 days
      1) Development of mature ovum under influence of FSH and LH
      2) Thecal cells of follicle cause endometrium to thicken and glands to secrete
   b. Progestational phase – 12 to 14 days
      1) Ovulation occurs on first day
      2) Corpus luteum forms and secretes progesterone; endometrium doubles in thickness
      3) Ischemia of coiled arteries on last day

c. Menstrual phase – 1 to 5 days
   1) Necrotic functionalis and accumulated blood form menstrual flow
   2) Near end of phase, blood level of FSH begins to rise; endometrial repair begins

4. Fertilization and Implantation
   a. Fertilization
      1) Ovum enters oviduct
      2) One sperm penetrates ovum
      3) Occurs in outer third of oviduct 12 to 24 hours after fertilization
      4) Ectopic fertilization may lead to ectopic pregnancy
   b. Implantation of blastocyst
      1) Occurs on sixth or seventh day after fertilization
      2) Trophoblast burrows into endometrium
   c. Germ layers and derivatives
      1) Ectoderm – certain epithelia and nervous tissue
      2) Mesoderm – certain epithelia, muscle tissue, connective tissue
      3) Entoderm – certain epithelia, respiratory and digestive structures
   d. Fetal membranes
      1) Amnion – amniotic fluid surrounds embryo
      2) Chorion – barrier between fetus and uterus
      3) Yolk sac
5. Pregnancy and the Placenta
   a. Development of placenta
      1) Fetal trophoblasts invade uterine mucosa
      2) Two parallel tissue layers formed with maternal blood spaces between them
      3) Umbilical cord mediates exchange of nutrients and waste products
   b. Functions of placenta – transfer, storage, and excretory
   c. Hormones in pregnancy – elaborated by placenta
      1) Chorionic gonadotropin – maintains corpus luteum
      2) Progesterone – prevents uterine contractions; prepares breasts for lactation
      3) Estrogens – growth of mother's reproductive organs; relaxes pelvis at birth

REVIEW QUESTIONS

1. What are the two main functions of the testes?
2. Describe the process of spermatogenesis.
3. What is contained in the spermatic cord? Can the term *ductus deferens* be substituted for *spermatic cord*? Explain.
4. How and where is the ejaculatory duct formed?
5. Describe the male urethra.
6. What is semen? What composes it?
7. What is meant by the term *erectile tissue*?
8. What is circumcision?
9. What is oögenesis? Describe it.
10. What does the term *atresia* refer to?
11. What is the final result of the two maturation divisions in oögenesis?

12. List the functions of estrogens and progesterone in the female.
13. What is the unusual structural feature of the uterine endometrium?
14. True or false: There is a direct, closed connection between the ovary and the uterine tube.
15. What is the clinical perineum?
16. Describe the events that occur in each phase of the menstrual cycle.
17. True or false: The placenta becomes functional as soon as implantation occurs.
18. What are the placental hormones and what are their functions?
19. What combines with mesoderm to form the chorion?
20. Name some body structures that develop from entoderm.
21. What special structural feature enables sperm to be highly motile?
22. What is an *ectopic* pregnancy?

ADDITIONAL READING

Csapo, A.: Progesterone. Sci. Amer. (April), 1958.
Marshall, F. H. A.: Physiology of Reproduction. 3rd ed. Edited by A. S. Parkes. New York, Longmans, Green, 1958.

GENERAL ANATOMY AND PHYSIOLOGY REFERENCES

Developmental Anatomy

Arey, L. B.: Developmental Anatomy. 7th ed. Philadelphia, W. B. Saunders Company, 1965.

Histology

Bailey's Textbook of Histology. 15th ed. (Revised by Copenhaven.) Baltimore, The Williams and Wilkins Company, 1964.

Bloom, W., and Fawcett, D. W.: A Textbook of Histology. 9th ed. Philadelphia, W. B. Saunders Company, 1968.

Gross Anatomy

Basmajian, J. V.: Primary Anatomy. 6th ed. Baltimore, The Williams and Wilkins Company, 1970.

Woodburne, R. T.: Essentials of Human Anatomy. 4th ed. New York, Oxford University Press, 1969.

General Physiology

Guyton, A. C.: Basic Human Physiology. Philadelphia, W. B. Saunders Company, 1971.

Tuttle, W. W., and Schottelius, B. A.: Textbook of Physiology. 16th ed. St. Louis, The C. V. Mosby Co., 1969.

Vander, A. J., Sherman, J. H., and Luciono, D. S.: Human Physiology. New York, McGraw-Hill Book Company, Inc., 1970.

DICTIONARIES AND ATLASES

Anson, B. J.: An Atlas of Human Anatomy. 2nd ed. Philadelphia, W. B. Saunders Company, 1963.

Blakiston's New Gould Medical Dictionary. 2nd ed. New York, McGraw-Hill Book Company, Inc., 1956.

Dorland's Illustrated Medical Dictionary. 24th ed. Philadelphia, W. B. Saunders Company, 1965.

Grant, J. C.: An Atlas of Anatomy. 5th ed. Baltimore, The Williams and Wilkins Company, 1962.

Lopez-Antunez, L.: Atlas of Human Anatomy. Philadelphia, W. B. Saunders Company, 1971.

GLOSSARY

absorption: the taking in and assimilation of material
acidosis: reduction of the body's alkaline reserve as a result of an excess of acid metabolites
afferent: carrying toward
alkalosis: an elevation of the body's alkaline reserve, especially of the bicarbonate content of the blood, either by ingesting large amounts of alkali or by the severe loss of body acid
alveolus: (1) a bony socket of a tooth; (2) an air cell of the lung
amino acid: any one of a large group of organic compounds containing nitrogen; used by the body to make proteins
amplitude: largeness; extent; range
ampulla: the dilated part of a canal or duct
anabolism: constructive metabolism; the changing of nutritive material into more complex living matter
antibody: a substance, natural or induced by exposure to an antigen, which has the capacity to react with the specific antigen
antigen: any substance which stimulates the production of antibodies or reacts with them
antiseptic: a compound which inhibits the growth of bacteria without necessarily killing them
antrum: a cavity or space
apex: summit or top of anything; point or extremity
aqueduct: a canal for the passage of fluid
autonomous: independent in origin, action, or function; self-governing
axis (pl. *axes*): an imaginary line passing through the center of a body; also the line about which a rotating body turns
bipartite: consisting of two parts separated by a septum
-blast: denoting a formative cell
cachexia: weakness and emaciation caused by some serious disease
catabolism: destructive phase of metabolism concerned with the breaking down of compounds by the body with the liberation of energy; opposite of anabolism
celiac: pertaining to the abdomen
chromosome: deeply staining body which arises from the nucleus of the cell during mitosis; carries the hereditary factors (genes). Man has 46 in each body cell, except for the mature ovum and sperm, each of which has 23
concave: having a curved, depressed surface
concha: a shell; a shell-like organ
convex: having a raised, rounded surface
coracoid: having the shape of a crow's beak
coronary: a wreath; a term applied to vessels or nerves that encircle a part or organ

cribriform: perforated like a sieve
cubic centimeter: a metric liquid measurement; 1 ounce equals approximately 30 cc.
-cyte: suffix denoting cell
deferens: carrying away or down
differentiation: an increase in complexity and organization of cells and tissues during development
dilatation: the state of being stretched; enlargement of a hollow part or organ
efferent: carrying away from
electrolyte: a substance which in solution is capable of conducting an electric current
embryo: the young organism from conception to the third month of pregnancy
emphysema: a condition in which there is overdistention of the air spaces in the lungs
encephalon: the brain
endocrine: secreting internally
erectile: capable of being dilated or erected
ethmoid: a sieve-like form
exocrine: secreting outwardly
fasciculus: a bundle of nerve, tendon, or muscle fibers
fetus: the unborn offspring from the end of the third month until birth
flexure: a bend or fold
follicle: a small secretory cavity or sac
fovea: a small pit or depression; especially the fovea centralis of the retina
fundus: the base of an organ; the part farthest from the opening of the organ
glossa: the tongue
groin: the depression between the abdomen and the thigh; the inguinal region
heparin: a substance occurring in tissues which has the property of prolonging the clotting time of blood
histamine: a chemical substance which stimulates smooth muscle, dilates capillaries, and stimulates secretions
humor: any fluid or semifluid
hypertonic: exceeding in strength or tension
incompatible: incapable of being used or put together because of antagonistic qualities
innervation: the distribution of nerves to a part
intromittent: conveying into or within, as into a cavity
ischemia: local lessening of the blood supply, caused by obstruction of inflow of arterial blood
isthmus: the neck or constricted part of an organ
labium (pl. *labia*)**:** lip
lacrimal: pertaining to the tears, or to the organs secreting the tears
lactation: milk secretion
lambdoidal: resembling the Greek letter lambda (λ)
luteo-: a combining form meaning "orange yellow"
macrophage: a phagocytic cell, not a leucocyte, which can store certain dyes in the form of granules in its cytoplasm
macula: a spot; the area of most acute vision of the retina

mammary: pertaining to the mammae (breasts)
manometer: an instrument for measuring the pressure of liquids and gases
meatus: an opening or passage
meningitis: inflammation of the meninges
metabolism: the process by which the body synthesizes food into complex tissue elements (anabolism) and complex substances into simple ones with the production of energy (catabolism)
mitosis: the indirect division of the nucleus and cytoplasm of a cell
nucleus: the differentiated part of the cell which controls growth, repair, and reproduction; a group of nerve cells within the central nervous system which have a particular function
orifice: an opening; an entrance to a cavity or tube
papilla (pl. *papillae*)**:** a small, nipple-like eminence
parenchyma: the essential or specialized part of an organ as distinguished from its supporting tissue
parturition: the act of giving birth
pectoral: pertaining to the chest
peduncle: a narrow part acting as a support
-phile: a combining form meaning "love of"; denotes a tendency toward
pitch (in sound)**:** the quality of a sound which depends on the rapidity of the vibrations that produce the sound
plasma cell: a fairly large mononuclear cell normally present in bone marrow and other tissues; they concentrate in chronically inflamed areas
pons: a bridge
porta: the hilum of an organ through which vessels enter
proliferation: multiplying, as by cellular division
protein: one of a group of complex substances containing nitrogen which are found in animals and plants and are characteristic of living matter
ramus: a branch, especially of an artery, vein, or nerve
refraction: the deviation of a light ray from a straight line when it passes from a medium of one density into a medium of a different density
renal: pertaining to the kidneys
saline: containing sodium chloride (salt)
sclero-: a combining form meaning "hard"
septum: a dividing wall between two cavities or spaces
serous: pertaining to, characterized by, or resembling serum
somatic: pertaining to the body, especially to the body framework
sphenoid: wedge-shaped
squamous: of the shape of a scale
stroma: the supporting framework of an organ, as opposed to the parenchyma
symphysis: a growing together
synergist: an agent coöperating with another
tegmentum: a covering; specifically part of the dorsal portion of the midbrain
temporal: pertaining to the temple
tension: the act or state of being stretched
torsion: a twisting
trochlea: a process having the nature of a pulley

trophic: of or pertaining to nutrition
-tropic: a combining form denoting a turning
tympanum: the middle ear
utricle: a delicate membranous sac in the inner ear
vagina: a sheath
vagus: wandering
varicose: descriptive of blood vessels that are dilated, knotted, and tortuous
vascular: pertaining to vessels, especially blood vessels
ventricle: a small cavity or pouch
vestibule: an approach; an entrance chamber
villus: a tiny, elongated projection from the surface of a membrane
volar: pertaining to the palm of the hand or the sole of the foot
zygoma: the cheek bone

METRIC SYSTEM MEASUREMENTS

MEASUREMENTS	*METRIC*	*LINEAR (approx. inches)*
meter	1.0 meter	40.0
decimeter	0.1 meter	4.0
centimeter	0.01 meter	0.4
millimeter	0.001 meter	0.04
micron (micrometer)	0.001 millimeter	1/25,000
millimicron (nanometer)	0.001 micron (micrometer)	1/25,000,000
Angstrom	0.1 millimicron (nanometer)	1/250,000,000

INDEX

Page numbers in *italics* indicate illustrations.